This book gives a comprehensive account of both experimental and theoretical aspects of electron microprobe analysis, and is an extensively updated version of the first edition, published in 1975, which was regarded as a standard work on the subject.

The design and operation of the instrument, including the electron column and both wavelength- and energy-dispersive X-ray spectrometers are covered in the first part of the book. Experimental procedures for qualitative analysis (including elemental mapping) and quantitative analysis, using both types of spectrometer, are then discussed. Matrix ('ZAF') corrections, as required for quantitative analysis, are treated in some detail from both theoretical and practical viewpoints. Special considerations applying to the analysis of 'light' elements (atomic numbers below 10) are covered in a separate chapter. The book also contains an appendix dealing with the physics of X-ray production, which includes a number of useful tables. The emphasis throughout is on a sound understanding of principles and the treatment is applicable equally to the electron microprobe in its 'classical' form and to scanning electron microscopes fitted with X-ray spectrometers.

This book will be of use to researchers, graduate students and industrial scientists using electron microprobe analysis including scanning electron microscopes with X-ray detection facilities.

**Electron microprobe analysis**

# Electron microprobe analysis
## SECOND EDITION

**S. J. B. Reed**
*University of Cambridge*

Published by the Press Syndicate of the University of Cambridge
The Pitt Building, Trumpington Street, Cambridge CB2 1RP
40 West 20th Street, New York, NY 10011-4211, USA
10 Stamford Road, Oakleigh, Victoria 3166, Australia

© Cambridge University Press 1975, 1993

First published 1975
Second edition 1993

Printed in Great Britain at the University Press, Cambridge

*A catalogue record for this book is available from the British Library*

*Library of Congress cataloguing in publication data*
Reed, S.J.B.
Electron microprobe analysis / S.J.B. Reed. – 2nd ed.
p. cm.
Includes bibliographical references and index.
ISBN 0–521–41956–5
1. Microprobe analysis. I. Title.
QD117.E42R43 1993
543′.08586–dc20 92–16263 CIP

ISBN 0 521 41956 5 hardback

# Contents

| | |
|---|---|
| *Acknowledgements* | xv |
| *Preface* | xvii |

**1 Introduction**    1
1.1 Historical notes    1
1.2 Principles of electron microprobe analysis    3
1.3 Characteristic X-ray spectra    4
1.4 Inner-shell ionisation    6
1.5 Fluorescence yield    7
1.6 Continuous X-ray spectrum    8
1.7 X-ray absorption    9
1.8 Fluorescence    10
1.9 Relationship between X-ray intensity and elemental concentration    10
1.10 Matrix corrections    11

**2 Essential features of the electron microprobe**    13
2.1 Probe-forming system    13
2.2 X-ray spectrometers    13
2.3 Specimen stage    15
2.4 Optical microscope    15
2.5 Scanning    17
2.6 Vacuum system    18
2.7 Related techniques    20
     2.7.1 Scanning electron microscopy    20
     2.7.2 Analytical electron microscopy    21
     2.7.3 Scanning transmission electron microscopy    21
     2.7.4 Auger analysis    22
     2.7.5 Proton probe analysis    22
     2.7.6 X-ray fluorescence analysis    22

## 3 Electron guns — 24
- 3.1 Conventional triode gun — 24
- 3.2 Gun alignment — 24
- 3.3 Electrical supplies — 26
  - 3.3.1 Filament heating current — 26
  - 3.3.2 High voltage — 26
  - 3.3.3 Bias — 26
- 3.4 Brightness — 28
- 3.5 Stability — 30
- 3.6 High brightness guns — 31
  - 3.6.1 Lanthanum hexaboride — 31
  - 3.6.2 Field emission sources — 32

## 4 The probe-forming system — 33
- 4.1 Introduction — 33
- 4.2 Magnetic lenses — 33
- 4.3 Final lens design — 35
  - 4.3.1 Mini-lenses — 37
- 4.4 Lens current stability — 38
- 4.5 Aberrations — 39
  - 4.5.1 Spherical aberration — 39
  - 4.5.2 Astigmatism — 40
  - 4.5.3 Chromatic aberration — 41
- 4.6 Apertures — 41
- 4.7 Column alignment — 42
- 4.8 Demagnification of the electron source — 42
- 4.9 Probe diameter and current — 44
- 4.10 Probe current monitoring and stabilisation — 46
- 4.11 Computer control — 48

## 5 Scanning — 49
- 5.1 Deflection systems — 49
- 5.2 Display systems — 49
- 5.3 X-ray images — 51
  - 5.3.1 Total X-ray images — 52
- 5.4 Colour images — 53
- 5.5 Electron images — 54
  - 5.5.1 Backscattered electron images — 55
  - 5.5.2 Secondary electron images — 56
  - 5.5.3 Specimen current images — 56
- 5.6 Electron detectors — 56
  - 5.6.1 Scintillation detectors — 56
  - 5.6.2 Solid-state detectors — 57

## Contents

**6 Wavelength-dispersive spectrometers** — 59
- 6.1 Bragg reflection — 59
  - 6.1.1 Reflection by perfect crystals — 60
  - 6.1.2 Reflection by imperfect crystals — 61
- 6.2 Crystals for w.d. spectrometers — 61
  - 6.2.1 High order reflections — 63
  - 6.2.2 X-ray refraction — 63
  - 6.2.3 The effect of temperature — 64
- 6.3 Diffracting structures for long wavelengths — 64
  - 6.3.1 Soap films — 65
  - 6.3.2 Evaporated multilayers — 65
  - 6.3.3 Diffraction gratings — 66
- 6.4 Focussing geometry — 66
  - 6.4.1 Doubly-curved crystal geometry — 68
- 6.5 Design of fully focussing spectrometers — 68
  - 6.5.1 Variable crystal curvature — 70
- 6.6 Semi-focussing geometry — 71
- 6.7 Line width and resolution — 71
  - 6.7.1 Slits — 74
- 6.8 Reflection efficiency — 75
- 6.9 Spectrometer performance — 75
- 6.10 Defocussing effects — 76
  - 6.10.1 $z$-axis defocussing — 77
  - 6.10.2 $x$- and $y$-axis defocussing — 77

**7 Proportional counters** — 79
- 7.1 Principles of operation — 79
- 7.2 Entrance window — 80
- 7.3 Counter gas — 81
- 7.4 Gas multiplication — 83
- 7.5 Pulse height depression — 85
- 7.6 Output pulses — 85
- 7.7 Ionisation statistics — 86
- 7.8 Anode wire — 88
- 7.9 Escape peaks — 89
- 7.10 Unorthodox proportional counters — 90

**8 Counting electronics** — 91
- 8.1 Counting systems for w.d. spectrometers — 91
- 8.2 Preamplifier — 92
- 8.3 Main amplifier — 93
  - 8.3.1 Operation at high count-rates — 95
- 8.4 Discriminator and pulse height analyser — 96
  - 8.4.1 Automatic pulse height analysis — 96

|  |  | Contents |  |
|---|---|---|---|

| | 8.5 | Ratemeter | 98 |
| | 8.6 | Scaler and timer | 99 |
| | 8.7 | Dead time | 100 |
| | | 8.7.1 Measurement of dead time | 101 |

**9 Lithium-drifted silicon detectors** — **103**

| | 9.1 | Principles of operation | 103 |
| | 9.2 | Construction | 105 |
| | 9.3 | Cryostat | 107 |
| | | 9.3.1 Alternative methods of cooling | 108 |
| | 9.4 | Entrance window | 108 |
| | | 9.4.1 X-ray collimation | 109 |
| | | 9.4.2 Electron trapping | 110 |
| | | 9.4.3 Removable windows | 110 |
| | | 9.4.4 Ice layers | 110 |
| | 9.5 | Energy resolution | 111 |
| | | 9.5.1 Comparison with other types of spectrometer | 111 |
| | 9.6 | Incomplete charge collection | 112 |
| | | 9.6.1 Trapping | 112 |
| | | 9.6.2 Silicon dead layer | 113 |
| | | 9.6.3 Non-linearity | 115 |
| | 9.7 | Escape peaks | 115 |
| | | 9.7.1 Internal fluorescence | 117 |
| | 9.8 | Efficiency | 118 |
| | | 9.8.1 Varying the collection efficiency | 120 |
| | | 9.8.2 Efficiency calibration | 121 |
| | | 9.8.3 Monitoring contamination layers | 122 |
| | 9.9 | Other detection media | 123 |
| | | 9.9.1 Germanium | 123 |
| | | 9.9.2 Mercuric iodide | 124 |
| | | 9.9.3 Ion-implanted silicon | 124 |

**10 Electronics for energy-dispersive systems** — **125**

| | 10.1 | Preamplifier | 125 |
| | | 10.1.1 Charge restoration | 126 |
| | 10.2 | Noise | 128 |
| | 10.3 | Main amplifier | 129 |
| | | 10.3.1 Pulse shaping | 129 |
| | | 10.3.2 Pole-zero cancellation | 131 |
| | | 10.3.3 Baseline restoration | 131 |
| | | 10.3.4 Resolution and count-rate | 132 |
| | 10.4 | Harwell pulse processor | 133 |
| | | 10.4.1 Strobed noise peak | 134 |
| | 10.5 | Multi-channel pulse height analysis | 135 |

|  |  |  |
|---|---|---|
| 10.6 | Pulse pile-up | 137 |
| 10.7 | Dead time | 139 |
| | 10.7.1 Pile-up losses | 140 |
| 10.8 | Beam switching | 140 |

## 11 Wavelength-dispersive analysis 142
| | | |
|---|---|---|
| 11.1 | Qualitative w.d. analysis | 142 |
| 11.2 | Principles of quantitative w.d. analysis | 143 |
| 11.3 | Peak selection | 145 |
| 11.4 | Pulse height analysis | 148 |
| 11.5 | Background corrections | 149 |
| | 11.5.1 Background 'holes' | 151 |
| 11.6 | Interferences | 152 |
| 11.7 | Counting strategy | 153 |
| 11.8 | Limits of detection | 154 |
| 11.9 | Choice of accelerating voltage | 155 |
| | 11.9.1 Measurement of accelerating voltage | 156 |
| 11.10 | Effect of conducting coating | 156 |
| | 11.10.1 Carbon contamination | 158 |
| 11.11 | Beam damage | 158 |
| 11.12 | Standards | 159 |

## 12 Energy-dispersive analysis 161
| | | |
|---|---|---|
| 12.1 | Introduction | 161 |
| | 12.1.1 Qualitative e.d. analysis | 161 |
| | 12.1.2 Mapping | 161 |
| | 12.1.3 Quantitative e.d. analysis | 162 |
| 12.2 | Energy calibration | 163 |
| | 12.2.1 Duane–Hunt limit | 164 |
| 12.3 | Peak integration | 165 |
| | 12.3.1 Peak shift and broadening | 165 |
| 12.4 | Peak overlap | 166 |
| 12.5 | Continuum modelling | 169 |
| | 12.5.1 Modifications to Kramers' law | 170 |
| | 12.5.2 Carbon reference spectrum | 172 |
| | 12.5.3 Absorption of the continuum | 172 |
| 12.6 | Digital filtering | 173 |
| 12.7 | Least squares fitting | 175 |
| | 12.7.1 Fitting multiple peaks | 176 |
| | 12.7.2 Fitting filtered spectra | 176 |
| | 12.7.3 Non-linear fitting | 177 |
| 12.8 | Filter-fit method – sources of error | 178 |
| | 12.8.1 Derivative references | 179 |
| 12.9 | Escape peak stripping | 180 |

|  |  | |
|---|---|---|
| 12.10 | Analysis at high count-rates | 180 |
| 12.11 | Peak to background ratio method | 181 |
| 12.12 | Standardless e.d. analysis | 182 |
| 12.13 | Precision and accuracy | 182 |
|  | 12.13.1 Detection limits | 183 |
| 12.14 | Energy- versus wavelength-dispersive analysis | 184 |
|  | 12.14.1 Combined e.d. and w.d. analysis | 185 |

## 13 X-ray generation and stopping power — 186

|  |  | |
|---|---|---|
| 13.1 | Introduction | 186 |
| 13.2 | Ionisation cross-section | 186 |
| 13.3 | Characteristic X-ray intensity | 188 |
| 13.4 | Bethe's stopping power expression | 190 |
|  | 13.4.1 Mean excitation energy | 190 |
|  | 13.4.2 Shell effects | 191 |
| 13.5 | Stopping power corrections | 193 |
|  | 13.5.1 Duncumb–Reed method | 194 |
|  | 13.5.2 Philibert–Tixier method | 195 |
|  | 13.5.3 Love–Cox–Scott method | 195 |
|  | 13.5.4 Pouchou–Pichoir method | 196 |
| 13.6 | Standardless analysis | 196 |
| 13.7 | Electron range | 197 |
| 13.8 | Lenard coefficient | 199 |
| 13.9 | Spatial resolution | 199 |

## 14 Electron backscattering — 202

|  |  | |
|---|---|---|
| 14.1 | Elastic scattering | 202 |
|  | 14.1.1 Single scattering | 202 |
|  | 14.1.2 Multiple scattering | 204 |
| 14.2 | Electron backscattering coefficient | 205 |
|  | 14.2.1 Empirical expressions for $\eta$ | 206 |
| 14.3 | Simplified backscattering models | 206 |
| 14.4 | Multiple reflection model | 207 |
| 14.5 | Transport equation | 207 |
| 14.6 | Monte-Carlo method | 208 |
|  | 14.6.1 Monte-Carlo models | 209 |
| 14.7 | Energy distribution of backscattered electrons | 210 |
| 14.8 | X-ray intensity loss | 211 |
|  | 14.8.1 Experimental determination of $\eta_x$ | 212 |
| 14.9 | Backscattering corrections | 212 |
|  | 14.9.1 Duncumb–Reed $R$ values | 212 |
|  | 14.9.2 Springer polynomial | 212 |
|  | 14.9.3 Love–Scott method | 213 |
|  | 14.9.4 Pouchou–Pichoir method | 214 |

|  |  |  |
|---|---|---|
| 14.10 | Non-normal electron incidence | 215 |
| 14.11 | Backscattered electron imaging and analysis | 216 |

## 15 Absorption corrections — 218
- 15.1 Introduction — 218
- 15.2 Experimental determination of $\phi(\rho z)$ and $f(\chi)$ — 220
- 15.3 Graphical correction methods — 222
- 15.4 'Phi-rho-$z$' models — 223
- 15.5 Philibert method — 224
  - 15.5.1 The effect of critical excitation potential — 226
  - 15.5.2 Variants of Philibert method — 226
  - 15.5.3 Other exponential models — 227
- 15.6 Rectilinear models — 227
- 15.7 Gaussian models — 229
- 15.8 Parabolic model — 231
- 15.9 Surface ionisation — 233
- 15.10 Non-normal electron incidence — 234
  - 15.10.1 Tilted specimens — 236
- 15.11 Surface roughness — 236
- 15.12 Mass attenuation coefficients — 237
- 15.13 Layered samples — 240

## 16 Fluorescence corrections — 242
- 16.1 Fluorescence excitation — 242
- 16.2 Theory of characteristic fluorescence corrections — 243
  - 16.2.1 The absorption term — 245
  - 16.2.2 Fluorescence excited by K$\beta$ radiation — 247
  - 16.2.3 Ternary and higher compounds — 248
- 16.3 K–K fluorescence corrections in practice — 248
- 16.4 Fluorescence involving L lines — 249
  - 16.4.1 L–L fluorescence corrections — 250
  - 16.4.2 K–L and L–K fluorescence corrections — 250
- 16.5 Continuum fluorescence — 251
  - 16.5.1 Continuum fluorescence in pure elements — 251
  - 16.5.2 Continuum fluorescence in compounds — 252
  - 16.5.3 Continuum fluorescence corrections in practice — 255
- 16.6 Fluorescence near phase boundaries — 257
  - 16.6.1 Fluorescence uncertainty — 261

## 17 Matrix corrections in practice — 262
- 17.1 Introduction — 262
- 17.2 Example of correction calculation — 263
  - 17.2.1 Absorption — 263
  - 17.2.2 Characteristic fluorescence — 264

## Contents

|  |  |  |
|---|---|---|
| | 17.2.3 Continuum fluorescence | 265 |
| | 17.2.4 Backscattering | 267 |
| | 17.2.5 Stopping power | 267 |
| | 17.2.6 Total correction factors | 268 |
| 17.3 | Iteration | 268 |
| | 17.3.1 Simple iteration | 268 |
| | 17.3.2 Hyperbolic iteration | 269 |
| | 17.3.3 Wegstein iteration | 270 |
| 17.4 | Alpha coefficients | 270 |
| 17.5 | Correction programs | 272 |

**18 Light element analysis** — **274**

| | | |
|---|---|---|
| 18.1 | Introduction | 274 |
| 18.2 | Long wavelength w.d. spectrometry | 276 |
| 18.3 | Low energy e.d. spectrometry | 278 |
| 18.4 | Peak intensity measurements | 280 |
| 18.5 | Background corrections | 281 |
| 18.6 | Conducting coating | 281 |
| 18.7 | Contamination | 282 |
| 18.8 | Absorption corrections | 283 |
| | 18.8.1 'Thin film' model | 285 |
| | 18.8.2 Monte-Carlo models | 286 |
| 18.9 | Specimen tilt | 287 |
| 18.10 | Mass attenuation coefficients | 287 |
| | 18.10.1 Experimental m.a.c. data | 287 |
| | 18.10.2 M.a.c. tables | 288 |
| | 18.10.3 Determining m.a.c.s from microprobe measurements | 288 |

**Appendix: Origin of characteristic X-rays** — **292**

| | | |
|---|---|---|
| A.1 | Atomic structure | 292 |
| A.2 | Characteristic X-ray emission | 293 |
| A.3 | Wavelengths and energies of X-ray lines | 296 |
| A.4 | Relative intensities | 300 |
| A.5 | Satellite lines | 300 |
| A.6 | Auger effect and fluorescence yield | 301 |
| A.7 | Coster–Kronig transitions | 301 |

*References* — 306

*Conference proceedings* — 319

*Index* — 322

# Acknowledgements

I am greatly indebted to P. Duncumb, A. Mackenzie and G. Love for constructive criticism of the manuscript. I am also very grateful to the following for assistance of various kinds: T. Bleser, A. Buckley and Meng Xing. Thanks are due to Oxford Instruments (Microanalysis Group) for sponsoring the colour plates. Finally I would like to express my gratitude to a host of colleagues and friends from whom I have learnt most of what is in this book, and especially to Jim Long, under whose guidance I started out many years ago.

# Preface

In 1951 Raymond Castaing, under the supervision of Professor A. Guinier of the University of Paris, wrote a doctoral thesis describing the successful development of a new instrument destined to have an impact in many fields of science and technology comparable with that of the electron microscope. The electron microprobe owed its existence to developments in electron optics making it possible to produce a focussed electron beam (or probe), which could be used to excite characteristic X-rays in solid specimens, with sufficient current in the beam to give useful X-ray intensities. This enabled chemical analysis to be carried out non-destructively with a spatial resolution of around 1 $\mu$m, as limited by the spreading of the beam in the sample.

The simplicity of characteristic X-ray spectra, their independence of physical and chemical state, and the regular dependence of the wavelength of the lines on atomic number make the identification of the elements present a straightforward matter. Quantitative analysis may be performed by comparing the line intensities with those from pure elements or standard compounds of known composition. Castaing developed a theoretical basis for the matrix (or 'ZAF') corrections required to allow for the dependence of the efficiency of X-ray production on composition. Matrix correction procedures have been subsequently refined and tested on a wide variety of samples. The accuracy of quantitative analysis is usually better than $\pm 2\%$ (relative) and detection limits are typically around 100 ppm in routine analysis, though 10 ppm can be approached in favourable circumstances. The volume analysed is of the order of a few cubic micrometres and the absolute elemental sensitivity is about $10^{-15}$ g.

This book provides comprehensive coverage of both theoretical and practical aspects of the subject for users in metallurgy and materials science, geology, biology, etc. The treatment emphasises physical principles, in relation to the design and operation of the instrument, analytical procedures and interpretation of results. Most of the material included is relevant not only to the conventional electron microprobe, but should also be of interest to users of scanning electron microscopes fitted with X-ray spectrometers for analytical purposes.

The physical basis of the technique is introduced in chapter 1. (A more detailed treatment of the physics of characteristic X-ray production is given in an appendix.) Chapters 2–5 are concerned with the design and operation of the electron column. X-ray spectrometers of the 'wavelength-dispersive' type are described in chapters 6–8, while the 'energy-dispersive' type is treated in chapters 9 and 10. Analytical procedures are covered in chapters 11 and 12, the former being devoted to w.d. analysis and the latter to e.d. analysis. Chapters 13–16 describe the physical basis for matrix effects and give methods for calculating 'ZAF' corrections. Chapter 17 covers computational aspects and includes a worked example of a correction calculation. Analytical procedures specific to 'light' elements (atomic numbers below 10) are covered in chapter 18.

Cambridge, February 1992                                                                 S.J.B.R.

# 1
# Introduction

## 1.1 Historical notes

Characteristic X-rays were discovered by Barkla and Sadler (1909), as a result of studies of fluorescence spectra, and Kaye (1909) who excited pure element spectra by electron bombardment, using filters to obtain wavelength discrimination. The invention of the crystal diffraction spectrometer (Bragg and Bragg, 1913) enabled Moseley (1913, 1914) to resolve the spectra better and establish the relationship between wavelength and atomic number known as Moseley's law. During the 1920s and 1930s X-ray spectroscopy attracted the attention of some of the leading physicists of the day, because of its relevance to atomic structure, and characteristic spectra were investigated in great detail. Also during this period chemical analysis by means of electron-excited X-ray spectra became a practical possibility.

Developments in electron optics which accompanied the evolution of the electron microscope in the 1930s enabled finely focussed electron beams to be produced. The principle of electron microprobe analysis, defined as chemical analysis using characteristic X-ray spectra excited by a focussed electron beam, was first described Hillier (1947).

The practical development of microprobe analysis was undertaken in 1948 as a Ph.D. project by R. Castaing, under the supervision of Professor A. Guinier at the University of Paris. Using a converted electron microscope Castaing obtained a current of a few nanoamps in a beam of 1 $\mu$m diameter. He realised that the resolution with a solid sample would not be better than about 1 $\mu$m, however small the beam diameter, because of the penetration and scattering of the incident electrons. At the time this seemed a serious restriction, but it has not, in fact, prevented the technique from finding a wide range of applications.

Castaing proceeded to fit an X-ray spectrometer, and in 1950 carried out experiments on diffusion couples to see if compositional differences could be detected. According to his own account (Castaing, 1967), the lack of an optical microscope and the instability of the instrument made it hard to tell if changes in X-ray intensity were really related to composition. However, on removing the specimen he found contamination spots produced by the electron beam clustered around the boundary of the couple. In his own words 'This was a revelation ... I realised that until then I had never really believed in electron probe microanalysis: henceforth I would believe in it.' Castaing also laid the foundations of the theory and practice of quantitative analysis in his Ph.D. thesis of 1951. An electron microprobe was also developed independently in the USSR, at about the same time (Borovskii, 1953).

The first commercial instrument, developed in France by the Cameca Company, appeared in 1958. Other early commercial instruments were produced by AEI Ltd in the UK (Page and Openshaw, 1960) and ARL Inc. in the USA (O'Brien, 1963). In 1960 the Cambridge Instrument Company produced the 'Microscan', an instrument with facilities for scanning an electron beam in a TV-like raster in order to produce element distribution images, based on the work of Cosslett and Duncumb (1956) and Duncumb and Melford (1960). Five years later the same company produced the first commercial scanning electron microscope ('Stereoscan'), though an experimental instrument had been constructed much earlier, by Knoll (1935). Soon afterwards, the JEOL company in Japan also produced a scanning electron microscope, followed by several other manufacturers.

A subsequent event of major importance was the development (in nuclear physics laboratories) of the solid-state lithium-drifted silicon, or Si(Li), detector. Fitzgerald, Keil and Heinrich (1968) reported attaching a Si(Li) detector to an electron microprobe and obtaining qualitative results, while Tenny (1968) demonstrated the feasibility of quantitative energy-dispersive (e.d.) analysis with such a detector, using least-squares peak fitting to overcome the poor resolution.

Dramatic improvements in resolution occurred over the next year or two and thinner entrance windows enabled low energy X-rays to be detected. The commercial availability of e.d. systems of good performance led to the rapid adoption of the technique from 1970 onwards. As a result of the development of Si(Li) detectors, either completely 'windowless' or having ultra-thin windows, the scope of e.d. analysis has been extended to include light elements such as O, C, B and even Be. It is interesting to note that e.d. analysis was first applied to light element analysis, using a thin-window gas-filled counter (Dolby and Cosslett, 1960).

## 1.2 Principles of electron microprobe analysis

In electron microprobe analysis, electron bombardment generates X-rays in the sample to be analysed. From the wavelength (or photon energy) and intensity of the lines in the X-ray spectrum, the elements present may be identified and their concentrations estimated. The use of a finely focussed electron beam enables a very small selected area to be analysed.

*Qualitative* analysis (identifying the elements present) entails recording the spectrum by means of an X-ray spectrometer, over the range of wavelengths or energies within which relevant lines may be present (§11.1). Lines are identifiable by reference to tables (see §A.3). In *quantitative* analysis the intensities of the X-ray lines from the specimen are compared with those from standards of known composition. The measured intensities require certain instrumental corrections, including subtraction of background, the chief source of which is the continuous X-ray spectrum (§1.6). The composition at the analysed point is calculated from the corrected intensities by applying 'matrix corrections' which take account of the various factors governing the relationship between intensity and composition (§1.10). These are commonly applied in the form of 'ZAF corrections', with separate correction factors dependent on atomic number, absorption and fluorescence (chapters 13–16).

The incident electrons typically have a kinetic energy of 10–30 keV (1 eV, or electron volt, is the energy associated with a change in potential of an electron of 1 volt), and penetrate the sample to a depth of the order of 1 $\mu$m, spreading out laterally to a similar distance. This imposes a lower limit to the analysed volume and hence the spatial resolution. Improving the resolution by reducing the electron energy is generally impracticable because the electrons must have enough energy for efficient X-ray excitation.

The essential features of the instrument are described in chapter 2, where related techniques (e.g. scanning electron microscopy) are also discussed. In chapters 3–10 the design and operation of the various parts of the instrument are treated in detail. Other topics covered are the use of energy-dispersive (e.d.) detectors for quantitative analysis (chapter 12) rather than wavelength-dispersive (w.d.) spectrometers (chapter 11), and the special considerations involved in analysing for 'light' elements with atomic number less than 10 (chapter 18).

An understanding of the physics of X-ray production is important for an appreciation of the limitations of quantitative analysis, and for the correct choice of instrumental operating conditions and analytical procedures.

## 1.3 Characteristic X-ray spectra

X-ray emission lines are produced by transitions between inner atomic electron energy levels. For such a transition to be possible, a vacancy must first be created by the ejection of an inner electron: in electron microprobe analysis the required inner level ionisation is produced by bombardment with electrons (fig. 1.1). If the initial ionisation is in the innermost atomic 'shell' (the K shell), the resulting X-ray emission is identified as K radiation. The K spectrum contains several lines due to transitions from different levels in the L, M, etc. shells, which contain electron orbits of increasing mean radius and decreasing binding energy. The L shell consists of three subshells ($L_{1-3}$), while the M shell includes five ($M_{1-5}$). The principal K

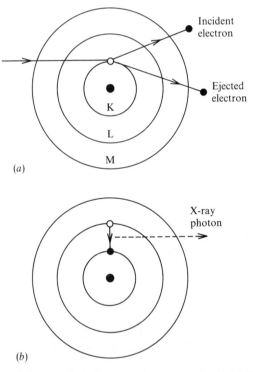

Fig. 1.1 Production of characteristic X-rays: (a) removal of inner (K shell) electron by electron bombardment; (b) emission of X-ray photon as result of electron transition between shells.

## 1.3 Characteristic X-ray spectra

line, designated $K\alpha_1$, is produced by an $L_3$–K transition, while the $K\alpha_2$ line results from an $L_2$–K transition. The only other important K line is $K\beta_1$ ($M_3$–K).

The quantum energy of the emitted radiation is equal to the difference between the potential energy of the atom in its initial and final states (e.g. ionised in the K and $L_3$ shells respectively in the case of the $K\alpha_1$ line). The relationship between the photon energy $E$ in electron volts and the wavelength $\lambda$ in Å (1 Å = $10^{-10}$ m) is: $E = 12\,396/\lambda$.

X-ray line emission is known as 'characteristic radiation' because the wavelength of the lines is specific to the emitting element. The energy of a particular line increases smoothly with the atomic number of the emitting atom, owing to the increasing binding energy of the inner levels. Since only inner electrons are involved, the wavelengths of characteristic lines are practically independent of the physical and chemical state of the emitter.

L and M spectra are more complicated than K spectra, as shown in fig. 1.2, because of the existence of subshells. The transitions giving rise to the more significant K, L, and M lines are given in §A.1, together with a more

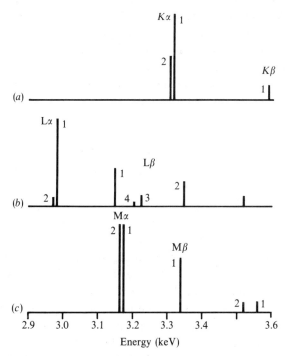

Fig. 1.2 Principal K, L and M lines of: (a) K ($Z=19$), (b) Ag ($Z=47$) and (c) U ($Z=92$), showing approximate relative intensities.

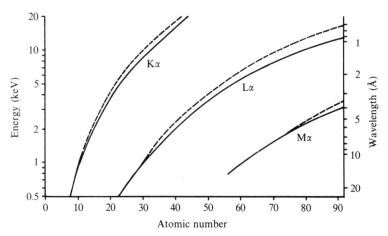

Fig. 1.3 Energy and wavelength of the most important characteristic X-ray lines as function of atomic number; dashed lines indicate critical excitation energies.

detailed account of the origin and nature of characteristic spectra. Tables of energies and wavelengths are given in §A.3.

Fig. 1.3 shows the variation of the energy and wavelength of the most important characteristic lines with atomic number. The X-ray spectrometers used in microprobe analysis operate mainly in the range 1–10 keV (1.2–12 Å), within which the Kα lines of elements of atomic number 11–32 occur. For higher atomic numbers L or M lines are used.

## 1.4 Inner-shell ionisation

Ionisation of inner electron shells may be caused by either electrons or X-rays with energy greater than the 'critical excitation energy' ($E_c$) of the shell concerned, which is the energy required to raise an electron from that shell to the first vacant energy level. Multiple shells (L, M, etc.) have separate critical excitation energies for each subshell. Fig. 1.3 shows the dependence on atomic number ($Z$) of $E_c$ for the K, $L_3$, and $M_5$ shells, ionisation of which gives rise to the Kα, Lα, and Mα lines respectively.

In the electron microprobe the incident electron energy ($E_0$) in electron volts is equal to the accelerating potential (in volts) applied to the electron gun. In order to excite a particular line, this energy must exceed the critical excitation energy of the appropriate shell. Furthermore, to generate a reasonable intensity, $E_0$ should preferably be at least twice $E_c$. To satisfy this condition an accelerating voltage of 10–30 kV is generally used.

The efficiency of ionisation by electron bombardment is rather low: typically only one electron per thousand produces a K shell vacancy. Most of the energy of the incident electron beam is dissipated in interactions which ultimately produce heat. However, despite this and the low collection efficiency of X-ray spectrometers, adequate X-ray photon emission rates are obtained because of the very large number of electrons arriving per second (a typical current of 10 nA corresponds to $6 \times 10^{10}$ electrons s$^{-1}$).

## 1.5 Fluorescence yield

The ionisation of an inner shell may be followed by a 'radiationless' transition rather than one resulting in the emission of an X-ray photon. This is known as the Auger effect. In such events the energy is used to eject another electron from the atom. The 'fluorescence yield' ($\omega$) is the probability of a radiative transition, or the fraction of ionisations of a specified shell that result in characteristic X-ray emission, and is independent of the method of ionisation. The fluorescence yield of the K shell ($\omega_K$) increases rapidly with atomic number, as shown in fig. 1.4. The effect of this is counteracted, however, by a decrease in the rate of production of K shell vacancies with increasing atomic number.

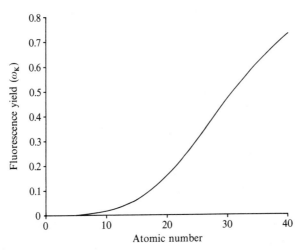

Fig. 1.4 Fluorescence yield of K shell ($\omega_K$) as function of atomic number (after Krause (1979)).

## 1.6 Continuous X-ray spectrum

Electron bombardment generates not only characteristic X-ray lines but also a 'continuous spectrum' or 'bremsstrahlung' ('braking radiation'), consisting of photons emitted by electrons decelerated in collisions with atoms. The continuous spectrum, or 'continuum', extends up to a limiting energy (the Duane–Hunt limit) equal to the incident electron energy.

The distribution of continuum photons as a function of energy is given approximately by the following expression derived by Kramers (1923) from classical theory:

$$N_c(E) = aZ(E_0 - E)/E, \qquad (1.1)$$

where $N_c(E)$ is the intensity in photons per second per unit energy interval, per incident electron. The constant $a$ has a value of about $2 \times 10^{-9}$ photons $s^{-1}$ $eV^{-1}$. According to this expression the intensity of the continuum is proportional to the atomic number ($Z$), but the shape is the same for all target materials. In practice the observed continuum intensity is reduced at low energies by absorption of the radiation emerging from the targets. This, together with falling spectrometer efficiency, causes a sharp decrease in intensity below about 1 keV (fig. 1.5), compared to the monotonic increase predicted by equation (1.1).

In electron-excited X-ray spectra, the continuum forms a 'background' upon which the characteristic lines are superimposed. Fig. 1.5 shows the general features of a typical X-ray spectrum plotted as a function of photon energy. When plotted as a function of wavelength, which is sometimes more appropriate, the spectrum appears as in fig. 1.6.

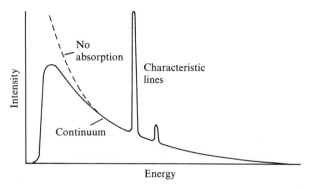

Fig. 1.5 X-ray spectrum as function of photon energy, showing effect of absorption at low energies.

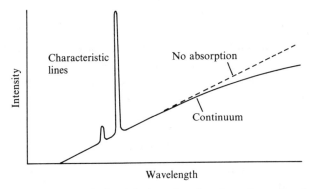

Fig. 1.6 Spectrum in fig. 1.5 plotted as function of wavelength.

## 1.7 X-ray absorption

By far the most important absorption mechanism for X-rays in the energy range of interest in microprobe analysis is the photoelectric effect, in which the whole energy of the incident X-ray photon is used to eject an atomic electron. Compton scattering, in which X-rays are scattered with loss of energy, is insignificant in the present context, though coherent (Rayleigh) scattering (involving a change of direction but no energy loss) is not necessarily negligible, especially for heavy elements. When only the undeflected beam is taken into consideration, coherent scattering causes the apparent absorption to be slightly greater than that due to photoelectric absorption alone.

The power of a material to absorb X-rays may be expressed in terms of the 'mass attenuation coefficient' ($\mu$), which is used in the expression:

$$I' = I \exp(-\mu\rho x), \tag{1.2}$$

where $I$ and $I'$ are respectively the intensity of a collimated X-ray beam before and after passage through a thickness $x$ of material of density $\rho$. The mass attenuation coefficient, or m.a.c., (usually expressed in units of cm² g⁻¹) is independent of the physical and chemical state of the absorber. This quantity is also commonly called the mass *absorption* coefficient. (In the earlier literature the symbol $\mu$ usually refers to the *linear* attenuation or absorption coefficient, the mass attenuation coefficient being written as $\mu/\rho$.) For a compound the m.a.c. may be calculated by summing the contributions of the constituent elements:

$$\mu = \sum C_i \mu_i, \tag{1.3}$$

where $C_i$ is the mass concentration of the $i$th element, with mass attenuation coefficient $\mu_i$.

Owing to the strong dependence of $\mu$ on atomic number, the absorption suffered by characteristic radiation emerging from a sample in the electron microprobe is highly dependent on composition. For quantitative analysis an absorption correction is therefore required.

## 1.8 Fluorescence

Fluorescence is the emission of characteristic X-rays from atoms which have absorbed X-rays by the photoelectric effect, as described above. X-ray absorption is thus an alternative mode of producing the inner-shell vacancies which precede characteristic X-ray emission. In microprobe analysis the absorption of the primary X-rays (both characteristic and continuum) produced by electron bombardment within the sample itself gives rise to the emission of secondary fluorescent X-rays. This causes enhancement of the measured X-ray intensity, which requires correction.

## 1.9 Relationship between X-ray intensity and elemental concentration

The reason that characteristic X-ray intensities are, to a first approximation, proportional to *mass* concentration (whereas *atomic* concentration might appear more reasonable) is related to the fact that incident electrons penetrate an approximately constant mass in materials of different composition. This is because these electrons lose their kinetic energy mainly through interactions with orbital electrons of the target atoms, the number of which is approximately proportional to atomic mass.

The consequences of this can be demonstrated as follows. Consider two elements, A and B, the latter being 'heavier' than the former. To determine the concentration of A in a sample containing a mixture of A and B, one compares the intensity of 'A' radiation emitted by the sample with that from pure A. For the sake of simplicity we assume that the sample contains equal numbers of A and B atoms. Fig. 1.7 shows diagrammatically the volumes excited in pure A (*a*) and the A–B compound (*b*), given that the masses excited are equal, as noted above. The number of atoms excited in pure A is greater than in the compound because of the presence of heavy B atoms in the latter. Consequently the ratio of the numbers of excited A atoms in sample and standard, and hence the X-ray intensity ratio, is less than 0.5 (the atomic concentration of A in the compound).

The following argument shows that this ratio is, in fact, equal to the mass

## 1.10 Matrix corrections

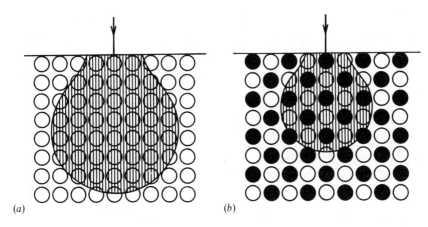

Fig. 1.7 Excited volumes in (a) pure A and (b) A–B compound containing equal numbers of A and B atoms (represented by open and filled circles respectively). The number of atoms excited in (b) is less than in (a) because the atomic weight of B is greater than A.

concentration (given the assumptions stated above). If the atomic concentration of A is $n_A$, then the mass concentration is given by $C_A = n_A A_A / [n_A A_A + (1 - n_A) A_B]$, where $A_A$ and $A_B$ are the atomic weights of A and B respectively. The number of atoms excited in the pure A standard is equal to $Nm/A_A$, where $N$ is Avogadro's number and $m$ the mass penetrated by the incident electrons. In the case of the compound, the number of A atoms in the excited volume is $n_A Nm/[n_A A_A + (1 - n_A) A_B]$. The X-ray intensity ratio (which is proportional to the ratio of the numbers of excited atoms) is given by this expression divided by $Nm/A_A$, which is equal to the expression given above for $C_A$. In reality the intensity ratio is not exactly equal to the mass concentration, firstly because $Z/A$ is not constant and secondly because the 'stopping power' of all bound electrons is not the same. This, together with other factors which affect the measured intensities, gives rise to the need for matrix corrections, as described below.

## 1.10 Matrix corrections

The relationship between the intensity of an X-ray line and the concentration of the element concerned depends on the composition of the sample. 'Matrix corrections' are used to convert specimen/standard intensity ratios into concentrations. The 'uncorrected concentration' ($C'$) of each element may be calculated, given the concentration ($C_0$) of that element in the standard, from the expression:

$$C' = C_0(I/I_0), \tag{1.4}$$

where $I$ and $I_0$ are the specimen and standard intensities respectively, corrected for instrumental effects. The ratio $I/I_0$ is commonly known as the '$k$ ratio'. Matrix corrections may be applied by multiplying the specimen intensity by a factor $F$ and the standard intensity by $F_0$. Hence from equation (1.4):

$$C = C'(F/F_0), \tag{1.5}$$

where $C$ is the true concentration in the specimen. The concentrations in the above discussion are mass concentrations (see §1.9).

The phenomena upon which matrix corrections depend are:

(1) absorption of characteristic X-rays emerging from the specimen;
(2) enhancement of the characteristic X-ray intensity due to fluorescence by other lines and the continuum;
(3) loss of X-ray intensity owing to incident electrons being backscattered out of the specimen;
(4) variation in the efficiency of X-ray production, as governed by the 'stopping power' of the specimen (a function of atomic number).

These effects may be represented by separate factors: $F_a$ (absorption), $F_f$ (fluorescence), $F_b$ (backscattering) and $F_s$ (stopping power). The overall correction factor $F$ in equation (1.5) is given by the product of these individual factors:

$$F = F_a F_f F_b F_s. \tag{1.6}$$

The combined backscattering and stopping power corrections (both of which are dependent on atomic number), together with the absorption and fluorescence factors, comprise the 'ZAF' correction (Philibert and Tixier, 1968a). Methods of calculating the four factors in equation (1.6) are described in chapters 13–16. Since they depend on composition, an iterative procedure is applied, whereby approximate concentrations are used to calculate correction factors, which are then used to derive improved estimates for the concentrations. The true composition is thus obtained by successive approximations.

# 2
# Essential features of the electron microprobe

## 2.1 Probe-forming system

The usual source of electrons is an electron gun in which a tungsten 'hairpin' filament is heated to about 2700 K in order to obtain thermionic emission. The filament is held at a negative potential (typically 10–30 kV) which accelerates the electrons through an aperture in the earthed anode plate. High vacuum ($10^{-4}$ Torr or better) is required in order to prevent oxidation of the hot tungsten filament, breakdown of the accelerating voltage, and scattering of the electrons in the beam.

Magnetic lenses are used to focus the beam into a fine probe incident on the surface of the specimen. The electron lenses produce a demagnified image of the source on the surface of the specimen (fig. 2.1). For microprobe analysis a probe diameter of 0.2–1 $\mu$m is typical, with a current of 1–100 nA. In order to obtain the required demagnification two lenses are needed, though three may be used if high resolution scanning images are required.

The current available in a probe of given diameter is dependent on the characteristics of the probe-forming system, especially the final lens. Optimum performance is obtained with the lens as close to the specimen as possible, but since this conflicts with optical viewing and paths to the X-ray spectrometers, a compromise is necessary.

## 2.2 X-ray spectrometers

In microprobe analysis the X-ray spectrum is recorded with either a 'wavelength-dispersive' (w.d.) or 'energy-dispersive' (e.d.) spectrometer (fig. 2.1). The former utilises a diffracting crystal which acts as a monochromator, selecting one wavelength at a time, depending on the angle of incidence of the X-rays. The crystal is curved so that the angle subtended at the point source is constant. Geometrical requirements

14    *2 Essential features of the electron microprobe*

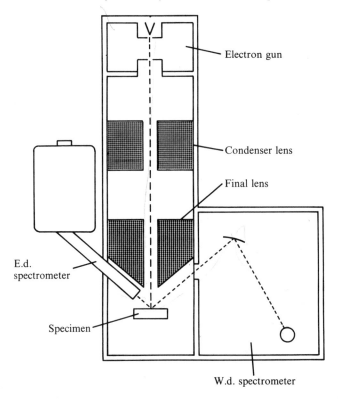

Fig. 2.1 Schematic diagram of electron microprobe.

dictate that the X-ray detector must be the same distance from the crystal as the crystal is from the X-ray source, at all angles of incidence. This necessitates a fairly complicated mechanism, which must be precise and reproducible. Most instruments have two or more spectrometers with crystals covering different wavelength ranges, contained in a chamber which is evacuated in order to prevent absorption of the X-rays in air (high vacuum is not essential – a pressure of about $10^{-1}$ Torr is sufficient). Each spectrometer has associated electronics for amplifying and counting the pulses from the detector. In present-day instruments spectrometer functions are controlled by a computer.

E.d. spectrometers employing solid-state X-ray detectors complement and to some extent have superseded w.d. spectrometers. The e.d. spectrometer records the whole spectrum simultaneously, electronic pulse height analysis being used to sort the pulses produced in the detector according to X-ray energy. This technique has many advantages, but the resolving power obtainable is substantially inferior to that of the w.d. spectrometer.

## 2.3 Specimen stage

The specimen stage usually holds several specimens and standards. The specimens are either round or rectangular with dimensions typically of the order of 2–3 cm. Standards may either be mounted individually in small mounts, or grouped in normal-sized mounts. For special purposes extra-large specimen stages are available: for example Matsuya *et al.* (1988) describe a holder with a working area of $30 \times 30$ cm. In this case the specimen chamber obviously must be much larger than normal. The specimen holder is usually electrically isolated and can be connected to a meter so that the current in the incident beam may be measured. The time taken to change specimens is reduced if they are introduced via an air-lock which can be isolated from the rest of the vacuum chamber.

For microprobe analysis flat polished specimens are required and the holder should be designed so that the front surface of the specimen is located in a fixed plane. (This contrasts with the usual arrangement in scanning electron microscopes, where the back of the specimen is attached to the holder and the level of the front surface depends on the thickness of the specimen.) Orthogonal $x$ and $y$ movements are required, with a means of reading the coordinates. It is usual to fix the focus of the optical microscope and use a fine mechanical adjustment of the specimen stage in the $z$ direction (normal to the surface) for focussing. This ensures that the position of the X-ray source is constant, which is especially important with w.d. spectrometers. In the present generation of instruments, computer control of the $x$, $y$ and $z$ movements is the rule. This enables large numbers of points to be analysed without intervention of the operator, using previously stored coordinates. Positional reproducibility to better than 1 $\mu$m is desirable.

In biology it is useful to be able to analyse samples in a frozen state in order to retain water. For this purpose special stages cooled with liquid nitrogen have been developed, as described for example by Taylor and Burgess (1977).

## 2.4 Optical microscope

The purpose of the optical microscope is to enable the operator to find areas of interest in the specimen and to relate the analysed points to the visible microstructure. It is also useful for observing cathodoluminescence (emission of light stimulated by electron bombardment), which gives a convenient indication of the location of the point of impact of the beam. For reconnaissance, a low magnification (less than 100) is preferable, whereas for observing fine detail a higher power (250–500) is required. For some kinds of specimen transmitted light illumination and polarised light facilities are useful. A television camera can be attached to the optical microscope in order to provide an image on a video screen, which has considerable advantages.

Since both light and electron lenses subtend a large solid angle and preferably operate at normal incidence, it is difficult to optimise both. One commonly used solution is to employ a reflecting microscope objective with a central hole to allow passage of the electron beam (fig. 2.2(*a*)). This has the advantage that optical viewing and analysis can be carried out simultaneously. However, the optical performance tends to be somewhat inferior to that of a conventional refracting microscope, and usually only a single magnification is available.

Normal incidence viewing with refracting optics while analysis is in progress is possible only if the electron beam is incident on the specimen at an inclined angle (fig. 2.2(*b*)). Interchangeable objectives of different power can easily be fitted, but objectives of the highest numerical aperture are

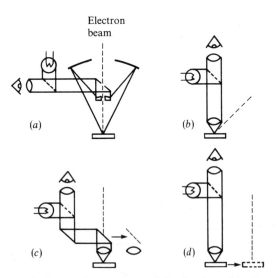

Fig. 2.2 Optical viewing systems: (*a*) coaxial reflecting objective; (*b*) refracting objective with inclined electron incidence: (*c*) removable refracting objective; (*d*) refracting objective with specimen translation.

incompatible with access to the specimen by the electron beam at a reasonable angle (not less than 45°), which somewhat limits the performance at high magnifications. A disadvantage of non-normal electron incidence is that correction methods for quantitative analysis are less well developed than for normal incidence, though there have been considerable advances in this regard in recent years.

Another way of using a refracting objective is to make it removable to allow the electron beam to reach the specimen (fig. 2.2(c)). Viewing during analysis is sacrificed, but if the mechanical translation of the objective is sufficiently reproducible this is not a serious drawback. Objectives of different powers are easily incorporated, and as there is no restriction on numerical aperture the performance at high magnifications is good. However, the extra space required between the final electron lens and the specimen to accommodate the optical objective is undesirable from an electron optical viewpoint.

An alternative system is specimen translation (fig. 2.2(d)), which permits ideal conditions for both light and electron optics. However, it is difficult to achieve the desired degree of reproducibility in the translation mechanism. As with the preceding system, viewing and analysis cannot be carried out simultaneously.

## 2.5 Scanning

By scanning the probe in a television-like raster and modulating the brightness of a cathode-ray tube with the X-ray spectrometer output, a scanning image showing the spatial distribution of a selected element may be produced (fig. 2.3). The beam is deflected by means of electromagnetic coils driven by a 'sawtooth' waveform generator which also supplies a synchronous signal to the display. The magnification of the scanning image is controlled by varying the amplitude of the raster scanned by the probe on the surface of the specimen.

The image is made up of spots, each one produced by a pulse occurring on the arrival of an X-ray photon in the counter. With a typical signal of up to a few thousand pulses per second the image is quite 'noisy'. In order to produce an image of reasonable quality the usual procedure is to photograph the screen with an exposure of several minutes. An alternative is to use a computer-based image store, which has the advantage that the image can be monitored during accumulation.

The image obtained by using the signal from an electron detector is much less noisy but does not give information on the distribution of specific

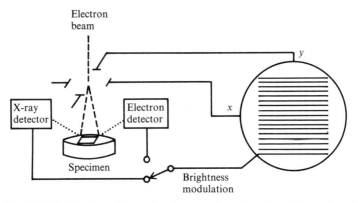

Fig. 2.3 Production of scanning image using electron or X-ray signal.

elements. In microprobe analysis electron images are employed mainly for inspection of the specimen prior to analysis and for correlating surface features with X-ray images. A backscattered electron (b.s.e.) image can be used to display differences in mean atomic number.

It is sometimes advantageous to make use of colour. For example, a composite X-ray image can be produced, in which a different colour is assigned to each of several elements. Alternatively different ranges of concentration of a single element may be assigned arbitrary ('false') colours, in order to enhance the visibility of compositional variations.

Sometimes it is useful to scan in one direction only. For example, a fast line-scan using the electron signal provides a convenient means of focusing the probe, the final electron lens being adjusted to give the maximum detail in the trace. A slow line-scan may be used to display the concentration profile of a selected element, using the signal from the X-ray spectrometer to deflect the spot in the vertical direction. Such a profile gives more detailed information than a scanning image, but only in one dimension.

## 2.6 Vacuum system

Conventional high vacuum technology is generally used in the electron microprobe. Rough evacuation is carried out with a rotary mechanical pump, and usually an oil diffusion pump backed by the rotary pump is used to achieve working pressure, which is typically $10^{-5}$ Torr or less. Other types of pump sometimes used include turbomolecular and ion pumps, which have the advantage of giving a 'cleaner' vacuum, with less contamination from hydrocarbons etc.

## 2.6 Vacuum system

Automatic operation of the vacuum system can be achieved by means of relays linked to the pressure gauges. Such a system should be designed to respond appropriately in the event of such contingencies as failure of the mains supply or cooling water for the diffusion pump. It is also desirable to have interlocks so that the electron gun supplies cannot be switched on unless the pressure is correct, and air cannot be admitted inadvertently with the gun on. These requirements are satisfied in present-day instruments by placing the vacuum system under the control of a suitably programmed microprocessor.

The point of impact of the electron beam on the specimen becomes contaminated by carbon from decomposed hydrocarbons, originating from diffusion pump oil etc. Backstreaming from the diffusion pump may be reduced by baffles between pump and vacuum chamber, at the expense of decreased pumping speed. Baffle efficiency can be enhanced by cooling with water or liquid refrigerant, or thermoelectrically. In the interests of 'clean' vacuum it is also desirable to have a trap in the backing line containing an adsorbent material.

A 'cold finger' cooled by liquid nitrogen placed close to the specimen is sometimes used to reduce contamination. Hydrocarbons condense on the finger and their partial pressure near the specimen is reduced. The finger must be warmed to room temperature if exposed to air, to avoid condensation of atmospheric water. Campbell and Gibbons (1966) observed that a copper cylinder cooled by liquid nitrogen and almost completely surrounding the specimen reduced the contamination rate by a factor of 1000, though in normal practice the solid angle subtended by the cold surface is considerably less and its efficiency is therefore lower.

Alternatives to hydrocarbon oils for use in diffusion pumps have become increasingly popular. Silicone oils are undesirable because of their tendency to produce non-conducting contamination layers and a spurious Si peak in the X-ray spectrum. A better choice is a perfluorinated polyether, which has the advantage of not polymerising under electron bombardment (Holland, Laurenson, Baker and Davis, 1972; Conru and Laberge, 1975). Such fluids are also resistant to oxidation in the event of accidental admittance of air to the pump when hot.

Contamination rates may be greatly reduced by using a turbomolecular pump, which relies on rapidly rotating vanes for its pumping action. Hydrocarbon levels associated with these pumps are very low. Vibration is potentially a problem in high resolution scanning instruments, but can be minimised by the use of a bellows mounting. Compared to the diffusion pump, initial and running costs (e.g. for bearing replacement) are higher.

The use of a cryogenic pump has been described by Venuti (1983). This has some advantages, including high pumping rate and low maintenance; also there is no need for continuous backing with a rotary pump. Compared to the diffusion pump, low specimen contamination rates are obtainable without the need for an additional cold trap.

For most purposes it is considered desirable to have as low a pressure as practicable in the specimen chamber. However, Robinson and Robinson (1978) have described the use of a pressure of around $10^{-1}$ Torr (obtainable with a rotary pump alone). The specimen chamber is separated from the column by a small aperture, enabling high vacuum to be maintained in the latter. This allows the analysis of uncoated insulators and materials with a relatively high vapour pressure. There is some loss of spatial resolution due to electron scattering, and only semi-quantitative X-ray analysis is possible, but this approach is useful where rapid sample throughput is important.

## 2.7 Related techniques

The following techniques are related to electron microprobe analysis in that they use either electron beams or X-ray spectrometry.

### *2.7.1 Scanning electron microscopy*

The scanning electron microscope (s.e.m.) shares many common features with the electron microprobe, but is designed primarily for producing high resolution topographic electron images, where its high resolution and depth of field give it a great advantage over the optical microscope.

In recent years electron microprobe and s.e.m. designs have tended to converge, with each instrument combining to a considerable degree the capabilities of the other. This has come about through the addition of X-ray spectrometers to the s.e.m. and by improving the imaging capabilities of the electron microprobe. The s.e.m. typically has a spatial resolution of better than 5 nm, but for X-ray analysis the resolution is limited to about 1 $\mu$m (for samples of average density) because of electron scattering in the specimen. The differences in design are a matter of emphasis: in the microprobe the X-ray spectrometers and optical microscope are given priority, while in the s.e.m. these take second place to imaging facilities. Also in the microprobe normal electron incidence is used as a rule, whereas in the s.e.m. the specimen is often tilted, in order to optimise the image contrast.

The conversion of scanning microscopes into analytical instruments has been encouraged by the advent of e.d. X-ray spectrometers, which have higher X-ray collection efficiency than w.d. spectrometers and hence can be used at lower probe currents. Furthermore, they are easier to fit.

For further details about scanning electron microscopy, see Goldstein *et al.* (1981) and Reimer (1986).

### 2.7.2 Analytical electron microscopy

The resolution limit imposed by the penetration and scattering of electrons in a solid sample may be overcome by using a thin specimen. Microchemical analysis of such specimens is made possible by adding an X-ray spectrometer to a transmission electron microscope (t.e.m.). E.d. detectors are ideal for this purpose on account of their high collection efficiency.

Transmission electron microscopes did not, at first, lend themselves readily to efficient X-ray detection, owing to the massive electron lens surrounding the specimen. Furthermore, they were not designed with a small probe diameter in view. These factors led Duncumb (1963) to develop a specialised instrument ('EMMA'), a version of which was produced commercially by AEI Ltd. In recent years t.e.m. manufacturers have produced instruments increasingly oriented towards analytical usage. These are described as 'analytical electron microscopes' and frequently have provision for other analytical techniques as well, e.g. electron energy loss spectroscopy.

Another possibility is to adapt a conventional electron microprobe to enable transmission electron images to be obtained, by adding lenses below the specimen (Rouberol, Tong, Conty and Deschamps, 1967). However, this approach has proved less popular.

For a detailed treatment of analytical electron microscopy, see Joy, Romig and Goldstein (1986).

### 2.7.3 Scanning transmission electron microscopy

An alternative technique for exploiting the high spatial resolution possible with thin specimens is scanning transmission electron microscopy (s.t.e.m.), in which a finely focussed beam is scanned as in the ordinary scanning microscope, but the image is produced by electrons transmitted through the sample. The potential of this technique has been enhanced by the development of intense field emission electron sources (§3.6.2), which enable images to be obtained with a probe less than 1 nm in diameter.

In scanning transmission microscopy the sample may be thicker than in ordinary transmission electron microscopy, because an image can be produced with electrons that have lost considerable amounts of energy, whereas in the t.e.m. this is impossible owing to chromatic aberration in the image-forming lenses. Energy analysis of the transmitted electrons enables different kinds of image to be produced. Scanning transmission microscopy can readily be combined with microanalysis by X-ray emission, as in the analytical t.e.m.

### 2.7.4 Auger analysis

A certain proportion of inner-shell ionisations result in the ejection of a bound electron, rather than the emission of a characteristic X-ray photon (§1.5). These 'Auger electrons' have discrete energies governed by the energy levels of the atom concerned, hence elemental analysis is possible with the aid of an electron energy spectrometer. Auger analysis is complementary to conventional electron probe analysis, being most sensitive for light elements, which have high Auger electron (and low X-ray) yields. It is essentially a surface analysis technique, since only those Auger electrons which originate from very shallow depths, and have suffered negligible energy loss in emerging from the sample, are recorded in the peaks. A related technique is photoelectron spectroscopy, in which the energy spectrum of electrons ejected as a result of irradiation by X-rays is recorded.

### 2.7.5 Proton probe analysis

The advantage of proton induced X-ray emission (p.i.x.e.) is the much lower continuum background (a consequence of the greater mass of the proton), which results in lower detection limits. Proton microbeams down to less than 1 $\mu$m in diameter are attainable. Energies in the range 1–5 MeV are necessary to obtain reasonable X-ray intensities. Such protons penetrate several tens of micrometres in solids, hence for thick samples the spatial resolution with respect to depth is relatively poor.

Further details about this technique can be found in the book by Johansson and Campbell (1988).

### 2.7.6 X-ray fluorescence analysis

In this technique characteristic X-rays are excited in the sample by exposure to an intense beam of X-rays. The emitted X-ray spectra are very similar to those produced by electron bombardment, but the background level is

## 2.7 Related techniques

much lower, owing to the absence of bremsstrahlung. Lower detection limits are thus attainable and the technique is widely used for both major and trace element analysis. With conventional X-ray sources, only bulk analysis is possible, but analysis on a microscale is feasible with very intense synchrotron sources, which can produce sufficient intensity in a beam only a few microns across.

Further information on X-ray fluorescence analysis can be found in the book by Williams (1987).

# 3
# Electron guns

### 3.1 Conventional triode gun

The triode gun generally used in electron probe instruments and electron microscopes consists of: (1) a heated filament constituting the cathode, which is at a high negative potential relative to earth, (2) a grid or 'wehnelt' at a negative potential of a few hundred volts relative to the cathode, and (3) an earthed anode plate. The filament is made of tungsten wire (usually 0.125 mm in diameter) bent into a 'hairpin' shape and heated to about 2700 K, which emits electrons thermionically. Thermionic emission increases rapidly with temperature and is strongly influenced by the work function of the emitter (the potential barrier at the surface). Tungsten is used because it has a fairly low work function (4.5 eV) and a high melting point (3643 K), which allows a high working temperature.

Electrons emitted from the filament are accelerated towards the anode, in which there is an aperture a few millimetres in diameter to allow the beam to pass. Emission from the filament is controlled by the grid. Between cathode and anode the electron trajectories form a 'crossover', which is the effective source of electrons (fig. 3.1). A demagnified image of the crossover is projected onto the surface of the specimen by means of the probe-forming lens system described in the next chapter.

### 3.2 Gun alignment

The filament must be accurately centred in the grid aperture, otherwise the beam will emerge at an angle to the axis. It is preferable to have provision for mechanical filament centring controlled from outside the vacuum chamber, so that adjustment can be carried out with the gun in operation. In the absence of such controls, other methods must be used to compensate for the effects of inaccurate centring and wandering of the filament tip in

## 3.2 Gun alignment

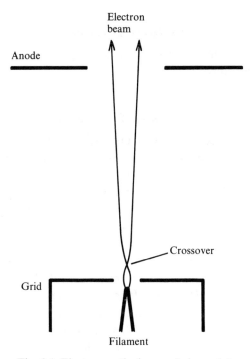

Fig. 3.1 Electron paths in gun (schematic).

operation. Various means of correction are possible, including moving the whole filament and grid assembly relative to the anode, or translating the anode only. Alternatively electromagnetic alignment coils may be used to deflect the beam after emerging from the gun, thereby compensating for the misorientation of the beam.

The position of the filament tip relative to the grid has important effects on the behaviour of the gun. The correct position is slightly below the plane of the grid (as shown in fig. 3.1). Moving the filament further from the grid decreases the maximum available gun current. If the filament–grid distance is reduced, a larger gun current can be obtained, but the effective source size increases.

The optimum grid–anode spacing varies with accelerating voltage. Ideally the spacing should be no more than the minimum necessary to avoid flashover: this can be achieved for different accelerating voltages by an adjustable or interchangeable anode, though for most purposes a single fixed spacing is satisfactory.

## 3.3 Electrical supplies

The electron gun requires various electrical supplies, which are described in the following sections.

### 3.3.1 Filament heating current

The filament is heated by a DC supply isolated from earth, the current required being in the region of 2–3 A for a tungsten filament of the usual type. This current should be reasonably stable, though in the saturated condition (§3.3.3) the emission from the gun is insensitive to small changes in temperature.

Tungsten filaments have a limited life owing to the sublimation of tungsten from the hottest region near the tip. As the filament thins during its lifetime, its resistance increases. If the internal resistance of the heating current supply is high (constant current source) the filament temperature rises as thinning proceeds, and its lifetime is shortened. On the other hand, with low internal resistance (constant voltage source) the temperature falls. Bloomer (1957) showed that the change of temperature with thinning is minimised if the source resistance is about three times that of the filament, which is approximately 1 Ω.

### 3.3.2 High voltage

The accelerating voltage applied to the gun is obtained from a high voltage generator, which for conventional microprobe analysis is required to cover the range 10–30 kV, though lower and higher voltages are useful in certain specialised applications. Drift and ripple should not exceed approximately 1 in $10^4$ for ordinary microprobe analysis, and 1 in $10^5$ for scanning electron microscopy. The voltage should be resettable to better than $\pm 100$ V for quantitative microprobe analysis. The current drawn from the supply is equal to that in the beam emitted by the gun ('gun current'), which is typically about 50 μA.

### 3.3.3 Bias

The commonest method of generating the required negative bias of the grid relative to the filament is to pass the gun current through a resistance of several megohms, as shown in fig. 3.2. The high voltage supply is connected to the grid, and the gun current ($i_g$) passing through the bias resistance ($R_B$)

## 3.3 Electrical supplies

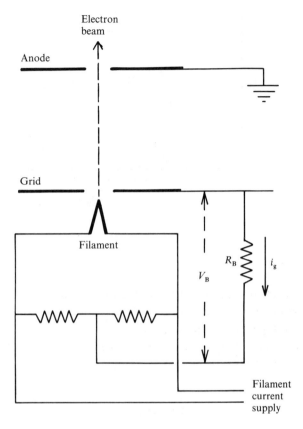

Fig. 3.2 Bias voltage ($V_B$) produced by gun current ($i_g$) flowing through bias resistor ($R_B$).

generates the bias voltage ($V_B$), which makes the filament positive relative to the grid. The gun current is stabilised by negative feedback: if the emission increases, the bias voltage increases, reducing the gun current.

Emission from the filament increases rapidly with temperature at first, but the gun current tends to level off above a certain temperature, the value obtained at a given temperature being dependent on $V_B$. The *probe* current shows more marked saturation, which is attributable to the effect of self-biassing on the area of the filament tip from which electrons are drawn. A false maximum is usually observed before the true plateau is reached (fig. 3.3). The correct operating temperature is just above the 'knee' of the curve of probe current versus temperature, which should occur at about 2600 K.

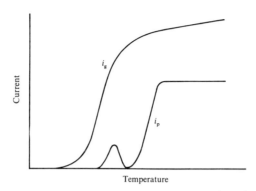

Fig. 3.3 Gun current ($i_g$) and probe current ($i_p$) as function of filament temperature.

This gives a reasonable compromise between high brightness and filament life.

Reducing the bias voltage increases the gun current by allowing emission from more of the filament tip to contribute to the beam, but if the emitting area is too large the curvature of the tip results in a hollow beam which cannot be focussed properly. This occurs at gun currents exceeding approximately 100 $\mu$A.

## 3.4 Brightness

The current per unit area ($j$) emitted by the filament increases rapidly with temperature, as shown in fig. 3.4, but the rate of thinning also increases steeply. According to Haine, Einstein and Borcherds (1958), the life in hours ($t$) of a 0.125 mm diameter filament and the current density ($j$) in A cm$^{-2}$ are related approximately by the expression: $t = 50/j$. A current density of 1 A cm$^{-2}$ obtained at 2640 K is thus about the highest compatible with a reasonable life (50 hours).

The beam emerging from the gun is focussed by electron lenses which produce an image of the crossover on the surface of the specimen. According to the electron-optical analogue of the Helmholtz–Lagrange law in light optics, the 'brightness' ($\beta$) of the beam (the current density in the focus divided by the solid angle of the focussed beam) is constant whatever the number and strength of the electron lenses. In order to obtain a probe of a given diameter it is necessary to limit the aperture of the final lens, because of spherical aberration, §4.5.1. This defines the solid angle, hence the current obtainable in a probe of a given diameter is proportional to $\beta$.

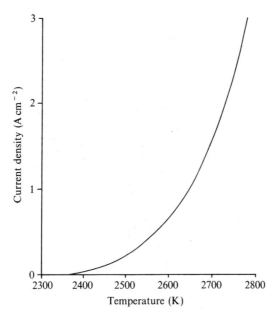

Fig. 3.4 Current density emitted by tungsten filament as function of temperature.

The electrons emitted by the filament are subject to thermal agitation determined by the absolute temperature ($T$) of the filament. The thermal velocity perpendicular to the axis determines the divergence of the beam and sets a fundamental limit to the brightness. The effective mean thermal energy is $kT$, where $k$ is Boltzmann's constant ($8.62 \times 10^{-5}\,\mathrm{eV\,K^{-1}}$). Langmuir (1937) showed that the resulting theoretical brightness limit is given (in $\mathrm{A\,cm^{-2}\,sterad^{-1}}$) by:

$$\beta = jeV_0/\pi kT,$$

where $e$ is the charge on the electron, and $V_0$ is the accelerating voltage. For example, with the typical values $j = 1\,\mathrm{A\,cm^{-2}}$, $T = 2640\,\mathrm{K}$, $V_0 = 20\,\mathrm{kV}$, the limiting brightness is $2.8 \times 10^4\,\mathrm{A\,cm^{-2}\,sterad^{-1}}$.

Haine and Einstein (1952) showed that $\beta$ is greatest at a bias voltage about 50 V less than that required to cut off the gun. The relationship between the cut-off bias voltage and the filament–grid spacing may be understood by considering the electrostatic field near the filament tip (fig. 3.5). Electrons can escape from the filament only if the equipotential corresponding to the filament potential ($-V_0 + V_B$) intersects the tip: if it does not the gun is cut off. With a large filament–grid distance ($h$), cut-off

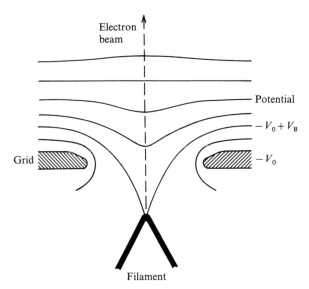

Fig. 3.5 Equipotentials in electron gun.

occurs at a small bias voltage, and as $h$ is decreased the bias voltage must be increased to maintain constant conditions at the tip.

## 3.5 Stability

With a saturated gun the main cause of probe current drift is movement of the tip of the filament. The angle of divergence of the beam emerging from the gun is usually only about 1°, and a small tilt can cause significant change in probe current. Haine and Einstein (1952) observed a beam tilt of nearly 10′ in a typical gun for a filament displacement of 25 μm (corresponding to less than one quarter of the wire diameter). The angle of divergence is strongly influenced by gun geometry and bias. Haine and Einstein showed that divergence increases rapidly as the bias voltage is reduced. Increased divergence makes the probe current less sensitive to movement of the filament, at the cost of some loss of brightness, though usually this is not serious.

Even if the gun parameters such as filament position and bias are selected to minimise drift, positive measures are needed to ensure that the current reaching the specimen (probe current) remains constant over long periods (§4.10). This requires a means of controlling the probe current. One method is to vary the grid potential, using an optical coupler to provide the necessary isolation (Baines, Dean and Wilson, 1975). However, more

## 3.6 High brightness guns

commonly the probe current is controlled by means of the condenser lens (§4.10).

### 3.6 High brightness guns

Improved performance has been claimed for tungsten filaments with finely pointed tips produced by grinding or electrolytic etching. The effective source diameter can be reduced to less than that of a conventional filament, and higher brightness is possible because of reduced space charge at the tip. Bowman and Hardie (1972) found that a pointed filament gave a factor of 10 increase in brightness, but the life was only 3 hours. Pointed filaments have the disadvantage of producing a beam of small divergence, making it harder to obtain a good probe current stability.

Alternative thermionic emitters with a lower work function than tungsten offer the possibility of higher brightness, partly by virtue of higher current density, but also through a reduction in working temperature (see §3.4). Oxide coated cathodes (as used in electronic valves) are susceptible to contamination. The most effective alternative to the conventional tungsten filament is the lanthanum hexaboride cathode described in the following section.

#### 3.6.1 Lanthanum hexaboride

The lower work function of $LaB_6$ (approximately 2.2 eV) gives a higher current density at a lower temperature than tungsten (Lafferty, 1951). The main difficulty with $LaB_6$ is its high chemical reactivity at elevated temperatures. In the gun designed by Broers (1967) this was overcome by attaching one end of the $LaB_6$ rod to a copper heat sink and heating the emitting end indirectly. A tungsten coil surrounded by a reflecting shield heated the rod by radiation and electron bombardment. Surplus heat was conducted along copper rods to an oil bath outside the vacuum chamber.

More recently $LaB_6$ cathodes which are interchangeable with conventional filaments have been developed. For example, Ramachandran (1975) used arc bonding to attach $LaB_6$ to rhenium wire, making a direct replacement for the usual filament. A method of embedding a small single crystal in a glassy carbon filament has been described by Futamoto et al. (1980). Direct replacement units for use in standard electron guns are now available commercially.

Early $LaB_6$ cathodes were made of sintered material, but oriented single crystals give higher brightness and more stable emission. A typical single crystal cathode has a conical form with a tip radius of 2–5 μm, though this may increase in use. The effective source diameter may be as little as 20 μm – considerably less than for a conventional tungsten filament.

At a temperature of 1900 K, $LaB_6$ gives a brightness of at least $10^6$ A cm$^{-2}$ sterad$^{-1}$ at 20 kV, which is more than ten times higher than is obtainable with a tungsten filament. Cathode life is dependent on the pressure in the gun: if this is kept below $10^{-1}$ Torr, a life of around 1000 hours can be expected. $LaB_6$ guns are used mainly in scanning electron microscopy, where brightness is more important than in microprobe analysis. This type of cathode has been reviewed by Crawford (1979) and Hohn (1985).

### 3.6.2 Field emission sources

Crewe, Eggenberger, Wall and Welter (1968) developed a field emission gun giving a brightness about $10^3$ times higher than the conventional gun, with an effective source diameter of only 3 nm. The cathode in this case was a tungsten point of about 0.1 μm radius. An external focus of about 25 nm was produced, which could be reduced to less than 1 nm by demagnifying lenses. The working pressure of below $10^{-9}$ Torr required for guns of this type necessitates the use of ultra high-vacuum techniques.

The field emission gun is superior to the conventional gun only for probe diameters of less than about 0.1 μm, hence it has no advantage for conventional microprobe analysis. Furthermore the beam current stability is relatively poor and the field emitter is expensive and rather fragile. The most significant application is in high resolution scanning transmission electron microscopy. It is less useful for ordinary scanning electron microscopy with solid specimens, where the resolution is limited by electron scattering.

Field emission guns have been reviewed by Van der Mast (1983).

# 4
# The probe-forming system

## 4.1 Introduction

The purpose of the electron optical system is to produce a demagnified image of the 'crossover' in the electron gun (§3.1) on the surface of the specimen. There are usually either two or three lenses in the electron optical column. The final lens is sometimes called the 'objective' by analogy with the comparable lens in the t.e.m., though this term is rather inappropriate in the case of the microprobe. The other lens is known as the 'condenser', also derived from electron microscope terminology (three-lens systems have two condensers). Magnetic lenses are used, their strength being controlled by varying the current in the coils. The condenser determines the size of the image of the source formed on the surface of the specimen and the current in the focussed beam. The final lens is adjusted to produce a sharp focus in the specimen plane.

Electron lenses are subject to aberrations analogous to those of ordinary (light-optical) lenses. Spherical aberration (§4.5.1) is of critical importance and plays a key role in determining the current available in a given beam diameter, as discussed in §4.9. Astigmatism is less serious and can be corrected (§4.5.2). Chromatic aberration (§4.5.3) can be neglected for most purposes, because the electrons are almost 'monochromatic' (i.e. they have a very small energy spread).

## 4.2 Magnetic lenses

Magnetic rather than electrostatic lenses are preferred because they are more convenient to use and have lower aberrations. The conventional magnetic lens consists of coils of copper wire wound symmetrically around the electron optical axis, surrounded by an iron shroud with a narrow gap across which the focussing field appears (fig. 4.1). The cylindrical iron parts

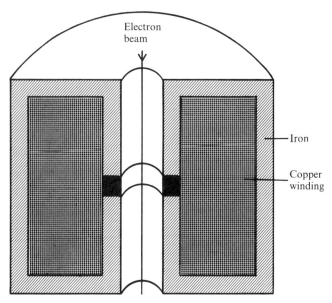

Fig. 4.1 Section through magnetic electron lens.

on each side of the gap are the 'polepieces'. The lens is energised by passing current through the windings – the magnitude of this current determines the strength of the lens.

The focussing action of a magnetic lens arises as follows. When an electron travelling initially parallel to the axis enters the field it experiences a tangential force due to the interaction of the axial velocity with the radial component of the field. The subsequent direction of travel being inclined to the axis, the tangential velocity component interacts with the axial component of the field to produce a radial force directed towards the axis (fig. 4.2). The resulting electron path is helical, though in a probe-forming lens only the converging effect of the lens need be considered. The focal length, $f$, of a magnetic lens is given by:

$$f = KV_0/(ni)^2, \tag{4.1}$$

where $K$ is a function of the lens geometry, $V_0$ is the accelerating voltage, $n$ is the number of turns in the winding and $i$ is the current.

It is usual for the condenser to be a symmetrical lens with polepieces of equal diameter, as in fig. 4.1, but other configurations are used for the final lens, as described in the next section.

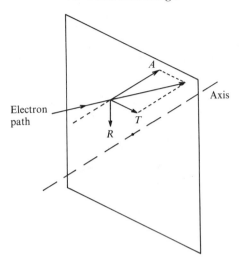

Fig. 4.2 Focussing action of magnetic field: interaction of tangential component ($T$) of velocity of electron with axial component ($A$) of field produces radial force ($R$).

## 4.3 Final lens design

It is desirable for the final lens to work at a short focal length, in order to obtain maximum demagnification and minimum spherical aberration. However, a short working distance may be incompatible with optical viewing requirements and the clearance required for X-ray paths to the spectrometers. Widely varying final lens designs are used, each a compromise between electron-optical and other considerations, as discussed below.

The final lens is often asymmetrical, with the smaller diameter polepiece on the specimen side. The 'pinhole' lens, with highly asymmetrical polepieces (fig. 4.3), has the advantage that the magnetic field at the specimen surface is very weak, which facilitates the collection of low energy secondary electrons used for scanning images (§5.5); also the large diameter of the back bore is useful for accommodating scanning coils (§5.1). Another advantage is that the effective centre of the lens is close to the front face, hence for a given focal length the space available between specimen and lens is maximised. Pinhole lenses have somewhat greater spherical aberration than the equivalent symmetrical lens, but the difference is not serious.

Final lens design and optical viewing arrangements are intimately linked. The only optical viewing system which does not affect lens design is that in which the specimen is translated from the optical microscope to the

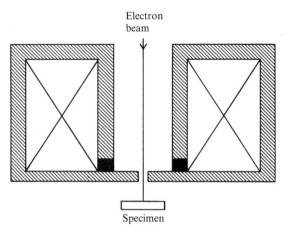

Fig. 4.3 'Pinhole' lens.

electron beam for analysis (fig. 2.2(*d*)), but this does not allow simultaneous viewing and analysis; also very high mechanical reproducibility is required, unless a scanning image is used for accurate location of points for analysis.

Fig. 4.4(*a*) shows a final lens design suitable for use with the coaxial reflecting type of microscope (fig. 2.2(*a*)). The removable refracting objective system (fig. 2.2(*c*)) may also be used, but suffers from the drawback that the large bulk of the lens makes it difficult to avoid a low X-ray take-off angle. This deficiency can be remedied by having X-ray paths going back through the bore of the lens (fig. 4.4(*b*)), allowing a high take-off angle (e.g. 75°). The polepieces require a rather large bore to provide the necessary clearance, and this has to be compensated by increasing the current in the lens windings. This type of lens can also be used with either coaxial reflecting or removable refracting light optics. The latter system takes up space between lens and specimen, thereby increasing the working distance, and hence the spherical aberration.

In both the preceding lens types the massive coil is an obstruction to the X-ray spectrometers and the light microscope. This can be avoided by moving the windings below the plane of the specimen (fig. 4.4(*c*)). The space above the specimen is then completely free for coaxial reflecting optics and X-ray spectrometers with high take-off angle. The disadvantage of this arrangement is that it imposes difficulties in arranging specimen changing and mechanical movements.

Simultaneous analysis and viewing with refracting optics (fig. 2.2(*b*)) can be obtained by tilting the specimen and mounting the microscope outside

## 4.3 Final lens design

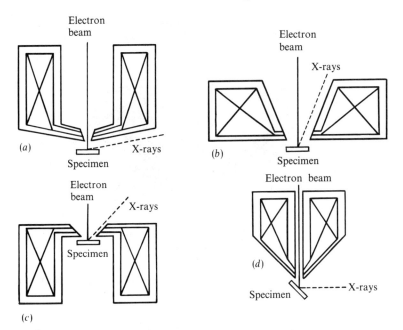

Fig. 4.4 Final lens configurations.

the lens (fig. 4.4(d)). This permits a reasonably high X-ray take-off angle, but non-normal electron incidence is somewhat undesirable for quantitative analysis.

### 4.3.1 Mini-lenses

The space occupied by the coil of a magnetic lens can be reduced if relatively few turns carrying a large current are used (Le Poole, 1964). 'Mini-lenses' of this type usually require liquid cooling. The lenses developed by Le Poole had no iron shroud, which in a conventional lens helps to smooth out irregularities in the magnetic field. The microprobe designed by Fontijn, Bok and Kornet (1969) used a mini-lens at 45° to the specimen plane, allowing a refracting optical microscope to be incorporated. There was space for several exit ports for X-ray spectrometers, electron detectors, etc., at a take-off angle of up to 45°. A mini-lens used in a current commercial electron microprobe is illustrated in fig. 4.5. In this case the lens has an iron shroud.

Bassett and Mulvey (1969) showed that in the Le Poole cylindrical solenoid lens, spherical aberration decreases as the ratio of length to

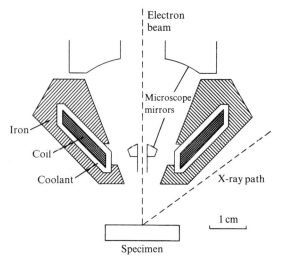

Fig. 4.5 'Mini-lens' as used in electron microprobe (by courtesy of JEOL Ltd.).

diameter is decreased, though heating becomes more intense. While a cylindrical solenoid can equal the performance of a conventional lens, it is difficult to obtain a significantly lower spherical aberration coefficient without encountering cooling problems. Similar considerations apply to conical solenoids, which permit some reduction in the solid angle subtended at the specimen by the lens, leaving more clearance for X-ray paths.

A novel type of mini-lens proposed by Huang and Lei (1981) consists of a cylindrical iron polepiece with a number of gaps of graded width throughout its length, giving a steadily increasing field along the axis with a sharp drop at the end near the focus. This configuration minimises spherical aberration.

## 4.4 Lens current stability

Differentiating equation (4.1) gives the change $\triangle f$ in focal length caused by a change $\triangle i$ in lens current: $\triangle f/f = -2\triangle i/i$. The effect of changing the focal length is to blur the image by an amount $\triangle d$, given by:

$$\triangle d = 4\alpha f \triangle i/i, \qquad (4.2)$$

where $\alpha$ is the semi-angle of the cone defining the focussed beam. In the electron microprobe, lens current stability should be such that

$\triangle d < 0.2\,\mu m$. For typical values: $f = 20$ mm, $\alpha = 20$ mrad, equation (4.2) indicates that current variations should not be exceed 1 in $10^4$. For high resolution imaging, about 100 times less defocussing is tolerable, but $\alpha$ is typically about four times smaller, hence the required lens current stability is a few parts in $10^6$. The relative change in accelerating voltage that can be tolerated is twice that of the lens current, as shown by the differentiation of equation (4.1) with respect to $V_0$.

## 4.5 Aberrations

Magnetic electron lenses suffer from inherent defects which affect their ability to produce a sharply focussed beam and are similar to the aberrations of lenses used in light optics. These are discussed in the following sections.

### 4.5.1 Spherical aberration

Magnetic lenses have marked spherical aberration, which causes off-axis electrons to be focussed more strongly than those close to the axis (fig. 4.6). This can generally be neglected in the condenser, but is important in the final lens. The diameter of the 'disc of least confusion' (the minimum diameter of the bundle of rays where it forms waist) in a focussed beam is given by:

$$d_s = C_s \alpha^3 / 2, \qquad (4.3)$$

where $C_s$ is the spherical aberration coefficient of the lens. Usually $C_s$ is about two or three times the focal length $f_1$, therefore a typical value for the electron microprobe is 50 mm. Hence $\alpha$ must not exceed 30 mrad if the disc of least confusion is to be smaller than $0.5\,\mu m$.

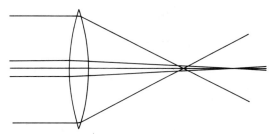

Fig. 4.6 Spherical aberration: effective focal length of lens is shorter for outer rays than for those near axis.

## 4 The probe-forming system

The diameter of the final lens aperture, which should be selected to give approximately the required $\alpha$, also governs the probe current. To achieve a high current in a probe of given diameter, it is therefore important to have a final lens with a small $C_s$. In principle, spherical aberration can be reduced by correction systems using quadrupole lenses, but no practical system has yet been developed. At present high brightness electron sources (§3.6) offer the best prospects for increasing the probe current.

### 4.5.2 Astigmatism

The other important aberration is astigmatism, which is caused mainly by imperfections in the polepieces of the final lens, especially ellipticity of the bores (Sturrock, 1951; Archard, 1953). Contamination of apertures in the column, and the presence of insulating foreign matter, can also cause astigmatism. Its effect is that instead of a single point focus, there are two line foci, the diameter of the disc of least confusion between these foci being greater than that of the ideal focus (fig. 4.7).

Astigmatism may be corrected by a 'stigmator', which in its simplest form consists of an electrostatic quadrupole with four electrodes arranged symmetrically about the beam. This produces astigmatism in the beam, and by suitable choice of voltage and orientation of the electrodes the original astigmatism can be cancelled out. The need for mechanical rotation may be obviated by using an 'octupole' stigmator, consisting of eight symmetrically disposed electrodes, which by varying the potentials can correct astigmatism in any orientation (Mulvey, 1959). A similar result may be achieved with a six-pole (hexapole) stigmator. By means of suitable circuitry, independent control of amplitude and orientation can be obtained. Usually electromagnetic coils are employed in preference to electrostatic electrodes. The stigmator is most commonly located in the bore of the final lens.

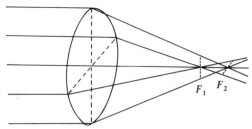

Fig. 4.7 Astigmatism: asymmetry in focussing action of lens produces two separate line foci ($F_1$, $F_2$).

### 4.5.3 Chromatic aberration

The focal length of an electron lens varies with the energy of the electrons: this effect is known a chromatic aberration, by analogy with the corresponding effect in light optics whereby the focal length of a lens varies with the wavelength of the light. The diameter of the disc of least confusion for a beam of electrons of mean energy $E$, with energy spread $\triangle E$, is given by:

$$d_c = (\triangle E/E)C_c \alpha,$$

where $C_c$ is the chromatic aberration coefficient of the lens and $\alpha$ is the semi-angle of the beam.

The thermal energy spread of electrons emitted by a conventional tungsten filament is about 0.3 eV, hence for a typical accelerating voltage, $\triangle E/E$ is approximately $10^{-5}$: thus, if $\alpha = 20$ mrad and $C_c = 10$ cm, $d_c = 0.02\ \mu$m. Chromatic aberration is therefore unimportant in relation to the spatial resolution of around 1 $\mu$m obtainable in microprobe analysis.

Variation in electron energy can also be caused by ripple or instability in the accelerating voltage. For the above values of $\alpha$ and $C_c$ it can be deduced that $d_c = 0.2\ \mu$m for $\triangle E/E = 10^{-4}$, hence it is desirable that the accelerating voltage should be constant to better than 1 in $10^4$.

## 4.6 Apertures

The solid angle of the beam subtended at the specimen is governed by a limiting aperture, ideally at or near the principal plane of the final lens. This consists of a hole in a flat disc made of platinum, molybdenum or tantalum. The choice of hole diameter is determined by the probe diameter and current required, and it is useful to have several apertures interchangeable under vacuum. This also permits a contaminated aperture to be replaced quickly. The effect of a small amount of contamination can often be corrected with the stigmator, but when distortion of the beam becomes excessive the aperture must be cleaned or replaced.

There are usually two or three additional aperture diaphragms in the column to prevent the divergent electron beam from striking lens bores etc. These 'spray apertures' typically have an internal diameter of 1-3 mm. They also need cleaning periodically, though the beam is less sensitive to contamination on them than on the final aperture. Some instruments have a column liner (made from a material of low atomic number in order to minimise electron scattering) which can be removed easily for cleaning. This reduces the need for spray apertures.

Platinum apertures may be cleaned by mounting on a platinum wire and holding in a flame, but they deteriorate fairly quickly with repeated cleaning. Alternative materials such as tantalum and molybdenum cannot be treated in this way because they oxidise in air, but can be cleaned by heating in a vacuum chamber. Chemical cleaning procedures can also be used. Apertures may be cleaned *in situ* in the column by means of a heating device built into the holder. Alternatively, special thin foil apertures can be heated by the impact of the electron beam. Such apertures in any case require less frequent cleaning because the surface area of the bore is very small.

## 4.7 Column alignment

The electron-optical column should be aligned so that the beam travels along the axis of each lens and passes through all the apertures in the system. Usually the main components are aligned when the instrument is assembled and residual misalignment is corrected by 'steering' the beam with electromagnetic coils. An external mechanical centring control is generally provided for the final lens aperture. This is adjusted by observing the point of impact of the beam on a luminescent specimen while the lens current is swept through focus, misalignment being indicated by lateral shift of the beam. Condenser alignment may be checked by observing the lateral movement of the probe on a luminescent specimen while the condenser current is changed. Gun alignment is discussed in §3.2.

## 4.8 Demagnification of the electron source

Geometrical optics may be applied to the calculation of the demagnification (source diameter/image diameter) of the probe-forming electron-optical system, given the simplifying assumption that the lenses are thin. For the two-lens system illustrated in fig. 4.8, the demagnification is given by:

$$D = (A/f_1) - B, \qquad (4.4)$$

where $A = ab/d$ and $B = (a+b)/d$, $f_1$ being the focal length of the condenser lens. When $f_1 = ab/(a+b)$, $D = 0$. In this condition the source is imaged by the condenser in the plane of the final lens and it is impossible to obtain a focus on the specimen. Also the current is very large because most of the electrons emitted by the gun pass through the final lens aperture. In normal operation a stronger condenser setting is used. If $c \ll b$, changes in $f_1$ have

## 4.8 Demagnification of the electron source

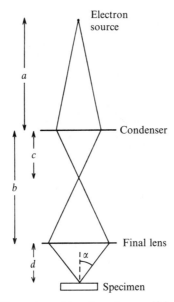

Fig. 4.8 Geometry of two-lens demagnifying system.

only a small effect on the setting of $f_2$ (the focal length of the finals lens) required to focus the beam on the specimen.

Substituting the following typical value in equation (4.4): $a = 200$ mm, $b = 300$ mm, $d = 20$ mm, $f_1 = 25$ mm, we obtain $D = 95$. Assuming a source diameter of 50 μm, the probe diameter will thus be 0.5 μm (neglecting the possible effect of aberrations), which is not unreasonable for microprobe analysis. A third lens is desirable if much higher demagnifications are required (to enable high resolution scanning images to be obtained), since there are practical limits to the strength of the condenser lens.

From equations (4.1) and (4.4) it follows that the demagnification for a two-lens system is related to the condenser excitation current $(i_1)$ by the expression:

$$D = (A/f'_1)(i_1/i'_1)^2 - B, \qquad (4.5)$$

where $f'_1$ is the condenser focal length required to focus the beam in the focal plane of the final lens, and $i'_1$ is the condenser current at which this occurs. Typical values of $A = 3000$, $B = 25$, $f'_1 = 120$ mm, have been used to plot $D$ versus $i_1/i'_1$ in fig. 4.9. This curve applies for all accelerating voltages, though the absolute value of $i'_1$ increases as $V_0^{1/2}$, and the value of $i_1$ required for a given $D$ increases proportionately.

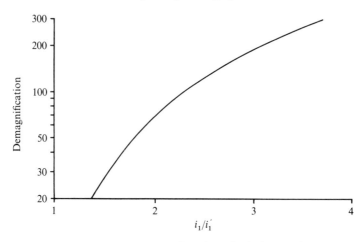

Fig. 4.9 Demagnification factor ($D$) as function of relative condenser current for typical two-lens system.

## 4.9 Probe diameter and current

The two fundamental limitations on the performance of electron probe instrument are gun brightness (§3.4) and spherical aberration in the final lens (§4.5.1). Other factors limiting the probe diameter, such as stray magnetic fields and vibration, can, in principle, be reduced to negligible levels. To this end the whole electron path needs to be magnetically screened, and locations with high levels of stray magnetic field and vibration should be avoided. Anti-vibration mountings for the column are desirable, and steps should be taken to avoid transmission of vibration from rotary vacuum pumps to the column. These precautions are especially important for the production of high resolution scanning images.

Neglecting spherical aberration, it follows from the definition of brightness that the maximum current in the focussed image of the electron source is given by:

$$i_p = (\pi/2)^2 d_i^2 \alpha^2 \beta, \tag{4.6}$$

where $d_i$ is the image diameter, and $\beta$ is the brightness. Assuming that the total probe diameter $d$ is given by $d^2 = d_i^2 + d_s^2$, where the diameter $d_s$ of the disc of least confusion due to spherical aberration is obtained from equation (4.2), then the optimum value for $\alpha$ (Mulvey, 1967) is given by:

$$\alpha = (d/C_s)^{1/3}, \tag{4.7}$$

## 4.9 Probe diameter and current

and the maximum current obtainable in the probe is:

$$i_p = (3/16)\pi^2 \beta C_s^{-2/3} d^{8/3}. \tag{4.8}$$

Substituting the typical values: $C_s = 50$ mm, $\beta = 2.8 \times 10^4$ A cm$^{-2}$ sterad$^{-1}$, and expressing $d$ in micrometres and $i_p$ in nanoamps, we have: $i_p = 380 d^{8/3}$. The optimum value of $d_i$ is $0.87d$. Also the optimum value of $\alpha$ in mrad, assuming $C_s = 50$ mm, is $27d^{1/3}$, where $d$ is in micrometres. Thus for a probe diameter of $0.5$ $\mu$m typical of microprobe analysis, $\alpha = 21$ mrad, whereas to obtain a probe diameter of 10 nm diameter for scanning electron microscopy, $\alpha$ would have to be reduced to about 6 mrad. The calculated probe currents in these examples are 73 nA and 0.8 pA respectively.

Optimisation for maximum current requires the final lens aperture to be changed for different probe diameters, which is somewhat inconvenient. This can be avoided by locating a fixed aperture between condenser and final lens, so that the effective final lens aperture automatically varies with the strength of the condenser, which is changed for different probe diameters in order to obtain the appropriate demagnification (see fig. 4.10).

The relationship between probe current and diameter for a typical instrument at 20 kV ($C_s = 50$ mm, $\beta = 2.8 \times 10^4$ A cm$^{-2}$ sterad$^{-1}$), obtained from equation (4.8), is plotted in fig. 4.11, together with the demagnifica-

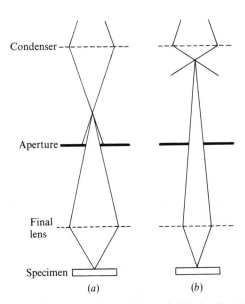

Fig. 4.10 Arrangement for automatic variation of effective final lens aperture: (a) low and (b) high demagnification.

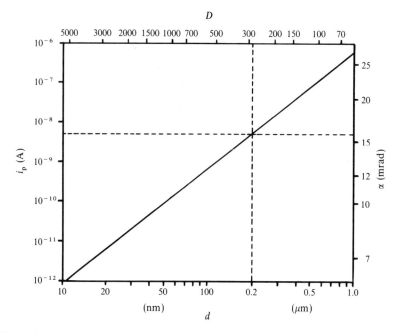

Fig. 4.11 Minimum probe diameter ($d$) as function of probe current ($i_p$), demagnification ($D$) and beam semi-angle ($\alpha$), for typical system.

tion (assuming a source diameter of 50 μm) and the value of $\alpha$ required. The dashed lines represent a typical case: the maximum current obtainable in a 0.2 μm diameter probe is nearly 5 nA, and the value of $\alpha$ required is about 16 mrad, with a demagnification of just below 300.

## 4.10 Probe current monitoring and stabilisation

In quantitative microprobe analysis it is important that the probe current should not change between X-ray intensity measurements. Causes of drift include variation in lens or gun supply voltages, electrostatic charging of insulating material in the column, aperture contamination, and inadequate conduction of the specimen, all of which can be corrected. On the other hand, drift due to movement of the filament is hard to avoid although it can be minimised by appropriate adjustment of the gun.

Various procedures can be adopted for dealing with drift, for example: (1) monitor probe current and normalise all X-ray measurements to a constant current; (2) monitor probe current and adjust the current to a

## 4.10 Probe current monitoring and stabilisation

constant value at intervals; (3) fit a device that stabilises probe current continuously. These methods all require a means of measuring probe current. It is insufficient to measure 'specimen current' (the current flowing from specimen to earth), since this is dependent on backscattering and secondary electron emission, which vary with composition.

The best way of measuring probe current is to put a 'Faraday cup' in such a position as to collect the whole current in the probe. The ideal Faraday cup allows no electrons to escape and can be realised sufficiently well by a hole of much greater depth than diameter, preferably in a material of low atomic number to minimise backscattering. The best arrangement is to have the cup on a movable arm located between the final lens aperture and the specimen. Alternatively it can be mounted on the specimen holder, which then has to be moved into the required position for current monitoring. In either case the current can obviously be measured only between analyses.

Probe current can be monitored indirectly using a double aperture in which the electrons passing through the first, larger, aperture but not the second are collected (fig. 4.12). Provided the current density distribution is uniform, the current collected by the inner aperture diaphragm can be

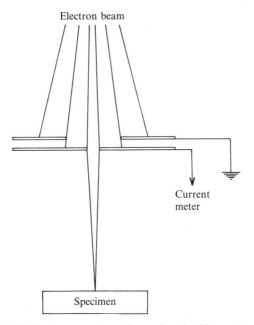

Fig. 4.12 Probe current monitor using double aperture.

taken to be proportional to the probe current. This method allows continuous monitoring during analysis. For returning the current to a predetermined value and maximising the current during gun alignment, an analogue meter is preferable to a digital display, though the latter is better when numerical readings are required for normalising X-ray intensities. An accuracy approaching 0.1% is desirable for quantitative analysis.

To correct probe current drift a means of changing the current is needed. Although the gun bias can be used, it is more usual to adjust the condenser lens current. This can be done automatically by means of a negative feedback system, using the beam current signal from a double aperture (Castaing, 1960). Correction should be restricted to a drift of the order of 10%, or the change in condenser current will cause the probe to go out of focus, and possibly shift.

### 4.11 Computer control

Computer control has been applied increasingly to electron-optical columns in recent years. This has several advantages: for example, the correct lens settings for a specified beam diameter and current can be recalled rapidly from the computer memory. Further, these settings can be corrected automatically for changes in accelerating voltage. Likewise, the magnification of scanning images can be kept constant regardless of changes in accelerating voltage. Automatic setting of the beam alignment controls is possible and probe current stabilisation can be carried out with the aid of the computer, enabling long measuring routines to be undertaken without risk of error due to drift.

# 5
# Scanning

## 5.1 Deflection systems

In order to scan the probe across the specimen, a means of deflecting the electrons is required. This may consist of either an electric or a magnetic field, but usually the latter. The deflection $\phi$ produced by circular coils close to the axis with $n$ turns in each, carrying a current $i$, is given approximately by:

$$\phi = 0.75 ni \, V_0^{-1/2}. \tag{5.1}$$

Thus for a 30 kV beam about 5 A turns are required for a deflection of 20 mrad. The magnetic flux is used more efficiently if the coils are wound on a ferrite former, which may take the form of a square with pairs of coils on opposite sides connected in series. However, they cannot be placed where there is significant magnetic field from the lenses.

Scanning coils may be located between specimen and final lens, as in fig. 5.1(*a*), but in this position occupy valuable space and increase the working distance of the lens. It is therefore common practice to use a double deflection system located between condenser and final lens (fig. 5.1(*b*)), the deflections by the first and second sets of coils being related in such a way that the beam always passes through the final lens aperture.

## 5.2 Display systems

'Sawtooth' waveform generators are used to supply the $x$ and $y$ deflections of both column and cathode-ray tube (c.r.t.). The magnification of the scanning image is equal to the ratio of the size of the raster on the c.r.t. to that on the specimen, the former usually being kept constant while the amplitude of the probe deflection waveform is varied. With magnetic deflection the current in the coils required for a given deflection is

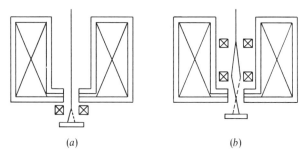

Fig. 5.1 Scanning by means of (a) coils located after final lens, (b) double deflection coils between condenser and final lens.

proportional to $V_0^{1/2}$ (see equation (5.1)), therefore the magnification control has to be calibrated for each accelerating voltage. In a computer-controlled system the magnification may be corrected automatically for changes in $V_0$. Magnification can be determined by observing the image of a grid of known spacing.

The frequency and amplitude of the line and frame scans are usually variable, either continuously or with selected fixed values. For some purposes it is useful to switch the frame scan off and scan a single line only. When a point analysis is required, both frame and line scans are switched off. The probe can then be moved manually with the aid of $x$ and $y$ shift controls.

It is useful to have more than one kind of display system. When the signal is noisy (e.g. for X-ray images and high resolution electron images) a c.r.t. with a long persistence phosphor is advantageous, since the effective integrating time is much longer than that of the eye (about 40 ms). It is appropriate to use a low frame-scan frequency (e.g. 0.3 Hz), matched to the decay time of the phosphor. Since this type of c.r.t. has a relatively large spot size, there is no benefit in using more than a limited number of lines (e.g. 500). For X-ray images, which require a long integrating time, photographic recording using a c.r.t. of higher resolution, or a computer-based image store, may be used.

Often it is more convenient to use a short persistence c.r.t. and a fast frame-scan speed (comparable to television rates). The image then responds immediately to changes when adjusting magnification or specimen position. However, this mode of operation is only practicable when a fairly strong electron signal is available, otherwise the image is unacceptably noisy. It also requires the amplifier to have a sufficiently fast response.

For photographic recording, a third type of display is preferable, utilising a high resolution c.r.t., with a 'single shot' slow frame scan. The advantage of the latter is that any instability during the exposure time merely causes slight distortion rather than blurring of the image. With this mode of recording, greater information content is possible than with the two other modes described above, and the number of lines can usefully be increased to 2000.

The preceding discussion refers to analogue scanning systems. The same basic considerations apply when digital techniques are employed, but a computer-based image store is much more powerful and flexible. The ability to modify images retrospectively is a useful feature of such systems. For example, the number of picture elements ('pixels') can be reduced after the image has been recorded, in order to decrease noise (though at the expense of resolution). Also contrast can be adjusted to emphasise features of interest. Integration by photographic recording can be replaced by the accumulation of image data in computer memory. Considerable capacity is required to match a high quality photographic image, which may limit the number of images which can be stored at the same time.

## 5.3 X-ray images

A commonly used type of X-ray image is the 'dot map' produced by modulating the brightness of the c.r.t. with pulses from a spectrometer set to detect the characteristic X-rays of a particular element. Each dot in the image thus corresponds to one X-ray photon: the density of dots is therefore approximately proportional to the elemental concentration.

A 500-line image contains $2.5 \times 10^5$ pixels, and with a frame-scan time of, say, 2.5 s, a typical X-ray signal of $10^4$ counts $s^{-1}$ gives only one count per ten pixels. Such an image is obviously very noisy and only gross concentration differences are detectable. Image quality can be improved by increasing the integrating time. Thus, in the above example, a 100-fold increase (to 250 s) will yield a mean of 10 counts per pixel: the statistical fluctuation from point to point is then approximately $\pm 30\%$. The noisiness of the image may also be reduced by using fewer pixels, though this obviously limits the spatial resolution obtainable.

According to Rose (1948) a difference in mean brightness between two areas in an image can just be detected when it exceeds five times the random signal fluctuation, which is equal to the square root of the number of counts. Hence, for areas of ten pixels, the minimum detectable difference in the above example would be 50%, decreasing to 15% for 100 pixels. The effective spatial resolution of an X-ray image is thus usually limited by statistics and is closely related to contrast.

Because of the presence of the continuous X-ray spectrum, dots appear in the image even where the element concerned is absent. Such background can be suppressed by discriminating against pulses separated by time intervals greater than a certain amount (Wakabayashi, Miyake, Date and Soezima, 1972; McCoy and Gutmacher, 1975), but this is not very effective for distinguishing low concentrations from background. Background is more significant when an e.d. spectrometer is used, because of the low peak to background ratio compared to the w.d. spectrometer.

An alternative to the 'dot map' form of image described above is to use the analogue output signal from a ratemeter (§8.5), which has the advantage of being more amenable to various forms of image enhancement. For example, the zero level can be suppressed so that background is eliminated and concentration differences are enhanced (Melford, 1962). An extension of this idea is to suppress the ratemeter signal when it falls outside selected upper and lower limits, thereby displaying the area within a certain concentration range (Heinrich, 1962, 1963). Another use of the ratemeter is to produce line scans, in which a profile of the distribution of a selected element is produced by scanning the probe along a single line. Noise is less troublesome in this case because the whole recording time is devoted to one line rather than a complete raster. Such analogue display modes have now generally been displaced by their digital equivalents.

### 5.3.1 Total X-ray images

An alternative to the conventional X-ray image is to use the total X-ray intensity of all energies, as recorded with a proportional counter (Duncumb, 1960), a solid-state e.d. detector (Ingram and Shelburne, 1980), or a scintillation detector (Reimer and Bernsen, 1984). Such images are less noisy than conventional X-ray images, but do not give element-specific information. The contrast in a total X-ray image is a function of mean atomic number, resembling a backscattered electron image (§5.5.1) in this respect. The variation in total X-ray intensity (continuum plus characteristic) as a function of atomic number (fig. 5.2) shows some distinctive features compared to that for b.s.e.s. (fig. 5.4), notably the reversal of slope around $Z = 30$. This is caused by the variation in the detected K and L intensities as a function of $Z$. Whether better contrast is obtained with one form of image or the other depends on the particular compositions involved.

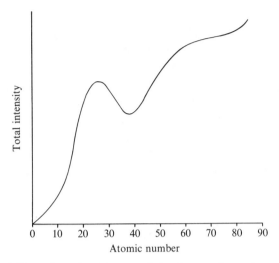

Fig. 5.2 Total X-ray intensity (characteristic plus continuum) as a function of atomic number Z (after Reimer and Bernsen (1984)).

## 5.4 Colour images

An ordinary monochrome X-ray scanning image represents the distribution of only one element. Often several elements are of interest, and with complex microstructures it can be quite difficult to correlate separate images. This problem may be solved by combining several element distributions in one colour image, in which each element is represented by a different colour. Examples of the use of colour are shown in plates 1–3.

One way to obtain such images is to use colour film to photograph a white c.r.t. screen through different coloured filters, while displaying scanning images of different elements in turn. However, the results are not easy to predict, and valuable instrument time may be wasted on trial exposures. In its simplest form this method allows three element distributions to be combined, using the primary colours (red, green, blue). By mixing colours, a larger number of elements can be represented; also phases containing different proportions of the same elements can be distinguished (Jones, Gavrilovic and Beaven, 1966).

On the whole, better results are obtainable by first taking monochrome photographs of the individual element distributions and then rephotographing them on colour film through filters (Duncumb and Cosslett, 1957; Yakowitz and Heinrich, 1969). Careful attention to the registration of the images is required, but this procedure is more flexible than direct photography of the screen.

A third alternative is to use a colour display system to produce a 'live' colour image, with three spectrometer outputs connected directly to the red, green and blue inputs of the c.r.t. Colour oscilloscopes used for this purpose in the past (Ordonez, 1971) have been displaced by colour television monitors (Pawley, Hayes and Falk, 1976).

The most flexible way of producing colour images is to use a computer-based image store with a colour display. If required, a colour printer can be used to provide a permanent record rather than photographing the screen. With such a system, a continuously updated 'live' image can be obtained. Once the image has been stored, the colours can be changed if desired, and different modes of image processing applied.

An alternative way of using colour, which is especially applicable to digital imaging systems, is the representation of intensities as 'false' colours, entailing the division of the intensity scale into bands and the assigning of a colour (or different shades of the same colour) to each. This enhances the visibility of intensity variations, as shown in plate 3(b), and can be applied to electron as well as X-ray images.

## 5.5 Electron images

Contrast in electron images depends on variations in the number of electrons emitted from the point of impact of the beam on the surface of the specimen. With appropriate choice of detection system, either compositional differences or topography may be emphasised.

Backscattered (or reflected) electrons are those that suffer collisions which result in their reemergence from the surface. Secondary electrons are those originally bound in the specimen that receive sufficient energy from

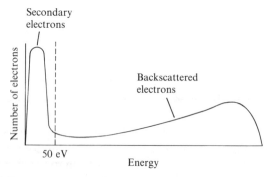

Fig. 5.3 Energy spectrum of secondary and backscattered electrons.

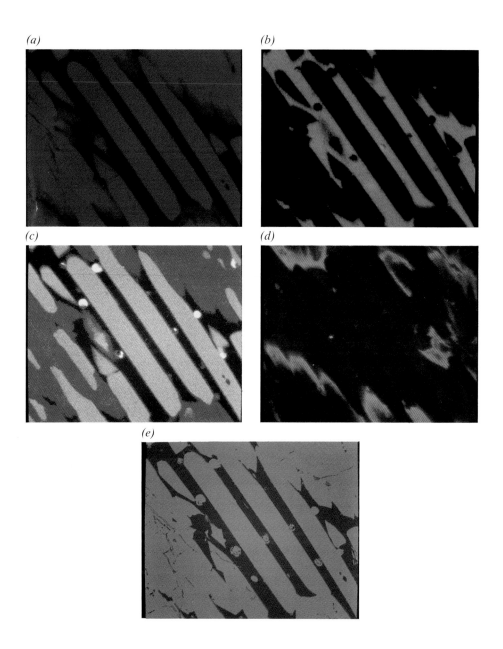

1. Images of microstructure in meteorite: monochrome elemental X-ray images of (*a*) Mg, (*b*) Al, (*c*) Fe, (*d*) Ca, and backscattered electron image (*e*) with enhanced contrast to emphasise atomic number differences.

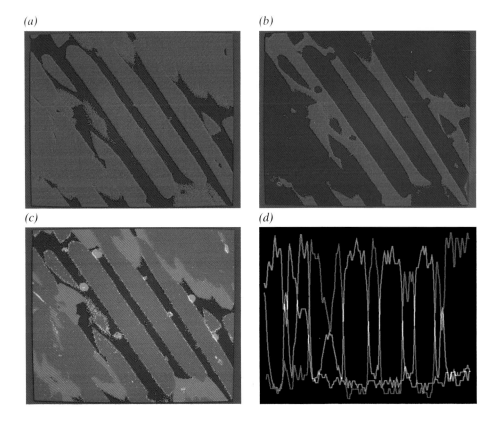

2. Multicolour images of same specimen as in plate 1: (a) 'binary' image of Mg showing areas where Mg X-ray intensity is below (blue) or above (red) threshold; (b) ditto for Al; (c) multiple image showing areas high in Mg (red), Al (blue), Ca (pink), Fe (green); (d) line scan profiles showing distributions of Mg (green), Al (red), and Ca (pink) across centre of scanned area.

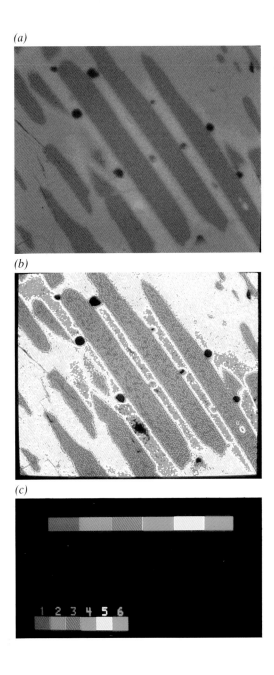

3. Images of Si distribution in same specimen as in plate 1: (*a*) monochrome, (*b*) false colours using colour scale shown in (*c*).

interactions with incident electrons to escape. Their distribution is strongly peaked at low energies, and all electrons below 50 eV are considered to be secondaries (fig. 5.3). The characteristics of the images produced by each of these electron signals are described below.

Electron line scanning is used mainly for focussing the probe by maximising the fine detail. A high scanning speed can be used because of the low noise level in the electron signal. In order to test for astigmatism, focussing is carried out with the probe scanning first in the $x$ direction and then in the $y$ direction: if astigmatism is present the optimum lens settings will differ for the two directions.

### 5.5.1 Backscattered electron images

The fraction of incident electrons backscattered is strongly dependent on atomic number, increasing from under 10% for C to over 50% for U (fig. 5.4). Images showing differences in average atomic number are therefore produced by a detector sensitive to backscattered electrons. Atomic number contrast can be enhanced by offsetting the zero of the electron signal. An image with less noise and higher resolution (about 0.1 $\mu$m) than an X-ray scanning image can be obtained, though absolute element discrimination is lacking (see plate 1($e$)).

The b.s.e. image is somewhat sensitive to surface topography, because the angle of the surface affects the fraction of the electrons backscattered; also the detector 'sees' the specimen only from one direction, and this causes shadowing effects. A method of separating compositional and topographic images described by Kimoto and Hashimoto (1966) uses two detectors on opposite sides of the specimen. A 'hill' or 'valley' causes an

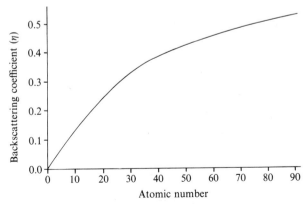

Fig. 5.4 Electron backscattering coefficient as function of atomic number.

increase in backscattering in one direction and a decrease in the other, therefore a topographic image is produced by subtracting one signal from another. If the signals are added, topographic effects are approximately cancelled out and the image shows atomic number differences. Greater flexibility can be achieved by using more than two detectors.

### 5.5.2 Secondary electron images

Higher resolution ($<10$ nm) is possible in a secondary electron image because the secondaries are emitted primarily from the point of impact of the incident electrons, whereas backscattered electrons originate from a finite depth where the beam has spread out significantly. The secondary electron yield does not vary monotonically with atomic number as in the case of the b.s.e. yield, but shows more complex behaviour. Secondary electron images are therefore used primarily for displaying topographic rather than compositional information.

### 5.5.3 Specimen current images

A different kind of image is obtained by amplifying the current flowing from specimen to earth. The 'specimen current' or 'absorbed current' image is influenced by both composition and topography. This form of imaging is unsuitable for use at very low current levels, owing to amplifier limitations.

Contrast in specimen current images is determined by the loss of electrons due to both backscattering and secondary emission. Compared with conventional b.s.e. and secondary electron images, the contrast is reversed and is less affected by shadowing, since electrons leaving the sample in all directions contribute equally. If a $+50$ V bias is applied to the specimen, the secondaries do not escape; contrast is then determined mainly by backscattering, and compositional differences are emphasised.

## 5.6 Electron detectors

There are different ways of detecting secondary and backscattered electrons. The two principal types of detector are described in the following sections.

### 5.6.1 Scintillation detectors

Scintillation detectors are widely used for electron detection. The light from the scintillator, which is usually plastic, is transmitted via a transparent rod

## 5.6 Electron detectors

or 'light pipe' to a photomultiplier outside the vacuum chamber (fig. 5.5). The front surface of the scintillator is aluminised in order to exclude light.

Scintillators are sensitive only to electrons of several keV energy, which includes most b.s.e.s. However, if a suitable accelerating potential is applied to the scintillator, secondary electrons (with initial energies below 50 eV) can also be detected. In the arrangement devised by Everhart and Thornley (1960), a grid in front of the scintillator controls whether or not secondary electrons are detected. A positive grid potential of, say, 200 V ensures that secondaries are attracted towards the grid, where, having passed through the spaces in the mesh, they are accelerated onto the scintillator. This rather bulky type of detector can be fitted into a scanning electron microscope with the usual tilted sample geometry, but not conveniently into electron microprobe instruments, in which solid-state detectors (see below) are generally preferrred.

Scintillation detectors of the Everhart–Thornley type have rather low efficiency for backscattered electrons, owing to the small solid angle subtended (given that high energy backscattered electrons travel in approximately straight lines). A large increase in efficiency can be obtained by using a wedge-shaped scintillator above the specimen with a hole for the beam to pass through (Robinson, 1974). Detectors of this type are especially advantageous for obtaining high quality atomic number images, since the low noise level allows expanded contrast to be used, enabling very small atomic number differences (less than 0.1) to be detected. However, it is difficult to combine this type of detector with a light microscope and paths to the X-ray spectrometers, as required in the electron microprobe.

### 5.6.2 Solid-state detectors

Solid-state electron detectors have some advantages for b.s.e. imaging, including low cost and compactness, though their performance may not always match that of the large area scintillator. Such detectors are only a

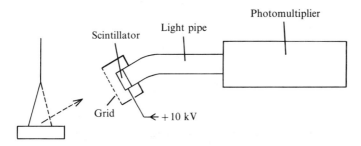

Fig. 5.5 Scintillation detector for electrons.

few millimetres thick and can be fitted into the specimen chamber of an electron microprobe relatively easily. Several detectors can be arranged around the specimen, thereby increasing the total solid angle. A similar result can be achieved with an annular detector (Stephen, Smith, Marshall and Wittam, 1975), though this may interfere with other facilities such as X-ray spectrometers. A split annular detector (Munden and Yeoman-Walker, 1973) allows contrast control by adding or subtracting the signals (fig. 5.6). A particularly cheap form of detector is the silicon solar cell (Wittry, 1972; Parker, 1980).

The signal from a solid-state detector requires amplification, which adds noise (the scintillator-photomultiplier detector has the advantage of essentially noise-free internal amplification); also it is liable to be more restricted in frequency response. Some, but not all, solid-state detectors are capable of TV-rate imaging (Gedcke, Ayers and DeNee, 1978). Frost, Harrowfield and Zuiderwyk (1981) noted that different types of detector are preferable for fast and slow scan rates: biassed Schottky barrier diodes of low capacitance are suitable for TV rates, whereas a better response at low scan rates is obtained with unbiassed photovoltaic diodes.

Solid-state electron detectors have been reviewed by Radzimski (1987).

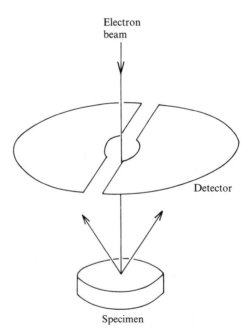

Fig. 5.6 Split annular solid-state electron detector.

# 6
# Wavelength-dispersive spectrometers

## 6.1 Bragg reflection

X-rays incident on a crystal are coherently scattered by the atomic electrons. At a certain glancing angle of incidence ($\theta$) X-rays of a given wavelength ($\lambda$) scattered by atoms in parallel planes are in phase and 'reflection' occurs (fig. 6.1). Bragg's law relating $\theta$, $\lambda$ and the interplanar spacing of the crystal ($d$) follows from simple geometrical considerations:

$$n\lambda = 2d \sin \theta, \tag{6.1}$$

where $n$ is the order of reflection, or the number of wavelengths corresponding to the path difference between rays scattered from successive layers.

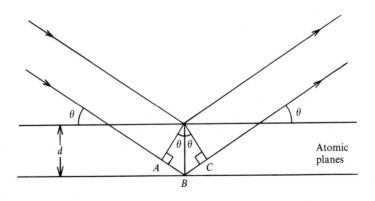

Fig. 6.1 Bragg 'reflection': when difference in path lengths ($ABC$) is equal to X-ray wavelength $\lambda$ (or multiple thereof), rays diffracted by successive planes of atoms are in phase.

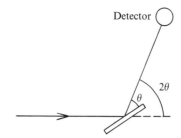

Fig. 6.2 Geometry of Bragg spectrometer.

A Bragg spectrometer can be constructed by mounting a crystal, with the diffracting planes parallel to the surface, on a rotating shaft. An X-ray detector is mounted on an arm geared to the shaft in the ratio 2:1 to maintain equal angles of incidence and reflection (fig. 6.2). Neglecting high order reflections ($n>1$), equation (6.1) is satisfied at a given angle for one wavelength only, and the detector records the intensity at that point in the spectrum: the spectrometer thus acts as a monochromator, different wavelengths being selected by changing $\theta$. This type of spectrometer is known as 'wavelength-dispersive' in order to distinguish it from the 'energy-dispersive' type, which employs electronic discrimination to resolve different X-ray energies (chapters 9 and 10).

Since usually only the first order reflection is used, and mechanical considerations impose a lower limit of 10–15° and an upper limit of 60–70° on $\theta$, the range of wavelengths covered is approximately $0.4d$–$1.8d$. Various crystals with different spacings are available (§6.2).

### 6.1.1 Reflection by perfect crystals

Darwin (1914) calculated the intensity reflected by a perfect crystal, close to the Bragg angle. The theoretical curve of reflectivity as a function of angle takes the form of a region of total reflection extending over a few seconds of arc, with very steep sides and approximately exponential tails. Absorption, neglected by Darwin, was shown by Prins (1930) to cause a reduction in reflectivity. Zachariasen (1967) gives a general treatment of this subject.

The reflecting characteristics of a crystal can be determined by plotting the reflected intensity of monochromatic radiation as the crystal is rotated through the Bragg angle. The area of the curve of reflectivity plotted against angle ('reflection curve' or 'rocking curve') is known as the 'integrated reflectivity' or 'integrated reflection' ($R_i$), and is measured in radians. For nearly perfect crystals $R_i$ is typically in the range $10^{-4}$–$10^{-5}$ rad. In focussing spectrometers employing curved crystal geometry (§6.4) the reflected intensity is dependent on the integrated reflectivity rather than the peak reflectivity, because the width of the reflection curve determines the area of the crystal which reflects (§6.8).

### 6.1.2 Reflection by imperfect crystals

The integrated reflectivity of some real crystals is much higher than for a perfect crystal, because they have a 'mosaic' structure consisting of slightly misoriented blocks or domains (fig. 6.3). The enhancement of $R_i$ is due to broadening of the reflection curve owing to the wider range of wavelengths satisfying the Bragg equation at a given angle of incidence. The theoretically very good resolution of a perfect crystal cannot be effectively utilised in practical spectrometers in any case (see §6.7), and mosaic crystals are preferable because of their higher integrated reflectivity.

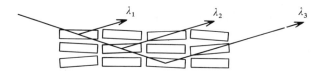

Fig. 6.3 Bragg reflection of slightly different wavelengths ($\lambda_1$ etc.) by misoriented blocks in mosaic crystal.

Bragg, James and Bosanquet (1921) demonstrated that a substantial increase in $R_i$ for rocksalt (NaCl) can be obtained by abrading the surface. This reduces the size of the mosaic blocks and increases the range of angles, resulting in an increase in the integrated reflectivity. Deformation of lithium fluoride (LiF) also causes a considerable increase in integrated reflectivity (Birks and Seal, 1957), but not all crystals respond in this way.

### 6.2 Crystals for w.d. spectrometers

The earliest crystals used in X-ray spectrometry were natural ones such as rocksalt, mica, calcite and quartz, but these have been displaced by synthetic alternatives. In choosing suitable crystals, physical properties are important. Thin plates of the required dimensions and orientation must be obtainable, either by cutting or cleaving, and need to be thin enough to be curved to the appropriate radius for focussing geometry. The response to

Table 6.1. *Crystals for X-ray spectrometry.*

| Name | Abbreviation | Formula | Reflecting plane | $2d(\text{Å})$ |
|---|---|---|---|---|
| Lithium fluoride | LIF | LiF | 200 | 4.027 |
| Quartz[1] | QTZ | $SiO_2$ | $10\bar{1}1$ | 6.686 |
| Pentaerythritol | PET | $C_5H_{12}O_4$ | 002 | 8.742 |
| Ammonium dihydrogen phosphate[1] | ADP | $NH_6PO_3$ | 101 | 10.64 |
| Mica (muscovite)[1] |  | $KAl_3Si_3O_{12}$ | 002 | 19.84 |
| Thallium acid phthalate[2] | TAP | $C_8H_5O_4Tl$ | $10\bar{1}0$ | 25.9 |

[1] No longer in common use in electron microprobes
[2] Rubidium and potassium acid phthalates have similar $2d$ values.

grinding is also relevant for Johansson geometry (see §6.4). The crystals chosen must be stable under vacuum and preferably not hygroscopic. Details of commonly used crystals are given in Table 6.1 and their wavelength coverage is shown in fig. 6.4. The most important of these are discussed in more detail below.

LiF is the standard crystal for short wavelengths ($<3$ Å). It is naturally mosaic but its reflectivity can be further enhanced by suitable mechanical treatment (see above). Crystals are quite easy to grow and can be cleaved along the required plane.

For intermediate wavelengths, pentaerythritol (PET) is the usual choice because of its high reflectivity. However, its physical properties are not ideal: it is rather soft and tends to deteriorate with time. Also it has a large thermal expansion coefficient, making it much more sensitive to temperature changes than other crystals (§6.2.3).

The acid phthalates are used for long wavelengths. The first to be introduced was potassium acid phthalate (KAP), but this has been supplanted by RAP and more recently by TAP (the Rb and Tl analogues of KAP), which give even higher reflectivity.

The crystals discussed above give complete coverage of the 'normal' wavelength range of 1–12 Å, allowing all elements from atomic number 11 (Na) upwards to be detected. The acid phthalates cover F K$\alpha$ and are just able to reach O K$\alpha$. For very long wavelengths, where the K peaks of the 'light' elements (B, C, N, O) are found, special multilayer diffracting structures (§6.3) are used.

It is desirable to have three separate spectrometers (or four if light elements are included) so that all elements can be covered without the need

## 6.2 Crystals for w.d. spectrometers

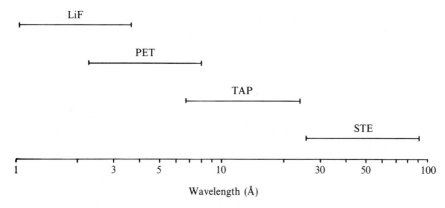

Fig. 6.4 Wavelength range of common crystals (first order reflection; $\theta = 15\text{--}65°$); also included is lead stearate (STE) – see §6.3.1.

for changing crystals. A further advantage of this arrangement is that the optimum type of counter (§§7.2, 7.3) for the wavelength range concerned can be fitted to each spectrometer.

### 6.2.1 High order reflections

So far only first order reflections have been considered, but the existence of higher order reflections (§6.1) must not be ignored. The relative intensity of successive orders varies with crystal structure: LiF has a fairly intense second order reflection (approximately 10% of the first order), whereas that of KAP is weak (about 2% of first order). Mica has particularly intense high order reflections.

There is hardly ever any reason for using high order reflections for analytical purposes, and their presence is important mainly as a possible cause of interferences. Pulse height analysis (§11.4) can be used to suppress high orders.

### 6.2.2 X-ray refraction

A small correction to Bragg's law is required to allow for the refraction of X-rays entering the crystal (Compton and Allison, 1935). The refractive index of a solid is less than unity by an amount $\delta$ which is typically of the order of $10^{-6}$. Bragg's law modified to allow for refraction takes the following form:

$$n\lambda = 2d[1 - (\delta \sin^2\theta)]\sin\theta.$$

It is convenient to substitute in equation (6.1) (Bragg's law) a modified value of the interplanar spacing $d$, given by:

$$d' = d[1 - (\delta \sin^2 \theta)].$$

Substituting for $\sin\theta$, this expression becomes:

$$d' = d[1 - (4d^2\delta/n^2\lambda^2)].$$

A simplification is possible because $\delta/\lambda^2$ is nearly constant, except close to absorption edges, hence:

$$d' = d[1 - (k/n^2)],$$

where $k$ is a constant for a given crystal and $n$ is the order of diffraction. Values of $k$ for commonly used crystals are as follows: LiF – $5.8 \times 10^{-5}$, PET – $1.44 \times 10^{-4}$, TAP – $2.18 \times 10^{-3}$.[1] The effect of refraction is thus greatest for long wavelengths.

### 6.2.3 The effect of temperature

Differentiation of equation (6.1) with respect to temperature gives the following expression (for $n=1$):

$$d\theta/dT = -(1/d)(dd/dT)\tan\theta = -\alpha \tan\theta,$$

where $\alpha$ is the linear expansion coefficient of the crystal (in the appropriate direction). For LiF (200 reflection), $\alpha = 3.4 \times 10^{-5}$: hence at $\theta = 45°$ the change in $\theta$ per degree is $3.4 \times 10^{-5}$ rad, or 7 seconds of arc. A temperature change of a few degrees may thus be significant. The effect is considerably more serious in the case of PET, for which $\alpha = 1.2 \times 10^{-4}$. Owing to the dependence of $d\theta/dT$ on $\tan\theta$, the sensitivity to temperature change is greatest at high $\theta$ values (about twice as great at 65° as at 45°).

### 6.3 Diffracting structures for long wavelengths

The lack of suitable crystals with large $2d$ values for long wavelength X-ray spectrometry has led to the development of synthetic multilayers in which widely spaced layers of heavy atoms act as diffracting planes. Diffraction gratings can also be used.

---

[1] By courtesy of Bureau de Recherches Géologiques et Miniéres, Orleans, France.

Table 6.2. *Soap film pseudo-crystals for X-ray spectrometry* ($x$ = number of $CH_2$ units).

| Name | $x$ | $2d$ (Å) (approx.) |
|---|---|---|
| Laurate (dodecanoate) | 12 | 70 |
| Myristate (tetradecanoate) | 14 | 80 |
| Palmitate (hexadecanoate) | 16 | 90 |
| Stearate (octadecanoate) | 18 | 100 |
| Lignocerate (tetracosanoate) | 24 | 125 |
| Melissate (triacontanoate) | 30 | 155 |

### 6.3.1 Soap films

Soaps are metal salts of straight-chained fatty acids, and multilayer soap films consisting of layers of heavy metal atoms separated by $CH_2$ chains can be used for X-ray spectrometry. The general formula is $(CH_2)_xO_4M$, where M is a divalent metal, usually Pb, and the number ($x$) of $CH_2$ units determines the $2d$ value (see table 6.2).

Soap film multilayers are made by the Blodgett–Langmuir technique (Blodgett, 1935; Blodgett and Langmuir, 1937): a fatty acid solution (e.g. stearic acid in benzene) is spread on the surface of water containing the required metal ions in solution, and soap molecules consisting of metal ions attached to fatty acid chains form a surface film. When this is compressed against a barrier the molecules become oriented with the $CH_2$ chains projecting from the surface. A suitable substrate repeatedly dipped vertically into the water picks up monolayers of soap, forming double layers of metal atoms separated by two fatty acid chain lengths.

The reflectivity of soap 'crystals' is strongly affected by the absorption edge of C in the $CH_2$ chains. Charles (1971) measured an integrated reflectivity of more than $4 \times 10^{-4}$ rad for C K$\alpha$ X-rays with 150 layers of lead stearate, compared with $< 10^{-4}$ rad for O K$\alpha$, which is on the high energy side of the edge and is therefore heavily absorbed. Typical peak reflectivities are of the order of a few per cent.

### 6.3.2 Evaporated multilayers

Another way of producing a layered structure for X-ray spectrometry is to deposit alternate layers of heavy and light elements on a suitable substrate by vacuum evaporation (DuMond and Youtz, 1935). X-rays are scattered by the heavy atoms (e.g. W or Mo), which are deposited in thin layers separated by thicker layers of light atoms (e.g. Si or C). Evaporated multilayers are available with a wide range of $2d$ values (40–200 Å) and can be substituted directly for 'true' crystals or soap film multilayers. Compared to the latter they are more stable and give more intensity, but have worse resolution. A useful characteristic is the very low intensity of reflections of order greater than two, which helps to minimise interferences from heavier elements. A theoretical model for the behaviour of these devices has been compared with experimental measurements by Henke, Uejio, Yamada and Tackaberry (1986).

Evaporated multilayers give broad, intense reflections. However, it is important to choose both spacing and composition correctly if optimum performance is to be obtained for a particular wavelength. For light element analysis it is therefore desirable to have several types available. For example, the following $2d$ values are appropriate for the elements indicated: 60 Å (O), 100 Å (C), 140 Å (Be, B). Reasonable (though not optimal) performance is obtained for the elements immediately adjacent to those given.

### 6.3.3 Diffraction gratings

The well known optical principle of diffraction by a grating consisting of a linear array of closely spaced reflecting elements can be applied in the long wavelength X-ray region. A low angle of incidence is necessary for X-rays in order to obtain efficient reflection, and special grating fabrication techniques are required (Sayce and Franks, 1964). The geometry of the grating spectrometer differs from that of the Bragg spectrometer, the angle of incidence being kept constant while the arm carrying the detector is moved to different angles to select the wavelength required. Such a spectrometer designed for use on an electron microprobe was developed by Nicholson and Hasler (1966), but in spite of performance comparing favourably with the Bragg spectrometer (using a stearate 'crystal'), grating spectrometers have not been widely adopted for microprobe analysis.

### 6.4 Focussing geometry

So far the discussion of Bragg reflection has assumed a flat crystal and parallel X-ray beam. With a point source, as in the electron microprobe, the angle of incidence varies across the surface of a flat crystal and Bragg

### 6.4 Focussing geometry

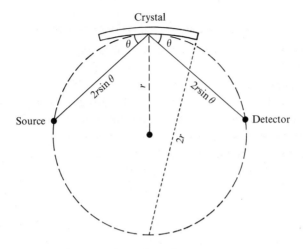

Fig. 6.5 Johann focussing geometry.

reflection occurs over a small area. Much improved efficiency can be obtained by using focussing geometry in which source, crystal and detector lie on the circumference of a 'Rowland circle' of radius $r$, the atomic planes of the crystal being curved to a radius $2r$. In 'Johann geometry' (Johann, 1931), shown in fig. 6.5, the surface of the crystal diverges from the Rowland circle and the angle of incidence is not strictly constant. This can be overcome by using 'Johansson geometry' (fig. 6.6), in which the crystal planes are curved to the same radius, but the surface is ground to the radius of the Rowland circle (Johansson, 1932, 1933). The angle of incidence is now constant at all points in the plane of the Rowland circle, though there is still an aberration for rays inclined to the plane. Owing to the difficulties associated with the grinding process, Johann geometry is generally used in practice.

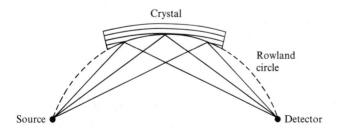

Fig. 6.6 Johansson focussing geometry.

### 6.4.1 Doubly-curved crystal geometry

In Johansson geometry, reflection occurs only along a strip down the centre of the crystal, since rays inclined to the plane of the Rowland circle have a slightly different angle of incidence (see §6.7). This limitation can be overcome by double curvature of the crystal (fig. 6.7). The appropriate radius about the line joining source and focus is given by $r' = 2r\sin^2\theta$, and is thus a function of $\theta$. In order to cover all wavelengths it is therefore necessary to use crystals with different values of $r'$, which is a major drawback. However, an alternative double-curved configuration has been shown by Wittry and Sun (1990a, b) to give high reflectivity (though somewhat less than in the ideal case) over a wide range of Bragg angles. There are practical problems in producing the doubly-curved shape with real crystal materials, though special techniques have been developed for this purpose (Golijanin, Wittry and Sun, 1989).

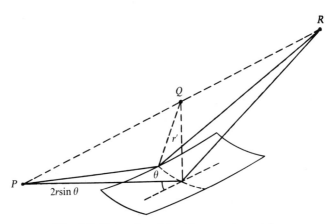

Fig. 6.7 Geometry of doubly-curved crystal.

Difficulties in fabricating doubly-curved crystals can be reduced by using a stepped profile approximating to the ideal shape (Wittry and Sun, 1991). The benefits of this compared to Johann or Johansson geometry are greatest for crystals with a narrow rocking curve, for which a useful increase in reflecting area is obtained.

## 6.5 Design of fully focussing spectrometers

Fully focussing geometry requires a complicated mechanism to maintain the correct angular and distance relationships. Since the source–crystal

## 6.5 Design of fully focussing spectrometers

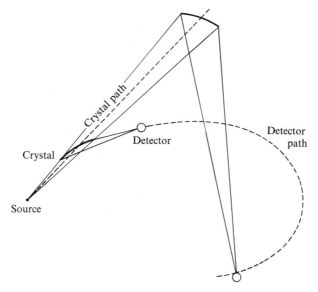

Fig. 6.8 W.d. spectrometer with linear crystal motion.

distance varies as $\sin\theta$, and hence is proportional to $\lambda$, linear wavelength calibration can be obtained by moving the crystal along a straight line (Sandstrom, 1952), which has the additional advantage that the X-ray take-off angle is constant (fig. 6.8).

The choice of Rowland circle radius is restricted by the minimum radius to which the crystal can be curved, and how close it can approach to the specimen. Too large a radius is inconvenient owing to the large chamber required to house the spectrometer; also the size of the crystals must be scaled proportionately in order to achieve a given solid angle for X-ray collection. In practice the radius is usually in the range 12–25 cm.

The spectrometer mechanism can be mounted with the plane of the Rowland circle vertical, allowing several spectrometers (e.g. up to five) to be fitted. However, in this orientation, measured intensities are susceptible to small differences in the position of the X-ray source in the $z$ direction (§6.10.1). This can be avoided by mounting the spectrometer horizontally, which is especially beneficial if the specimen surface cannot be accurately located in the $z$ direction (e.g. in an instrument lacking an optical microscope). In the electron microprobe the specimen is usually horizontal: the spectrometer plane must therefore be inclined in order to obtain the required take-off angle (fig. 6.9), but this does not materially affect the immunity to $z$ shift effects.

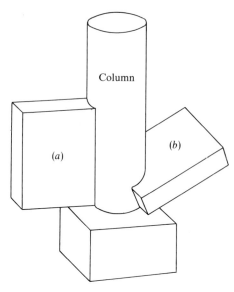

Fig. 6.9 W.d. spectrometers mounted (a) vertically, (b) inclined to the horizontal.

Analogue servo systems used in the past for controlling spectrometers have been superseded by digital systems. A stepping motor is often used to drive the spectrometer. In the case of the linear mechanism described above, the motor controls the source–crystal distance (by means of a lead-screw for example), with each step representing a constant increment of $\sin\theta$. The system is first initialised by reference to a known peak and thereafter $\theta$ is changed by applying the appropriate number of pulses to the stepping motor, as calculated by the computer. The speed of stepping motors is somewhat limited: a faster response can be obtained by driving with an ordinary motor and sensing the crystal position with an optical encoder.

Most spectrometers have a crystal-changing device. Some designs allow the crystal to be changed at any Bragg angle, whereas in others it is necessary to move the spectrometer to a particular angle. An optional feature is a changeable detector slit, which gives some control over the resolution (§6.7.1).

### 6.5.1 Variable crystal curvature

Elion and Ogilvie (1962) designed a spectrometer with a flexible crystal of variable curvature and a fixed axis of rotation. Focussing was maintained by adjusting the curvature automatically to keep the radius equal to $a\operatorname{cosec}\theta$ (where $a$ is the source–crystal distance) as required by Rowland circle geometry. Mica on which a soap-film multilayer (§6.3.1) was deposited gave complete wavelength coverage, using high order reflections for short wavelengths.

## 6.6 Semi-focussing geometry

Fully focussing geometry requires a fairly complicated mechanism, as discussed above. It is much simpler to have a fixed crystal axis and to mount the detector on an arm of fixed length (Long and Cosslett, 1957). A crystal of given radius then focusses correctly at only one wavelength. In practice the resolution and intensity are acceptable over a range of wavelengths and it is sufficient to have a limited number of interchangeable crystals of different radii. If the source–crystal distance is $c$, the correct radius of curvature for wavelength $\lambda$ is $2cd/\lambda$.

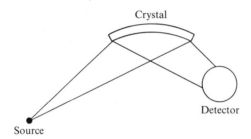

Fig. 6.10 Semi-focussing spectrometer.

Since focussing is inexact at most wavelengths it is not necessary to maintain the correct distance between detector and crystal, hence space can be saved by moving the detector closer to the crystal (fig. 6.10). The semi-focussing spectrometer is simple and cheap to produce, though performance is sacrificed to some extent by comparison with the fully focussing type.

## 6.7 Line width and resolution

The angle of incidence of the X-rays is not constant over the whole crystal in focussing spectrometers, except when a doubly-curved crystal is used

# 6 Wavelength-dispersive spectrometers

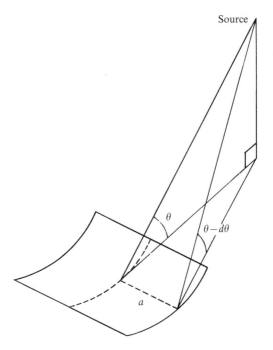

Fig. 6.11 Bragg angle 'error' for divergent ray.

(§6.4.1). Divergent rays are incident at a lower angle than those lying in the plane of the Rowland circle. The maximum difference, $\triangle\theta$, for rays incident at the edge (fig. 6.11) is given (in radians) by:

$$\triangle\theta = (a/r)^2/4\sin2\theta, \qquad (6.2)$$

where $a$ is the half-width of the crystal and $r$ is the radius of the Rowland circle. The function $\sin2\theta$ has a maximum value of 1 at $\theta = 45°$, falling to 0.5 at $\theta = 15°$ (typically the lowest Bragg angle of interest). Thus if $a = 10$ mm and $r = 150$ mm, then $\triangle\theta = 0.06°$ when $\theta = 45°$, and $0.12°$ when $\theta = 15°$.

Usually the width of the rocking curve of the crystal (§6.1.1) is considerably smaller than $\triangle\theta$, hence Bragg reflection occurs only over a limited part of the crystal. In the case of Johansson geometry, reflection occurs in a band lying along the centre of the crystal when $\theta = \theta_\lambda$ (the Bragg angle for wavelength $\lambda$). As $\theta$ is increased the reflecting area splits into two symmetrical bands which reach the edge of the crystal when $\theta = \theta_\lambda + \triangle\theta$. The line width in terms of angle is thus equal to $\triangle\theta$. The rocking curve also contributes to the line width, but as a rule the geometrical factor is predominant. Consequently the resolution is governed primarily by the

## 6.7 Line width and resolution

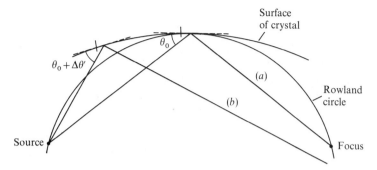

Fig. 6.12 Bragg angle 'error' in Rowland circle plane, for Johann geometry: (a) ray reflected at centre of crystal, (b) ray reflected at off-centre point which differs in Bragg angle by $\Delta\theta'$ ands does not pass through focus.

ratio $a/r$ rather than the properties of the crystal (though this is not necessarily true for multilayers with broad rocking curves).

In focussing spectrometers employing Johann geometry (§6.4) an additional cause of line broadening is the departure of the curved surface of the crystal from the Rowland circle. The resulting angular 'error' ($\Delta\theta$) is positive (fig. 6.12) and for a crystal of half-length $b$ the maximum value, occurring at the edges, is given by:

$$\Delta\theta' = (b/r)^2/4\tan\theta. \qquad (6.3)$$

In the direction perpendicular to the plane of the Rowland circle $\Delta\theta$ is the same as for Johansson geometry (see above) and is negative. It follows that the errors cancel out at points which lie on a cross, as shown in fig. 6.13 (Ditsman, 1960). Further, the total range of angles (and hence the line width) is given by adding the expressions in equations (6.2) and (6.3). The resolution is therefore worse for Johann than Johansson geometry for a given crystal size. Substituting $b = 10$ mm and $r = 150$ mm in equation (6.3), we find $\Delta\theta = 0.06°$ for $\theta = 45°$ and $0.24°$ for $15°$. Combining these with the figures for $\Delta\theta$ calculated above, the angular line width is $0.12°$ at $\theta = 45°$ and $0.36°$ at $15°$. The difference in resolution between Johann and Johansson geometries is thus greatest at low Bragg angles. This tendency can be mitigated by reducing the length of the crystal and by the use of a slit at the focus (see below), but only at the expense of intensity.

It is sometimes more useful to express the line width in terms of wavelength ($\Delta\lambda$), as obtained by differentiating equation (6.1):

$$\Delta\lambda/\lambda = (\cot\theta)\Delta\theta. \qquad (6.4)$$

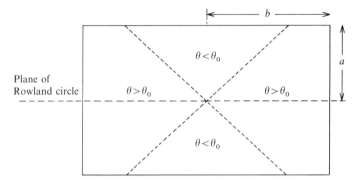

Fig. 6.13 Variation in Bragg angle 'error' with location on surface of crystal (Johann geometry); cross is locus of points for which $\theta$ is 'correct'.

In the case of Johansson geometry, $\triangle\theta$ is relatively constant and the variation in $\triangle\lambda/\lambda$ with $\theta$ is mainly determined by the $\cot\theta$ term in equation (6.4), which varies from 3.7 for $\theta=15°$ to 0.47 for $\theta=65°$. Wavelength resolution is thus quite strongly dependent on $\theta$ and, as discussed above, is relatively independent of the properties of the crystal. (For Johann geometry the dependence on $\theta$ is even more marked.) It follows that the resolution for Ti K$\alpha$, for example, is considerably better with LiF ($\theta=43°$) than PET ($\theta=18°$), though the intensity is lower (see §6.8).

### 6.7.1 Slits

The length of the line focus with either form of focussing geometry is twice the width of the crystal. Resolution can be improved by restricting the length, thereby excluding rays reflected from the outermost parts of the crystal, which is equivalent to reducing $a$ in equation (6.2). With Johansson geometry the line focus is ideally infinitely narrow, but in practice has finite width owing to imperfections in the crystal. A narrow slit improves resolution by eliminating reflections from parts of the crystal not at the correct angle. In the case of Johann geometry there is a geometrical broadening of the focus (see fig. 6.12) and a narrow slit has the effect of excluding rays reflected from the extremities of the crystal, thereby improving the resolution. From a practical viewpoint it is not usually necessary to have the highest possible resolution and since there are drawbacks (e.g. loss of intensity and increased sensitivity to the position of the source) the use of narrow slits is to be avoided as far as possible.

## 6.8 Reflection efficiency

For the purpose of calculating spectrometer efficiency it is convenient to use the integrated reflection ($R_i$) to describe the reflecting properties of the crystal (§6.1.1). Reflection can be considered as total over an angular range equal to $R_i$. In the case of Johansson geometry the angle of incidence varies as $(x/r)^2/4\sin2\theta$ with distance $x$ from the plane of the Rowland circle (§6.7). The distance over which total reflection can be assumed to occur is thus $2r(R_i\sin2\theta)^{1/2}$. Since there are equal reflecting areas on either side of the plane of the Rowland circle, the total reflecting area is $4lr(R_i\sin2\theta)^{1/2}$, where $l$ is the length of the crystal. The projection of this area perpendicular to the X-ray path is $4lr(R_i\sin2\theta)^{1/2}\sin\theta$, and the distance to the source is $2r\sin\theta$, hence the solid angle, $\Omega$, subtended at the source is given by:

$$\Omega = 4lr(R_i\sin2\theta)^{1/2}\sin\theta/4r^2\sin^2\theta = 2^{1/2}(l/r)(R_i\cot\theta)^{1/2}.$$

This solid angle is a measure of the collection efficiency of the spectrometer (though the detection efficiency of the counter is also relevant). Substituting typical values: $r = 150$ mm, $l = 20$ mm, and $R_i = 10^{-4}$, $\Omega$ ranges from $7 \times 10^{-4}$ for $\theta = 65°$ to $2 \times 10^{-3}$ for $\theta = 15°$. (Note, however, that this assumes that $R_i$ is constant, which is not true in general.) Given that characteristic X-rays are emitted isotropically over a solid angle of $4\pi$, the fraction collected is of the order of 0.01%. The efficiency is similar for Johann geometry.

The collection efficiency can be increased by using a longer crystal, but this is restricted by practical considerations. Improved efficiency can also be achieved by increasing $R_i$, for example by suitable mechanical treatment in the case of mosaic crystals (§6.1.2). With a doubly-curved crystal, reflection occurs over the whole area and an order of magnitude improvement in efficiency is possible in principle, though there are practical drawbacks to doubly-curved crystals (§6.4.1).

## 6.9 Spectrometer performance

The performance of a crystal spectrometer is usually measured in terms of the count-rate obtained for a given element with a specified probe current. Efficiency defined in terms of count-rate includes the effect of the variation of the efficiency of the detector with wavelength. Resolving power should be taken into account: the quality of a spectrometer is indicated by the extent to which high intensity and high resolution are combined.

Peak to background ratio is related to resolution and is often quoted as

Table 6.3. *Typical performance data for wavelength-dispersive spectrometers (count-rates and peak to background ratios for Kα lines of pure elements).*

| Element | Accelerating voltage (kV) | Crystal | Counts $s^{-1} \mu A^{-1}$ | Peak/bg |
|---|---|---|---|---|
| C | 10 | Pb stearate | $2 \times 10^5$ | 100 |
| Al | 20 | TAP | $3.5 \times 10^6$ | 900 |
| Ti | 25 | PET | $3.5 \times 10^6$ | 500 |
| Fe | 30 | LiF | $1.5 \times 10^6$ | 500 |

an index of spectrometer performance. Background is due principally to the continuous X-ray spectrum: contributions from electronic noise, scattered X-rays, etc. are generally small. A misleadingly high peak to background ratio is obtained if the background is measured on the short wavelength side of the absorption edge of the emitting element, where the intensity of the continuous spectrum is reduced by absorption in the specimen.

Typical performance figures are given in Table 6.3. When comparing data recorded under different conditions it should be noted that intensity and (to a lesser extent) peak to background ratio are dependent on accelerating voltage. The X-ray take-off angle of the instrument also has some effect. Sometimes intensities are quoted for a given 'absorbed' current: the difference between this and the true probe current varies up to approximately a factor of 2 for heavy elements (due to backscattering).

## 6.10 Defocussing effects

Focussing geometry requires the X-ray source to be precisely located relative to crystal and counter (§6.4). When spectrometers are installed, they are adjusted on the assumption that the source is at the intersection of the electron optical axis with the plane in which the specimen surface is intended to lie. Any departure from the correct position of the source may cause a loss of X-ray intensity as a result of 'defocussing'. The main cause of intensity loss is the change in Bragg angle resulting from displacement of the source. Change in source–crystal distance has much less effect. The source may shift along the $z$ axis (i.e. parallel to the electron optical axis in the case of normal incidence) because of differences in specimen height. Movement in the $x$ and $y$ directions, caused for instance by scanning, must also be considered. The consequences of such variations in source position, for both vertical and inclined spectrometers, are discussed below.

## 6.10 Defocussing effects

### 6.10.1 z-axis defocussing

The effect of displacement of the source in the $z$ direction, when the spectrometer is mounted vertically, is shown in fig. 6.14(a). For a small displacement $\triangle z$, the change $\triangle \theta$ in Bragg angle is given (in radians) by:

$$\triangle \theta = \triangle z \cos\psi / 2r\sin\theta.$$

Assuming that the maximum value of $\triangle \theta$ permissible for quantitative analysis is 25 $\mu$rad (5 seconds of arc), then for typical $r$ and $\psi$ (150 mm and 40° respectively), the corresponding value of $\triangle z$ ranges from 2.5 $\mu$m at $\theta = 15°$ to 9 $\mu$m at 65°. The depth of field of a typical optical microscope as fitted to an electron microprobe is sufficiently small to satisfy this requirement provided care is taken with focussing.

In the case of an inclined spectrometer (§6.5) there is no change in Bragg angle with displacement of the source in the $z$ direction. A small focussing error arises owing to the change in the source–crystal distance, but its effect can be neglected for practical purposes.

### 6.10.2 x- and y-axis defocussing

The effect of horizontal displacement of the source is shown in fig. 6.14(b). In the case of a vertically mounted spectrometer, a displacement of $\triangle x$ (where the $x$-axis lies along the intersection of the plane of the spectrometer and the surface of the specimen) causes a change $\triangle \theta$ in the Bragg angle, given by:

$$\triangle \theta = \triangle x \sec\psi / 2r\sin\theta.$$

Using the same assumptions as in §6.10.1, the maximum displacement allowable for quantitative analysis ranges from 1.5 $\mu$m at $\theta = 15°$ to 5 $\mu$m at

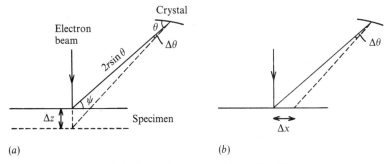

Fig. 6.14 Defocussing effects: (a) $z$-axis, (b) $x$-axis.

65°. Thus, it is very important that the beam should not be moved in the $x$ direction. (In w.d. analysis only the specimen and not the beam should be moved when selecting points for analysis.)

Loss of intensity due to change in $\theta$ is also important in scanning, though larger effects are tolerable. Thus, a change in $\theta$ of 0.001 rad, corresponding to a movement of 60–200 $\mu$m (depending on $\theta$) in the $x$ direction, may be permissible. Displacement in the orthogonal ($y$) direction does not change the Bragg angle and has a negligible effect on the intensity: an appropriately oriented line scan is thus immune from defocusing effects. Similar considerations apply in the case of a spectrometer mounted in the inclined mode, except that displacement in the $x$ direction as defined above has no effect, whereas the Bragg angle is very sensitive to movement of the source in the $y$ direction.

Defocussing effects in scanning images can be avoided by varying $\theta$ in synchronism with the appropriate deflection waveform. This requires the raster to be oriented appropriately relative to the spectrometer, which is somewhat inconvenient when images of several elements, using different spectrometers, are required. An alternative approach is to apply a correction to the intensity at each point, derived from a previously recorded profile of the spectrometer response as a function of the displacement of the spot (Marinenko, Myklebust, Bright and Newbury, 1989). Element distribution maps can also be produced by moving the specimen only, with a static beam, which avoids defocussing effects.

# 7
# Proportional counters

## 7.1 Principles of operation

The proportional counter consists of a metal cylinder containing a suitable gas, with a concentric anode wire mounted on insulators at each end (fig. 7.1). X-rays enter through an aperture covered with window material strong enough to withstand the pressure difference between the gas and the surrounding vacuum chamber, and are detected by virtue of their ability to ionise the gas. Free electrons generated by incident X-rays move to the anode wire, while the positive ions drift outwards to the cathode. The resulting negative signal at the anode is amplified electronically.

X-ray absorption in the counter gas takes place by the photoelectric effect, in which an inner electron is ejected from a gas atom. There follows the emission of either an Auger electron or a fluorescent X-ray photon which is itself absorbed by another atom producing another photoelectron, and so on. The Auger and photoelectrons are rapidly slowed down in the gas by collisions which produce more free electrons. One 'ion pair' (an electron and a positive ion) is generated for each 25–30 eV of incident X-ray energy: thus a 1 keV photon produces about 40 ion pairs. The amplitude of the pulse at the anode is proportional to the X-ray energy, but is subject to statistical variance owing to the fairly small number of ion pairs involved.

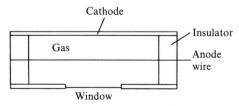

Fig. 7.1 Proportional counter.

To produce an output pulse of reasonable amplitude it is necessary to use a high enough anode potential for the free electrons to be accelerated between collisions and cause further ionisation, generating an 'avalanche' of secondary electrons. Since each primary electron generates the same mean number of secondaries, the output pulse height is still proportional to the incident X-ray energy. The internal amplification produced by 'gas multiplication' in the avalanche is known as the 'gas gain', and increases rapidly with anode voltage. When the avalanche spreads along the whole anode wire, the output is constant regardless of the X-ray energy: this applies when a counter is operated in the 'Geiger' mode. The counter then not only lacks energy resolution, but also has an excessively long dead time and is unsuitable for use in microprobe analysis.

## 7.2 Entrance window

The window is required to transmit X-rays while preventing leakage of counter gas into the surrounding vacuum chamber. A 'sealed counter' is permanently sealed, and its useful life depends on the impermeability of the window. This is usually made of Be, which has good X-ray transmission because of its low atomic number (4). Fig. 7.2 shows the transmission of such a window (thickness 127 μm) as a function of X-ray wavelength.

For long wavelength X-rays a 'flow counter' with a thin window is used. A continuous slow stream of gas passing through the counter makes up for any loss through the window, which must still be strong enough to withstand the pressure of the gas. A window material commonly used

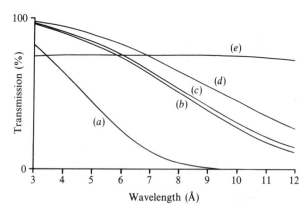

Fig. 7.2 Transmission of counter windows: (a) 127 μm (5 mil) Be, (b) 6 μm Mylar, (c) 25 μm (1 mil) Be, (d) 3.75 μm Mylar, (e) 0.1 μm formvar with 75% transmission grid.

in flow counters is 'Mylar' or 'Melinex' (polyethylene terephthalate, $C_{10}H_8O_4$), which is available in a standard thickness of 6 $\mu$m, and is very strong. Thinner Mylar is also available and gives significantly better transmission at long wavelengths (fig. 7.2).

Another useful material is polycarbonate ($C_{16}H_{14}O_3$), available as thin sheet, which will withstand atmospheric pressure over small areas. Thin windows can also be made from stretched polypropylene (($CH_2)_n$) sheet. Caruso and Kim (1968) described a procedure in which 26 $\mu$m sheet is clamped over an opening in a chamber pumped out to give the required amount of stretching. Polypropylene has the advantage for some applications that it contains no O and therefore has no absorption edge at 23.32 Å (532 eV). Materials containing O are relatively transparent to the O K$\alpha$ line itself, but heavily absorb X-rays just on the short wavelength side of the edge.

'Ultra-thin' windows ($< 1$ $\mu$m) need mechanical support in the form of a grid on the outside of the window. Nickel grids produced by etching or electroforming are generally used. Holes typically occupy 70–80% of the area, the remainder being practically opaque to X-rays. The resulting intensity loss is more than compensated by the high transmission of the window material. Suitable materials for ultra-thin windows include collodion (cellulose nitrate, $C_{12}H_{11}O_{22}N_6$) and formvar (polyvinyl formaldehyde, $C_5H_7O_2$). Films of about 0.1 $\mu$m thickness can be produced by evaporating the solvent from drops of solution spread on a water surface (Henke, 1965). The transmission of 0.1 $\mu$m formvar with a 75% transmission grid is plotted in fig. 7.2. Ultra-thin windows are liable to rupture occasionally and the vacuum system should be protected against this eventuality.

Apart from beryllium, most window materials are electrically insulating. In order to avoid disturbance of the field in the counter, the inside surface should be coated with a conducting layer connected to the body of the counter. This coating may eventually be eroded by ion bombardment, necessitating recoating or replacement. The front surface may also need to be coated to prevent charging up due to electrons scattered from the specimen; vacuum evaporated aluminium is generally used for this purpose.

## 7.3 Counter gas

In principle, any gas can be used in a proportional counter, but electronegative gases such as oxygen are undesirable because negative ions formed by electrons becoming attached to gas atoms have low mobility and are susceptible to recombination. This does not occur in the noble (rare) gases, which also have the advantage that adequate gas gain can be obtained at a lower anode voltage than with other gases. Argon is commonly used, but is unsuitable in the pure state, mainly because its transparency to UV radiation allows photons emitted by excited gas atoms to reach the counter wall, where they release electrons which contribute to the avalanche, causing the early onset of Geiger operation. The addition of a few per cent of a polyatomic gas to absorb UV radiation enables the counter to be operated at much higher gain. The most commonly used gas mixture is 90% argon–10% methane (known as 'P10').

For high counting efficiency all or most of the incident radiation should be absorbed in the counter gas. Fig. 7.3 shows absorption by 2 cm of argon at atmospheric pressure as a function of X-ray wavelength. Argon is suitable for wavelengths above about 5 Å, and is generally used in flow counters for soft X-ray detection. An exit window opposite the entrance window prevents high energy X-rays from striking the back wall of the counter and producing spurious pulses.

Better efficiency for short wavelengths is obtained by using xenon, as shown in fig. 7.3. Another way of increasing absorption of short wavelength X-rays is to use argon at high pressure (e.g. 2–3 atm). In a crystal spectrometer, flow and sealed counters can be mounted in tandem, low energy X-rays being absorbed in the flow counter, while high energy X-rays pass through to the sealed counter behind.

Very soft X-rays ($> 10$ Å) are so heavily absorbed in ordinary counter gases that primary ionisation takes place very close to the window, where the field is weak and may be inadequate to prevent recombination.

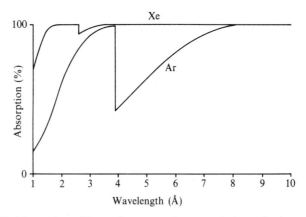

Fig. 7.3 Absorption of 2 cm of argon and xenon at atmospheric pressure.

Performance is improved by decreasing the gas density, either by changing the composition or reducing the pressure. Baun (1969) described a method of operating at reduced pressure, whereby the gas enters the counter through a control valve, with the outlet connected to the rough vacuum line via a capillary tube. For soft X-rays the methane content of the argon–methane mixture may be increased to 50% or more in order to reduce the density, but a higher anode voltage is then necessary. Neon is sometimes used because it absorbs less strongly than argon, owing to its lower atomic number.

To obtain optimum performance the counter gas must be of high purity, especially with regard to oxygen and water vapour contamination. Commercial gases are generally adequately pure, but it is advisable to discard the cylinder before it is completely exhausted, because the concentration of impurities tends to increase as the cylinder empties. Sealed counters stored in air eventually deteriorate due to the slight permeability of the window, and should preferably be freshly filled shortly before installation.

The internal gain of a proportional counter is a function of the density of the gas. In the case of a flow counter with the gas vented directly to the atmosphere, the density varies with ambient temperature and pressure, causing variations in pulse height, which typically changes by 1% for a pressure change of 1 mbar and 3% for a temperature change of 1°. This problem can be solved by employing a gas density stabiliser such as that described by Deslattes, Simson and La Villa (1966), incorporating an aneroid barostat to control the gas flow. If the temperature is constant, it is merely necessary to stabilise the gas pressure.

## 7.4 Gas multiplication

The electrostatic field, $X$, in a cylindrical counter at a distance $r$ from the axis is given by $X = V_1/r\ln(r_2/r_1)$, where $r_1$ and $r_2$ are the radii of the anode wire and cathode respectively, and $V_1$ is the anode potential. As $V_1$ is increased, the critical field ($X_c$) required for gas multiplication is reached at the surface of the anode with an anode potential $V_1'$ given by the expression: $X_c = V_1'/r_1\ln(r_2/r_1)$. When $V_1 > V_1'$ the radial distance $r'$ at which gas multiplication starts is given by: $r' = r_1(V_1/V_1')$. Assuming $r_1 = 25\,\mu\text{m}$ and $r_2 = 10\,\text{mm}$, and substituting $X_c = 2\,\text{V}/\mu\text{m}$ for argon (Sharpe, 1964), then $V_1' = 300\,\text{V}$ and for $V_1 = 1500\,\text{V}$, $r' = 125\,\mu\text{m}$. In this typical case, gas multiplication thus occurs only within 100 μm of the surface of the anode wire.

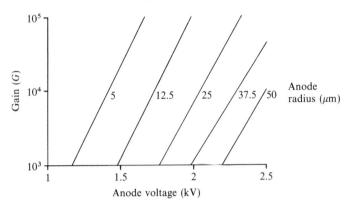

Fig. 7.4 Gas gain of proportional counter (cathode radius 2.8 cm) containing P10 gas at atmospheric pressure (after Charles (1972)).

Charles (1972) showed that many earlier gas gain measurements were inaccurate due to neglect of the effect of pulse shaping in the main amplifier. He found that corrected experimental measurements of gain ($G$) using P10 gas agreed to within $\pm 10\%$ with the expression attributed to Diethorn:

$$\ln G = [AV_1/\ln(r_2/r_1)]\ln[V_1/Bpr_1\ln(r_2/r_1)], \tag{7.1}$$

where $p$ is the gas pressure in atmospheres, and the constants $A$ and $B$ depend on the gas. Charles derived the empirical values: $A = 0.0228$, $B = 2.84$ ($r_1$ in micrometres). Hendricks (1972) also found that equation (7.1) fitted experimental measurements of $G$, and determined empirical values: $A = 0.0204$, $B = 3.62$ (for the same gas mixture). Fig. 7.4 shows $G$ as a function of $V_1$ for various anode radii, as measured by Charles. For a given $r_1$, $G$ increases exponentially with $V_1$, necessitating a voltage supply stable to a few parts in $10^4$ for a gas gain stability of better than 1%.

High gas gain is advantageous in that less gain is required in the amplifier, resulting in a better signal to noise ratio, but at excessively high gas gain, $G$ decreases at high count-rates ('pulse height depression' – see §7.5). Proportionality between pulse height and incident X-ray energy is maintained over a wide range of gas gain, but non-proportional behaviour sets in at high gain, close to the Geiger region. Hanna, Kirkwood and Pontecorvo (1949) observed that in a typical counter, non-proportionality occurs when the product of gas gain and incident X-ray energy (in keV) exceeds $10^5$.

## 7.5 Pulse height depression

It is a common observation that at high count-rates the size of the output pulses decreases, which is troublesome when pulse height analysis is used. Screening by positive ions forming a cloud around the anode wire is a possible explanation for this effect (Hendricks, 1969). However, numerous studies (e.g. Burkhalter, Brown and Myklebust, 1966; Spielberg, 1966, 1967a,b; Mahesh, 1976) have shown that pulse height depression occurs at count-rates too low for screening to be significant and various other mechanisms have been proposed. Contamination of the anode wire probably plays a significant role (den Boggende, Brinkman and de Graaff, 1969; Bawdekar, 1975; Spielberg and Tsamas, 1975).

Regardless of the cause, pulse height depression can be reduced to an acceptable level by using a sufficiently large anode wire diameter or low voltage. The loss of gas gain must be compensated by increasing the electronic gain. With a low noise, high gain preamplifier, a gas gain of only a few hundred can be used.

## 7.6 Output pulses

Counter output pulses are produced by the induction of charge on the electrodes resulting from the motion of electrons and ions in the electric field. The total charge generated by an ion pair is $e$ (the charge on the electron), but the contribution of each electron or positive ion is dependent on the difference in its potential before and after collection. By far the majority of ion pairs are produced close to the anode, where the avalanche occurs (§7.4). The charge induced by electrons moving to the anode is therefore very small, and the output pulse is produced almost entirely by the outward motion of positive ions. The shape of the output pulse is as shown in fig. 7.5.

The total amplitude of the pulse is equal to $Q/C$ volts, where $Q$ is the charge generated and $C$ is the capacitance of the counter and associated components. For $10^5$ ion pairs, $Q = 1.6 \times 10^{-14}$ C and if $C = 5$ pF the pulse amplitude is 3.2 mV. Since the amplitude is inversely proportional to $C$, it is desirable to minimise cable capacitance by locating the preamplifier close to the counter. The preamplifier provides a low impedance output so that the signal can be transmitted to the main amplifier through a long coaxial cable.

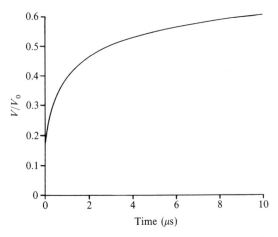

Fig. 7.5 Response of proportional counter to arrival of X-ray photon.

## 7.7 Ionisation statistics

The numbers of ion pairs produced both by primary ionisation and in the avalanche are subject to statistical fluctuations, and mono-energetic X-rays therefore produce pulses of varying amplitude. The pulse height distribution is very close to the gaussian function:

$$y = a \exp[-(x-b)^2/2\sigma^2],$$

where $x$ is the pulse height, $y$ the number of pulses, and $a$, $b$, and $\sigma$ are constants (fig. 7.6). The law of proportionality between X-ray energy and pulse height (§7.1) applies to the *mean* pulse height. For X-rays below about 1 keV the distribution becomes slightly skew, with an extended tail on the high energy side.

The resolution of the counter is usually defined as the full width at half maximum (FWHM) of the pulse height distribution. The FWHM is 2.355 times the standard deviation ($\sigma$) of the gaussian function. Since pulse height is equivalent to X-ray energy, the FWHM can be expressed in terms of energy ($\triangle E$).

If $\varepsilon$ is the mean energy required to produce an ion pair, then the mean number ($\bar{N}$) of primary ion pairs produced by an X-ray photon of energy $E$ is given by: $\bar{N} = E/\varepsilon$. Some of the photon energy (about half in the case of argon) is dissipated in non-ionising collisions. If ionising collisions were a small fraction of all collisions they would occur randomly, and the standard deviation ($\sigma$) of $\bar{N}$ would be $\bar{N}^{1/2}$. On the other hand, if all the photon

## 7.7 Ionisation statistics

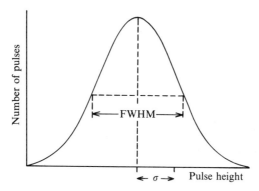

Fig. 7.6 Gaussian pulse height distribution.

energy were devoted exclusively to ionisation there would be no fluctuation in the number of ion pairs. Since the actual situation lies between these extremes, $\sigma < \bar{N}^{1/2}$. This can be taken into account by means of a factor $F$, defined by the relationship: $\sigma = (F\bar{N})^{1/2}$ (Fano, 1947):

$$\sigma = (FE/\varepsilon)^{1/2}. \qquad (7.2)$$

The production of secondary ion pairs in the avalanche causes additional statistical spread in the size of the output pulses, which can be taken into account by using an effective 'Fano factor' ($F$) which is larger than the value for primary ionisation alone.

Fig. 7.7 shows the calculated resolution as a function of X-ray energy for $\varepsilon = 26\,\text{eV}$ and $F = 0.8$, as determined for argon by Charles and Cooke (1968). In practice it is usually somewhat worse than this because of imperfections in the anode wire giving rise to local variations in gas gain. The resolution is also liable to deteriorate at very low and very high gas gain. Burek and Blake (1973) found that the best resolution with P10 gas occurred with a gain of 500–1000.

Improved resolution is obtainable in principle if the contribution of gas multiplication can be eliminated. However, in the conventional mode of operation the output pulses would then be so small that the resolution would be seriously degraded by electronic noise. An alternative possibility is to make use of the light emitted by the gas, which can be detected without such degradation. Thus, Alves, Policarpo and Santos (1973) obtained a resolution of 470 eV for 5.9 keV X-rays using a 'gas scintillation counter' operating on this principle. However, the improvement over the conventional proportional counter is modest and would offer no appreciable advantage in w.d. spectrometry.

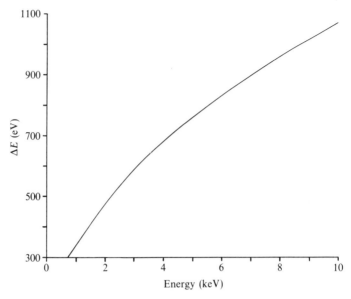

Fig. 7.7 Energy resolution ($\Delta E$) of typical proportional counter as function of X-ray energy.

## 7.8 Anode wire

The most common anode material is tungsten wire. Vogel and Fergason (1966) suggested using tungsten alloyed with 3% rhenium, because it can be heated to 3000 K for cleaning without loss of pliability, which occurs in pure tungsten at 1300 K. Burek and Blake (1973) found nichrome to be better than tungsten because of its smoother surface.

The wire should be under tension to prevent it from sagging: this can be achieved by prestressing it on fitting, or by attaching one end to a spring. The gas gain may change within about one cathode diameter of each end of the wire owing to disturbance of the field by the insulators: the wire should therefore extend well beyond the active region of the counter.

For constant gas gain the wire should be accurately round, with a uniform diameter and a smooth finish. Charles and Cooke (1968) measured diameter variations of 1% along the length of ordinary 50 $\mu$m diameter tungsten anode wire, and obtained a significant improvement in resolution by changing to wire uniform in diameter to better than 0.25% with a smoother finish. O'Boyle (1965) showed that heating tungsten to 1850 K for several hours under vacuum while applying an alternating electric field removes surface roughness.

Dust on the wire can cause serious loss of resolution, and entry of dust into flow counters via the gas supply should be avoided. If a flow counter wire becomes dusty the window may be removed and the wire cleaned by wiping with a fine brush wetted with acetone. With prolonged use the wire acquires a contamination layer that is not so easy to remove: the wire may then need to be cleaned by heating to a high temperature in vacuum. However, sometimes it is necessary to replace the wire to maintain good resolution. When the performance of a sealed counter with respect to resolution or background count-rate deteriorates the whole counter is replaced.

## 7.9 Escape peaks

In an argon-filled counter about 90% of incident X-rays of energy greater than 3.20 keV (the energy of the Ar K absorption edge) are absorbed by photoelectric ionisation of the Ar K shell (as distinct from other shells). The K-shell fluorescence yield of argon is 0.12, hence 11% ($0.9 \times 0.12$) of all ionisations result in the emission of Ar K photons. These are much more penetrating than the Auger electrons emitted by the majority of the ionised atoms, and an Ar K photon may escape absorption in the gas. If this happens, the energy deposited in the counter is equal to the energy of the photoelectron (the incident X-ray energy minus 2.96 keV) and the amplitude of the output pulse is correspondingly reduced. This gives rise to a satellite of the main peak in the pulse height distribution, known as the 'escape peak' (fig. 7.8).

If all the Ar K X-rays escape, the escape peak will contain about 11% of all recorded pulses. However, some are absorbed in the counter gas, resulting in normal output pulses. Typically, the escape peak intensity is about 5% of the main peak. X-rays of energy below 3.20 keV do not produce an escape peak (in principle they produce an Ar L escape peak but this is negligible).

The K absorption edge of xenon is well above the range of X-ray energies encountered in microprobe analysis, but the L edges lie in the range 4.78–5.45 keV. The L-shell fluorescence yield is 0.08, but as only about 50% of absorption is due to L-shell ionisation, the escape peak cannot exceed 4% of the main peak. In fact the ratio is much lower since Xe L X-rays are relatively heavily absorbed in xenon. The escape peak in a xenon-filled counter is therefore very small and can usually be ignored.

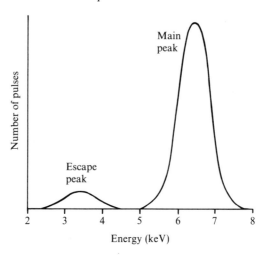

Fig. 7.8 Argon escape peak for Fe Kα X-rays.

## 7.10 Unorthodox proportional counters

In proportional counters the geometry of the anode wire is very important, but the shape of the cathode is uncritical. Therefore when the orthodox shape is inconvenient it is possible to use different configurations with little loss of resolution. Thus, for the purpose of e.d. spectrometry an 'end window' counter may be placed closer to the source than a conventional side window counter and is therefore more efficient. Such a counter with a point anode (Duncumb, 1960; Ranzetta and Scott, 1967) gives reasonably satisfactory resolution, though it is more susceptible to pulse height depression (§7.5) than conventional counters. An end window counter described by Cairns, Desborough and Holloway (1970) used a short anode wire parallel to the window, thereby avoiding the disadvantages of the point anode. By adjusting the anode position, 'tuning' for high or low energy X-rays was achieved.

Armigliato, Bentini and Ruffini (1976) detected the Be K peak at 109 eV and even Si L at 92 eV with such a counter. These peaks are difficult to detect with a windowless Si(Li) detector (§9.4.3) on account of the noise tail, which commonly extends beyond 100 eV. (The proportional counter suffers less from noise because of its internal gain.) However, other than at these very low energies, the windowless Si(Li) detector is preferable for e.d. analysis, on account of its better resolution.

# 8
# Counting electronics

## 8.1 Counting systems for w.d. spectrometers

The electronic units used in conjunction with the proportional counter amplify the counter pulses to a voltage suitable for processing (e.g. 1–10 V), and convert them into signals of various kinds representing count-rate. A preamplifier placed as close as possible to the counter provides a low impedance output suitable for driving the coaxial cable to the main amplifier, which provides most of the gain and also contains pulse shaping circuits. A discriminator following the main amplifier produces a standard output pulse when the input voltage rises above a preset threshold voltage, thereby eliminating low amplitude noise pulses. Unwanted high amplitude pulses may also be excluded by using a pulse height analyser, with an upper voltage level defining a 'window' of acceptable pulse heights. The discriminator/p.h.a. output is connected to a ratemeter which provides a visual count-rate reading on a meter, as well as an output voltage that can be displayed on a c.r.t. or used for plotting count-rate variations with a pen recorder or other device.

For quantitative analysis the spectrometer is set to a selected line and the X-ray intensity measured by counting the pulses with a scaler controlled by a timer. The number of counts recorded may be displayed visually, and can also be printed out. With modern computerised instruments the timers and scalers are under the control of the computer and the output is transferred to the computer on completion of each counting period.

An understanding of the function and behaviour of the various units is essential for their correct operation and the diagnosis of malfunctions. A block diagram of a standard counting system is shown in fig. 8.1 and the individual units are discussed in the following sections. In some current instruments these units and their controls are inaccessible to the user and can only be addressed via a computer keyboard; the operating principles

92                    8 Counting electronics

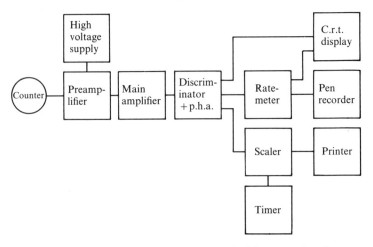

Fig. 8.1 Block diagram of electronics associated with proportional counter.

are nevertheless the same. This chapter is concerned with counting systems for w.d. spectrometers: those for e.d. spectrometers are treated in chapter 10.

## 8.2 Preamplifier

The preamplifier is usually operated in the 'charge sensitive' mode, the signal from the counter being integrated on the feedback capacitor (fig. 8.2). The output is then proportional to the *charge* in the pulse, rather than the *voltage*, which is affected by the capacitance of the counter, connecting wires, etc. Such capacitance still influences the noise level, however, and the preamplifier should be placed as close as possible to the counter. Preamplifiers have a fast rise time (e.g. 20 ns), in order to avoid distortion of the leading edge of the counter pulse. Usually a differentiation time constant of about 50 $\mu$s is incorporated, producing 'tail pulses' with a steep leading edge and a relatively long exponential tail (fig. 8.3). This differentiation prevents pile-up at high count-rates.

The noise generated in the preamplifier can be defined in terms of the corresponding fluctuation at the input, which may be expressed as the equivalent number of ion pairs (as if produced in the counter). Assuming optimum pulse shaping in the main amplifier, a preamplifier might have a noise level equivalent to about 300 ion pairs. Since X-rays of a few keV energy produce a comparable number of primary ion pairs in the counter, it

## 8.3 Main amplifier

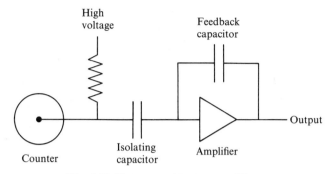

Fig. 8.2 Charge sensitive preamplifier.

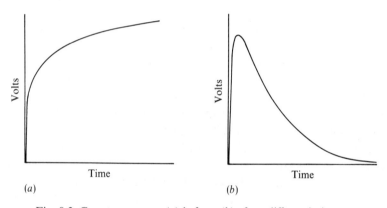

Fig. 8.3 Counter output: (a) before, (b) after, differentiation.

is only necessary to have a gas gain of a few hundred for the noise to be negligible compared to the signal, in which case there is no significant broadening of the pulse height distribution. The use of a low noise preamplifier has the advantage of allowing low gas gain in the counter and thereby avoiding pulse height depression at high count-rates (§7.5).

### 8.3 Main amplifier

The main amplifier serves to increase the amplitude of the counter pulses to a voltage suitable for the discriminator/p.h.a. The gain required is of the order of several hundred, and a typical output consists of pulses of a few volts amplitude. Another function is pulse shaping, which is used to

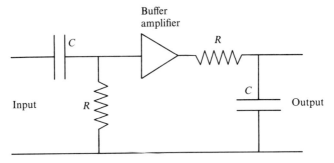

Fig. 8.4 *RC* differentiation and integration.

maximise the signal to noise ratio and minimise pulse overlap at high count-rates.

Noise, which determines the lowest X-ray energy that can be detected, and broadens the pulse height distribution by virtue of the random fluctuation of the baseline, originates mainly in the preamplifier. Pulse shaping circuits are essentially frequency filters that pass only the frequency band necessary for the transmission of the pulses, and not other frequencies carrying noise. The simplest shaping circuit consists of *CR* differentiating and *RC* integrating stages (fig. 8.4). The product *RC* is the 'time constant' and optimum performance is obtained with equal integration and differentiation time constants. The frequency response of an *RC* circuit can be calculated as a function of the characteristic frequency $f_0$ ($=1/2\pi RC$). Fig. 8.5 shows the calculated frequency response for the *CR–RC* circuit in fig. 8.4. For a typical time constant of 1 μs the peak response occurs at about 100 kHz.

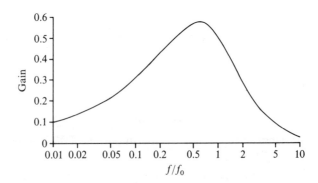

Fig. 8.5 Frequency response of *RC* differentiating and integrating circuit (fig. 8.4).

## 8.3 Main amplifier

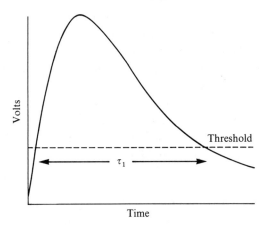

Fig. 8.6 Pulse shape produced by *RC* differentiation and integration, with effective pulse width ($\tau_1$) determined by discriminator threshold.

Fig. 8.6 shows the waveform produced by *CR–RC* shaping. At high count-rates closely spaced pairs of pulses can be resolved only if the waveform of the first pulse falls below the threshold of the discriminator before the arrival of the second. Pulse pair resolution is therefore determined by the effective pulse width ($\tau_1$). Even after falling below the threshold, the tail of the first pulse may still influence the apparent amplitude of the next.

Single *CR* differentiation and *RC* integration is the simplest but not the best form of pulse shaping. Modest improvements in signal to noise ratio are obtainable with other forms of shaping, but the choice is not critical in the context of w.d. analysis.

### 8.3.1 Operation at high count-rates

The choice of about 1 µs for the time constants is governed by signal to noise considerations. However, with low noise preamplifiers it is possible to use shorter time constants, thereby reducing dead time without serious broadening of the pulse height distribution due to noise. Variation in the rise time of the counter pulses also causes broadening, which increases as *RC* is decreased. However, some broadening is not necessarily unacceptable, since pulse height resolution is usually of secondary importance.

Reduction in *RC* increases the attenuation of the pulses, but this is not serious: pulse amplitude only decreases by a factor of about 2 when *RC* is decreased from 1 µs to 0.1 µs, allowing an order of magnitude increase in count-rate. At $10^5$ counts s$^{-1}$, a statistical precision of better than $\pm 1\%$ can be obtained in a counting time of only 1 s.

Pulse height depression at high count-rates may be controlled by suitable choice of anode wire and by using a low gas gain (§7.5). Electronic effects may also appear at high count-rates: for instance in an AC-coupled system the positive pulses are balanced by depression of the baseline between pulses, causing an apparent reduction in pulse height. This can be counteracted by 'baseline restoration', which is discussed later in the context of e.d. spectrometers (§10.3.3).

### 8.4 Discriminator and pulse height analyser

The purpose of the discriminator is two-fold: to stop unwanted low energy pulses being counted, and to convert wanted pulses into a standard form for counting, or for modulating a c.r.t. to produce X-ray scanning images. The discriminator compares the input signal with the voltage level set by the 'threshold' potentiometer. When the threshold voltage is exceeded a standard output pulse is produced.

The discriminator is used to exclude electronic noise. A greater reduction in background can be obtained with pulse height analysis, whereby only pulses between the 'lower level' (threshold) and 'upper level' voltages are accepted. The pulse height analyser consists of two discriminators and an anti-coincidence circuit. An output pulse is generated only if the waveform, after crossing the lower level, does not cross the upper level before again falling below the lower level.

The principal advantage of pulse height analysis is that second and higher order reflections from the spectrometer crystal (§6.2.1) can be suppressed by setting the upper level between the pulse amplitudes of first and second orders. The background contribution from scattered X-rays is also reduced. A narrow window gives the highest peak to background ratio, but if the window is too narrow some of the peak is lost; also the apparent count-rate is very sensitive to pulse height depression, electronic drift, and fluctuations in counter gas density (§7.3).

#### 8.4.1 Automatic pulse height analysis

To obviate the necessity for continually resetting the p.h.a. threshold for different X-ray lines, it can be varied automatically by means of potentiometers connected to the spectrometer mechanism. The relationship between X-ray energy ($E$) and Bragg angle is: $E \propto \mathrm{cosec}\theta$, and ideally the

## 8.4 Discriminator and p.h.a.

window width ($\triangle E$) should vary as $E^{1/2}$ to match the width of the pulse height distribution (§7.7). The threshold is then required to vary in such a manner that the voltage of the centre of the window is proportional to $\mathrm{cosec}\theta$. However, as shown in fig. 8.7 the use of a fixed window width is a reasonable approximation.

In a computer-controlled system there is no need to make physical provision for varying the threshold and window width with Bragg angle: the required pulse height analysis conditions can be generated in the software.

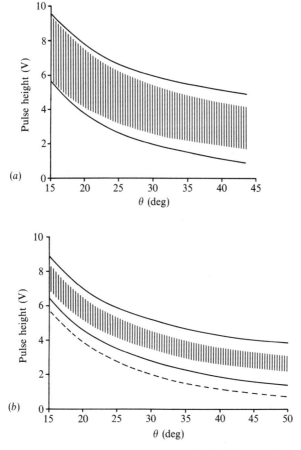

Fig. 8.7 Pulse height analysis for (a) TAP and (b) LiF crystals: shaded areas represent $\pm 2\sigma$ of pulse height distribution (solid lines – upper and lower limits of p.h.a. window with constant window width; dashed line – pulse height of Ar escape peak).

## 8.5 Ratemeter

The purpose of the ratemeter is to produce an output voltage proportional to the input count-rate. For each pulse a fixed increment of charge is transferred to a 'tank' capacitor across which a resistor provides a leakage path (fig. 8.8). The voltage on the capacitor is determined by the relative rates of inward and outward flow of charge and is proportional to the rate of arrival of the pulses. The time constant $RC$ governs the speed of response of the output voltage to changes in count-rate. The response to an abrupt change is exponential: for a 'step function' input, 90% of the change takes place in a time interval of $2.3RC$. Random variations in the intervals between pulses cause the output voltage to fluctuate with a relative standard deviation equal to $100(2nRC)^{-1/2}$ per cent, where $n$ is the count-rate.

In selecting a suitable time constant the operator must balance noise against response speed (fig. 8.9). For example, a time constant of 1 s gives a standard deviation (s.d.) of only 0.7% for a count-rate of $10^4$ counts s$^{-1}$, but the 90% response time is 2.3 s, which is excessively long for some purposes, e.g. line scans. With a time constant of 0.1 s the response time is reduced to 0.23 s and the s.d. increases to 2.2%. While this may be acceptable for many purposes, the fluctuations become large at relatively low count-rates: for example at 100 counts s$^{-1}$ the s.d. is 22%.

Other uses of the ratemeter include setting the spectrometer on a peak by maximising the count-rate, and monitoring a selected element while moving the specimen, in order to locate a particular phase. For the latter purpose an audible output is useful so that the operator can look down the microscope or at a scanning image while listening to the ratemeter signal. A further application is for qualitative analysis, in which the spectrometer is scanned through a range of wavelengths in order to record the lines in the

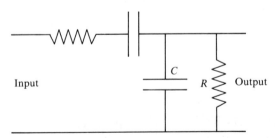

Fig. 8.8 Ratemeter circuit: input pulses charge capacitor $C$, which continuously discharges through resistor $R$, producing DC voltage proportional to count-rate.

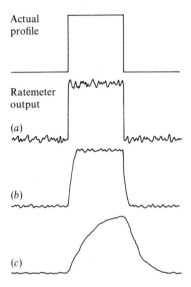

Fig. 8.9 Ratemeter line scan: effect of time constant: (a) short, (b) medium, (c) long.

spectrum (§12.1.1). To minimise noise the time constant should be the longest compatible with adequate response to the peaks. It is convenient to use a logarithmic ratemeter scale when large and small peaks coexist. coexist.

The present tendency is for the analogue ratemeter, as described above, to be replaced by a digital version which essentially carries out a rapidly repeated counting operation. A visual analogue output is still desirable and can be provided in the form of a 'bar-graph' display on a v.d.u. screen.

## 8.6 Scaler and timer

The basic element in electronic counting circuits is the 'bistable' circuit, which is stable in either of two states, changing from one to the other on the arrival of a pulse. By generating an output pulse every time it is switched to one of its two states, it acts as a 'scale of 2', i.e. the number of input pulses is scaled down by a factor of 2. Four bistables connected together can count up to $2^4$ ($=16$), but usually they are arranged to count only up to 10. By connecting 'scale of 10' circuits in series, decimal counting can be performed.

In order to measure count-rates the scaler must count for a known time interval. Time is measured by a separate scaler counting pulses generated by an electronic 'clock'. In 'preset time' mode the timer terminates the count after recording a preselected number of clock pulses, while in the 'preset count' mode the scaler terminates the count after accumulating a given number of counts. The advantage of the latter mode of operation is that the error from counting statistics is constant, but it is usually more convenient to count for a fixed time.

## 8.7 Dead time

A counting system has a characteristic 'dead time' ($\tau$), which is the interval after the arrival of a pulse during which the system does not respond to further pulses. The dead time may be 'extendable' or 'non-extendable', as defined by Ruark and Brammer (1937) and illustrated in fig. 8.10. In the latter case the arrival of a pulse during the inoperative period does not extend that period. In each second such a system is dead for $n'\tau$ seconds, where $n'$ is the recorded count-rate. Hence the 'live time' is $1 - n'\tau$, and the true count-rate $n$ is given by:

$$n = n'/(1 - n'\tau). \tag{8.1}$$

A different relationship applies to systems with extendable dead time, where any pulse following another within an interval $\tau$ is lost. The intervals between random pulses follow the Poisson distribution, according to which the fraction of intervals greater than $\tau$ is $e^{-n\tau}$, hence:

$$n'/n = e^{-n\tau}. \tag{8.2}$$

For $n\tau \ll 1$, equations (8.1) and (8.2) are equivalent, but the difference becomes significant at high count-rates.

Actual counting systems consist of several components, of which the only

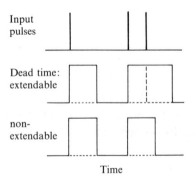

Fig. 8.10 Extendable and non-extendable dead time.

ones usually relevant in connection with dead time are the main amplifier and the discriminator/p.h.a. The amplifier has no inherent dead time, but in combination with the discriminator it effectively has an extendable dead time determined by the pulse width $\tau_1$ (fig. 8.6), since once the pulse wave form has crossed the discriminator threshold no further pulses are recorded until the voltage falls below the threshold again. The discriminator has a non-extenable dead time related to the output pulse width ($\tau_2$).

The overall behaviour of the amplifier and discriminator depends on the relative magnitude of $\tau_1$ and $\tau_2$. If $\tau_1 > \tau_2$, the discriminator has no effect on dead time, and the system has an extendable dead time determined by the amplifier. If $\tau_2 \gg \tau_1$, the system behaves as though it had a non-extendable dead time of $\tau_2$, and equation (8.1) may be applied. The intermediate case where $\tau_2$ is only somewhat larger than $\tau_1$ is more complicated. This and other combinations of extendable and non-extendable dead time have been investigated theoretically and experimentally by Beaman, Isasi, Birnbaum and Lewis (1972). In practice the dead time of counting systems associated with w.d. spectrometers can be considered to be non-extendable.

If pulse height analysis is used, the apparent dead time may be enhanced because of pulses being displaced out of the 'window', due to pulse height depression at high count-rates. To avoid this effect the upper and lower levels should be set far enough apart to allow for some change in pulse height without significant loss of counts, and the counter should be operated in such a way as to minimise pulse height depression (§7.5).

### 8.7.1 Measurement of dead time

It is desirable to measure the dead time of a counting system rather than relying on a nominal value given by the manufacturer. Methods used in other fields, involving placing filters in front of the counter, for example, are not always applicable to the electron microprobe, because of the enclosure of the spectrometer inside a vacuum chamber. Heinrich, Vieth and Yakowitz (1966) suggested measuring the apparent count-rate as a function of probe current. The true count-rate ($n$) is proportional to the current ($i$), hence from equation (8.1):

$$n'/i = k(1 - n'\tau), \tag{8.3}$$

where $k$ is a constant. If the system is characterised by an ideal non-extendable dead time, a plot of $n'/i$ versus $n'$ will be a straight line, whereas if the dead time is extendable, the experimental points will fall increasingly below the straight line with increasing $n'$ (fig. 8.11). In the

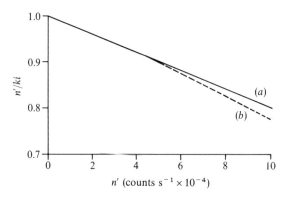

Fig. 8.11 Dead time plot for (a) non-extendable, (b) extendable, dead time of 2 μs.

non-extendable case the slope of the line is constant at $-k\tau$ and the intercept on the $n'/i$ axis is $k$, hence $\tau$ can be calculated from the graph. Alternatively it can be calculated from two measurements of $n'$ and $i$ using the expression:

$$\tau = [1/(i_2 - i_1)][(i_2/n'_2) - (i_1/n'_1)].$$

A different approach is the 'ratio method', whereby the apparent ratio of two X-ray intensities (their actual ratio being constant) is measured as a function of count-rate. This procedure was applied by Short (1960) using X-ray diffraction equipment, an absorbing foil being moved in and out of beams of different intensity.

Adapting this idea to the electron microprobe, where access to the X-ray paths is more difficult, Heinrich et al. (1966) measured the count-rates $n'_1$ and $n'_2$ from two spectrometers tuned to peaks of different intensity (actually Cu K$\alpha$ and K$\beta$, though the choice of peaks is unimportant), while varying the probe current. The intercept of a plot of $n'_1/n'_2$ against $n'_1$ is equal to $C$, the true intensity ratio, and the slope $B$ is a function of the dead time. In order to obtain independent values for the dead time of each spectrometer it is necessary to carry out a second set of measurements with the spectrometers tuned to the opposite peaks. The dead times are then given by:

$$\tau_1 = (B - B')/(C' - C), \quad \tau_2 = (C'B - CB')/(C' - C),$$

where $B$ and $B'$ are the slopes obtained from the first and second plots, while $C$ and $C'$ are the respective intercepts. Though more time consuming, this method avoids the necessity for accurate probe current measurements.

# 9
# Lithium-drifted silicon detectors

## 9.1 Principles of operation

Like the proportional counter, the Si(Li) detector relies on ionisation for the detection of X-rays, but this takes place in a solid instead of a gaseous medium. Wave mechanics predicts that in crystals electrons can only possess energies lying within certain bands. A semi-conductor such as Si has a fully occupied 'valence band' and a largely unoccupied 'conduction band' at a higher energy, separated by an energy gap (fig. 9.1). In a Si(Li) detector, some of the energy of incident radiation is applied to raising electrons from the valence to the conduction band, where they are free to move through the lattice. The 'holes' left in the valence band behave like free positive charges. A bias voltage is applied across the detector so that the charge carriers (electrons and holes) move to opposite electrodes, producing a signal which enables the X-rays to be detected.

As in the proportional counter, the energy of incident X-rays is first transferred to photo- and Auger electrons, which in turn dissipate their energy in the detection medium, generating charge carriers as described above (fig. 9.2). The minimum energy necessary to create an electron–hole pair is equal to the energy gap (1.1 eV in silicon). However, some energy is 'wasted' in exciting lattice vibrations, etc., and the average energy ($\varepsilon$) absorbed for each electron–hole pair produced is 3.8 eV. This is much lower than the 25 eV typical of a gas-filled counter, and, owing to the larger number of charge carriers generated, the statistical fluctuations in pulse height are smaller, hence the energy resolution is better.

Noise considerations demand a low DC leakage current in the detector, therefore silicon of high resistivity is desirable. At low temperatures the conductivity of perfectly pure silicon is low, being caused solely by thermally excited electron–hole pairs. However, even with the purest material available, the actual conductivity is considerably enhanced by

# 9 Lithium-drifted silicon detectors

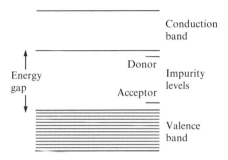

Fig. 9.1 Electronic band structure of a semi-conductor.

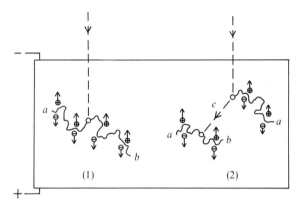

Fig. 9.2 X-ray detection in Si(Li) detector: production of electrons (−) and holes (+) by Auger electrons (*a*) and photoelectrons (*b*): Si atom ionised by incident X-ray produces photoelectron and either (1) Auger electron, or (2) Si Kα photon (*c*), which is absorbed by another Si atom.

impurities. The most important of these is boron, which gives silicon 'p-type' properties, with the conductivity caused mainly by 'acceptor' energy levels lying just above the valence band (fig. 9.1). Electrons are readily raised by thermal excitation from the valence band to occupy such levels, leaving holes which act as positive charge carriers.

The resistivity of p-type silicon may be greatly increased by adding 'donor' impurity atoms to neutralise the acceptors. Exact 'compensation' of p-type silicon by donors results in high resistivity, comparable to that of ideally pure ('intrinsic') silicon. Lithium is used for compensation because its small ionic radius enables it to diffuse easily into silicon; also it is an efficient electron donor. The required lithium distribution is obtained by 'drifting' under an applied electric field, resulting in the 'lithium-drifted silicon', or Si(Li), detector (Pell, 1960).

The Si(Li) detector, unlike the proportional counter, has no internal gain, and a typical X-ray photon produces around one thousand electron–hole pairs, corresponding to a charge of only about $10^{-16}$ C. For such a small signal, external amplification with low noise and high gain is required. The first stage of amplification is provided by a preamplifier connected directly to the detector in order to minimise stray capacitance. A field effect transistor (FET) is used because it offers the lowest noise. The detector and FET are cooled to about 100 K with liquid nitrogen to reduce noise and prevent diffusion of lithium in the detector.

Further gain is provided by a main amplifier similar to that used with a proportional counter (§8.3). The spectrum is recorded with the aid of a multichannel p.h.a., which sorts the pulses according to height (and hence X-ray energy). The number of counts received in each channel may be displayed on a v.d.u., plotted on an $x$–$y$ recorder, or printed out.

## 9.2 Construction

Si(Li) detectors are fabricated from wafers of high purity silicon. The first stage of the lithium drifting process consists of applying lithium to one face of the silicon wafer and heating for a few minutes at about 700 K in vacuum or an inert atmosphere. Lithium diffuses into the silicon, resulting in the gradient shown in fig. 9.3. A high lithium concentration produces n-type

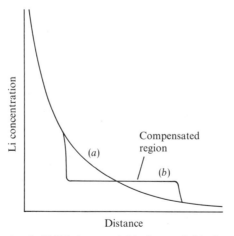

Fig. 9.3 Li distribution in Si(Li) detector (*a*) before and (*b*) after 'drifting' process produces region in which impurities are compensated by Li.

material in which electrons from donor atoms are the predominant carriers, while below a certain lithium concentration the silicon retains its original p-type properties. The resulting p–n junction has a narrow 'depletion layer' between the p and n regions, which lacks free charge carriers.

Drifting is carried out at about 400 K with a reverse bias (n region positive relative to p region) of several hundred volts. The strong field existing in the depletion layer because of its high resistivity causes positive lithium ions to migrate from the n to the p region. The result is an expanding depletion layer in which the silicon is exactly compensated. Drifting is carried on for several hours until the 'intrinsic' region reaches the required thickness. Further details of the drifting process have been given by Goulding (1966).

Surplus p-type silicon is removed to expose the intrinsic material, onto which a thin layer of gold is evaporated forming a 'surface barrier contact', which enables an electrical connection to the detector to be made. The gold plated surface is at the front of the detector and is connected to the negative bias voltage supply. The back electrode is connected to the FET.

Surface leakage current, which adversely affects resolution by adding noise, occurs because the silicon acquires n-type properties at the sides,

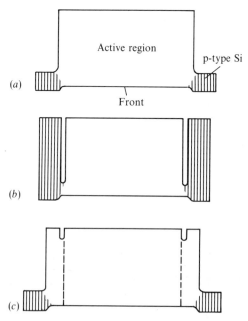

Fig. 9.4 Types of Si(Li) detector: (a) 'top hat', (b) 'grooved', (c) 'guard ring'.

## 9.3 Cryostat

with the result that conductivity is enhanced. The 'grooved' form of detector has a longer surface leakage path between electrodes (fig. 9.4(b)). In the 'guard ring' type (fig. 9.4(c)) the sensitive region is bounded not by an external surface but by electric field lines in the bulk material (Landis, Goulding and Jarrett, 1972). Other details regarding the construction of Si(Li) detectors have been given by Trammell (1978).

### 9.3 Cryostat

The detector–FET assembly is mounted on a copper rod, the other end of which is immersed in liquid nitrogen at 77 K (fig. 9.5). The mounting is designed so that the heat generated internally and leaking from the surroundings just suffices to maintain the detector and FET at the optimum temperature (about 100 K). Sometimes a small additional heat source is incorporated for this purpose. The assembly needs to be highly resistant to vibration, since very small changes in capacitance between components can give rise to significant electronic noise, thereby degrading the energy resolution.

The liquid nitrogen is contained in a dewar, which requires refilling every few days. The detector should preferably be kept cold permanently, though no damage is done by warming up to room temperature occasionally, provided the bias voltage is off (if this is on, however, the effect is catastrophic). Ice crystals may accumulate inside the dewar, where they can cause vibrations which degrade the energy resolution, and may have to

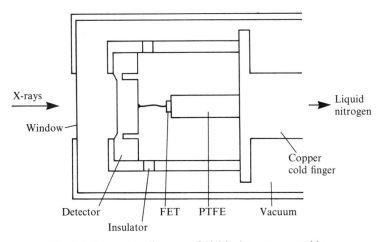

Fig. 9.5 Schematic diagram of Si(Li) detector assembly.

be removed from time to time. The manufacturer's instructions for this operation should be followed.

Since the performance of the detector is extremely sensitive to surface contamination, the cryostat must be permanently evacuated, and is usually sealed for life. Adsorbent material (e.g. zeolite) or a small ion pump is incorporated in order to deal with slow leaks and outgassing. If the pressure rises significantly, for example due to a leak, the noise level increases, resulting in broadening of the peaks in the spectrum. Also the rate of consumption of liquid nitrogen may rise perceptibly.

### 9.3.1 Alternative methods of cooling

Cooling with liquid nitrogen entails some inconvenience and expense, hence there is an incentive to consider alternatives. One possibility is to make use of the Joule–Thomson effect, whereby cooling occurs as a result of the expansion of pressurised gas passing through a small orifice (Alberti, Clerici and Zambra, 1979). A temperature comparable to that obtained with liquid nitrogen can be achieved, but it is difficult to avoid some loss of detector resolution due to vibration.

Other refrigeration techniques can be employed, such as a modified Solvay cycle system using helium (Stone, Barkley and Fleming, 1986). Again, careful design is required to minimise vibration. Another approach is to use the thermoelectric effect, whereby cooling occurs on passing an electric current through a junction between two dissimilar metals (Madden, Jaklevic, Walton and Wiegand, 1979; Madden, Hanepen and Clark, 1986). In early versions detector resolution was somewhat degraded owing to the temperature being higher than with liquid nitrogen cooling but this drawback has been eliminated in currently available systems.

A further possibility is to use a different type of detector, capable of operating at room temperature (§§9.9.2, 9.9.3), though the resolution obtainable is relatively poor.

### 9.4 Entrance window

X-rays reach the detector via a thin 'window', which has to withstand atmospheric pressure while having the highest possible X-ray transmission. Beryllium foil about 8 $\mu$m thick is commonly used and gives reasonable transmission for X-ray energies above about 1 keV.

Vacuum pump oil deposited on the window (which is colder than its surroundings) may significantly reduce the detection sensitivity for low

## 9.4 Entrance window

energy X-rays. Contamination originating from plastic materials in the specimen chamber can also occur (Love and Scott, 1980). Warming the window with a small lamp reduces the contamination rate (Steele, Smith, Pluth and Solberg, 1975). It may be necessary to clean the window from time to time, in which case the manufacturer's instructions should be followed. Beryllium windows are fragile and should be treated with care, since breakage entails an expensive repair and possibly irreversible loss of energy resolution.

Windows with much higher transmission for low energy X-rays, but strong enough to withstand atmospheric pressure, have been developed, offering an alternative to removing the beryllium window (§9.4.3) for light element analysis. One such proprietary window material consists predominantly of boron and is 0.25 $\mu$m thick. Other thin window materials are also used, but in some cases require a supporting grid in order to sustain a pressure of one atmosphere.

### 9.4.1 X-ray collimation

A Si(Li) detector is sensitive to X-rays entering the window over a considerable range of angles and lacks the selectivity possessed by the w.d. spectrometer on account of its focussing geometry. To avoid spurious X-ray peaks caused by electrons backscattered onto the polepiece of the final lens and other objects near the specimen, it is desirable to restrict the angle of acceptance of the detector by placing a collimator in front of the window (fig. 9.6). The best material for the inside surface of the collimator is carbon, since this has a low electron backscattering coefficient and does not emit X-rays in the energy range usually of interest.

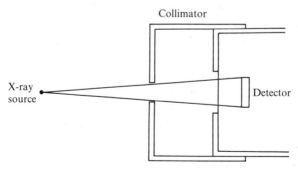

Fig. 9.6 X-ray collimator in front of Si(Li) detector.

### 9.4.2 Electron trapping

Scattered electrons with sufficient energy ($>20$ keV) may penetrate the beryllium window, causing spurious pulses in the detector. This effect is not usually serious in conventional microprobe analysis, where the electron energies are fairly low. However, in t.e.m.s. an accelerating voltage of 100 kV is usual and scattered electrons penetrating the window can cause a large increase in background. A magnet in front of the detector may be used to stop such electrons reaching the window (Hall, 1977). Details of the design of a magnetic electron trap have been given by Neumann, Reimer and Wellmans (1978). Statham (1979) observed that electrons of moderate energy can be absorbed by a Mylar film 4 $\mu$m thick placed over the beryllium window, the additional X-ray absorption being minimal.

### 9.4.3 Removable windows

The beryllium window is the main obstacle to the detection of X-rays below 1 keV. For light element analysis ($Z<11$), it is therefore appropriate either to remove the beryllium window, or replace it by one causing much less absorption (Jaklevic and Goulding, 1971).

A completely windowless detector is at risk from contaminants in the vacuum system of the instrument, hence it is preferable to retain a thin organic film ('ultra-thin window') to protect the detector. This should be aluminised to exclude light, which is liable to affect the detector. The turret mechanism needed to change windows occupies extra space, which is a disadvantage in some applications.

### 9.4.4 Ice layers

Windowless (or ultra-thin window) detectors are vulnerable to ice formation on the detector surface, which can seriously reduce the sensitivity to low energy X-rays. This can also occur with permanently sealed beryllium windows, due to slight leakage through the window. Methods of monitoring the thickness of such ice layers are discussed in §9.8.3.

Ice layers can be removed by warming the detector up to room temperature with the cryostat connected to a pumping system (Musket, 1986), but this is not practicable for the normal user. Some detectors are provided with a small heater which can be switched on in order to sublimate the ice, the water vapour produced being condensed elsewhere in the cryostat.

## 9.5 Energy resolution

The number of electron–hole pairs generated by X-rays of a given energy is subject to statistical fluctuations which give a gaussian pulse height distribution (fig. 7.6). The relative s.d. of the number of electron–hole pairs is given by the same equation (7.2) as for the proportional counter. Fano factor values in the region of 0.12 are typical.

Sources of noise include fluctuations in the DC leakage current in the detector and thermal noise in the FET and its associated circuits (§10.1). Noise causes random pulse height variations superimposed upon those resulting from ionisation statistics. Its effect can be expressed by a constant term $\triangle E_n$ added in quadrature to the contribution of ionisation statistics to the FWHM ($\triangle E$):

$$\triangle E^2 = \triangle E_n^2 + 2.355^2 \varepsilon F E. \quad (9.1)$$

Substituting $\varepsilon = 3.8\,\text{eV}$ this becomes:

$$\triangle E^2 = \triangle E_n^2 + 21.1 F E. \quad (9.2)$$

Fig. 9.7(a) shows $\triangle E$ for a typical system, increasing from 95 eV at 1 keV to 178 eV at 10 keV. Detector resolution is conventionally defined as that for 5.89 keV (the energy of the Mn K$\alpha$ peak, which is readily available from radioactive $^{55}$Fe for bench testing purposes). This is 146 eV in the above example. It seems unlikely that $F$ can be reduced below 0.1, though further reduction in $\triangle E_n$ is possible. The resolution that would be obtained with zero noise is shown in fig. 9.7(b): the advantage is greatest at low energies.

The preamplifier signal to noise ratio is affected by the capacitance of the detector (§10.2). Detectors of small area and hence small capacitance give the best performance. Unless high collection efficiency is of paramount importance an area of about 10 mm² is generally used.

### 9.5.1 Comparison with other types of spectrometer

Fig. 9.8 shows the energy resolution of a typical Si(Li) detector compared with a proportional counter and a w.d. spectrometer using various crystals. The proportional counter has much worse resolution except at very low energies, owing to the unavoidable effect of ionisation statistics, and cannot be considered seriously for e.d. spectrometry. The w.d. spectrometer has a resolution that varies considerably with Bragg angle, but is always better than the Si(Li) detector by a large margin. Only at high energies ($>20\,\text{keV}$) which are not usually of interest in electron microprobe analysis does the Si(Li) detector approach the w.d. spectrometer in resolution.

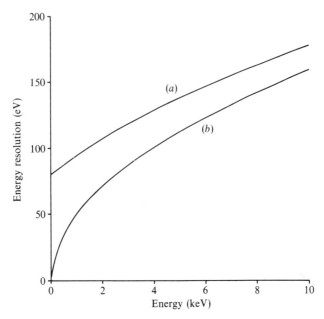

Fig. 9.7 Energy resolution ($\Delta E$) of Si(Li) detector as function of X-ray energy ($E$), calculated from equation (9.2), assuming $F = 0.12$ and $\Delta E_n = 80$ eV (a). Curve (b) shows resolution for hypothetical case of $\Delta E_n = 0$.

## 9.6 Incomplete charge collection

Ideally all the charge carriers generated by an incident X-ray photon are collected and contribute to the output. However, in reality there are various ways in which carriers can be lost, making the output pulse smaller than it should be, so that it appears in the wrong place in the energy spectrum. This phenomenon is known as incomplete charge collection.

### 9.6.1 Trapping

Charge carriers (electrons and holes) moving through the detector can be 'trapped' in energy levels lying between valence and conduction bands, caused by impurities and defects. Such carriers may be released, or 'detrapped', by thermal excitation, but their contribution to the output pulse is lost if this does not occur promptly. Also holes and electrons may recombine at trapping sites and hence be lost. Trapping causes asymmetrical broadening of the pulse height distribution.

The quality of the silicon determines the density of the traps in the bulk

## 9.6 Incomplete charge collection

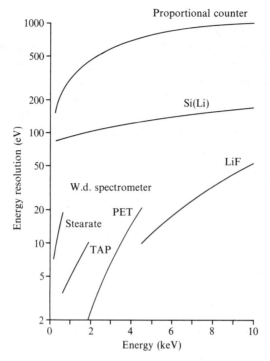

Fig. 9.8 Energy resolution ($\Delta E$) of Si(Li) detector compared to proportional counter and w.d. spectrometer using different crystals.

material. The probability of trapping is reduced by rapid charge collection, hence a high bias voltage is desirable, so far as the need for a low leakage current allows. If the detector is too cold, trapping effects are worsened, both because the thermal detrapping rate is lower and because impurity levels have a lower probability of being filled by thermal excitation from the valence band. In practice bulk trapping effects in Si(Li) detectors are small: most of the observable incomplete charge collection effects are associated with the 'dead layer' discussed in the next section.

### 9.6.2 Silicon dead layer

The existence of a layer of silicon which is 'dead', in the sense that X-rays absorbed within it are not detected, is suggested by the jump in the detection efficiency at 1.84 keV (the Si K absorption energy), which is observable as a step in the continuum. The nature of this 'dead layer' is somewhat controversial. It could be caused by a high density of near-surface defects, giving a high trapping probability. However, this would be expected to vary from one detector to another, whereas observed dead-layer thicknesses appear fairly uniform at around 0.2 μm (Goulding, 1977). This observation favours an alternative explanation, according to which some of the electrons produced by X-ray absorption near the surface diffuse to the front electrode rather than moving to the rear electrode under the influence of the bias voltage (Llacer, Haller and Cordi, 1977). A similar 'electron escape' process can also occur at the sides of the detector, but this effect is usually less important and can be eliminated by means of an aperture in front of the detector.

Incomplete charge collection in the dead layer causes a 'shelf' to appear on the low energy side of the main gaussian peak produced by mono-energetic X-rays (fig. 9.9). This is most prominent for X-rays strongly absorbed by silicon, including those of low energy (especially below 1 keV) and just above 1.84 keV, e.g. P Kα. The size of the shelf is dependent on the

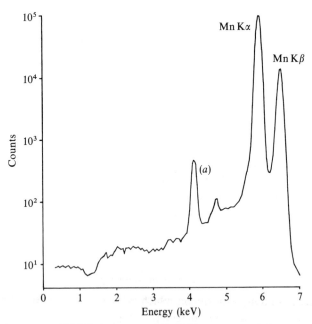

Fig. 9.9 Response of Si(Li) detector to Mn K X-rays: recorded counts (log scale) versus energy. Features on low energy side of peaks are caused by incomplete charge collection. For explanation of escape peaks (a) see §9.7.

## 9.7 Escape peaks

fraction of absorbed photons for which incomplete charge collection occurs. Since the shelf extends over a wide energy range, it is difficult to observe in an electron-excited spectrum, owing to the presence of the continuum. With a background-free radioactive source such as $^{55}$Fe, however, features of the detector response related to incomplete charge collection can be seen easily. Typically the number of counts per channel at an energy of, say, 1 keV, caused by incomplete charge collection, is several thousand times smaller than the Mn K$\alpha$ peak height.

### 9.6.3 Non-linearity

If incomplete charge collection affects a large proportion of the detected X-ray photons, the mean pulse height is reduced and the peak appears to be shifted to a lower energy. Since this effect is energy dependent, the relationship between peak position and energy becomes non-linear. In practice this behaviour is observed only in the low energy region which becomes accessible when the beryllium window is removed (§9.4.3). Thus the K peaks of B, C and N may be significantly displaced, though improvements in detector fabrication techniques have minimised this effect.

## 9.7 Escape peaks

Photoelectric absorption of an incident photon in the detector is followed by the emission of either an Auger electron or a Si K X-ray photon. Auger electrons are absorbed in a short distance, but X-rays travel further: a Si K$\alpha$ photon has a 10% probability of penetrating 30 $\mu$m. If such a photon escapes from the detector (fig. 9.10(a)), the pulse recorded is equivalent to that produced by a photon of energy $E - E_{Si}$, where $E$ is the incident photon energy and $E_{Si}$ the Si K$\alpha$ photon energy (1.739 keV). The result is an 'escape

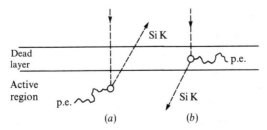

Fig. 9.10 Mechanism for escape peak (a) and internal Si fluorescence (b): in (a) photoelectron stays in active region of detector but Si K photon escapes; in (b) Si K photon is absorbed in active region, but photoelectron remains within 'dead layer'.

Table 9.1. *Silicon escape peaks.*

| Detected Kα peak | Energy of escape peak (keV) | Kα peaks within 100 eV of escape peak |
|---|---|---|
| P | 0.274 | B, C |
| S | 0.568 | O |
| Cl | 0.882 | |
| K | 1.573 | Al |
| Ca | 1.951 | P |
| Sc | 2.349 | S |
| Ti | 2.769 | |
| V | 3.210 | |
| Cr | 3.672 | Ca |
| Mn | 4.155 | Sc |
| Fe | 4.659 | |
| Co | 5.185 | |
| Ni | 5.732 | |
| Cu | 6.301 | Fe |
| Zn | 6.891 | Co |

peak' similar to that produced by a proportional counter (§7.9). The Si K$\beta$ escape peak is negligibly small (about 2% of the main escape peak) because of the low probability of Si K$\beta$ photon emission. Only X-rays of higher energy than the Si K absorption edge (1.841 keV) produce escape peaks. Table 9.1 gives escape peak energies and lists Kα peaks for which interferences may occur.

Equations for the escape peak intensity derived for NaI scintillation detectors (Axel, 1954) and for Ge(Li) detectors (Fioratti and Piermattei, 1971) can be adapted to the Si(Li) detector (Reed and Ware, 1972). The fraction of all pulses appearing in the escape peak instead of the main peak is given by the product of the following factors:

(1) the proportion of primary ionisations of Si that involve the K shell, which is equal to $(r-1)/r$, where $r$ is the Si K absorption edge jump ratio (the ratio of the mass attenuation coefficient on the high energy side of the edge to that on the low energy side);
(2) the proportion of K-shell ionisations followed by Si K emission, which is equal to the fluorescence yield, $\omega_K$;
(3) the fraction of generated Si K photons that escape, which for an infinite plane detector is $x/2$, where:

$$x = 1 - (\mu_{Si}/\mu_i)\ln[1 + (\mu_i/\mu_{Si})],$$

## 9.7 Escape peaks

in which $\mu_i$ and $\mu_{Si}$ are the mass attenuation coefficients of Si for the incident radiation and Si K$\alpha$ radiation respectively.

The fraction of the total intensity that appears in the escape peak is given by the product of the above factors. The ratio $r_e$ of escape peak to parent is:

$$r_e = 0.023x/(1 - 0.023x), \qquad (9.3)$$

assuming $(r-1)/r = 0.92$ and $\omega_K = 0.050$. For energies above about 20 keV an additional term is needed to allow for Si K photons escaping through the back of the detector (Dyson, 1974).

For non-normal X-ray incidence, the escape probability is higher, due to primary absorption occurring nearer the surface. Statham (1976a) derived a modified formula taking this into account, and incorporated an expression for $\mu_i$ in terms of the energy $E$ of the incident X-rays, giving:

$$r_e = 0.023/[1 + (mE + b)E^2], \qquad (9.4)$$

where $m = 0.01517 \cos\theta - 0.000803$ and $b = 0.0455 \cos\theta + 0.01238$, $\theta$ being the angle between the X-ray beam and the normal to the detector surface, and $E$ the X-ray energy in keV. (The constant in equation (9.4) is scaled to the same value as $\omega_K$ as used above.) This is more convenient than equation (9.3) for escape peak stripping (§12.9). Fig. 9.11 shows the variation in $r_e$ with $E$. Owing to the small size of escape peaks (usually less than 1% of their parent), they do not present serious difficulties in e.d. analysis.

It is to be expected that the width of an escape peak will be the same as for a 'real' peak of the same energy, and hence somewhat narrower than the parent peak. This has been confirmed experimentally by Van Espen, Nullens and Adams (1980), who also noted that escape peaks occur at slightly lower energies than theory suggests (probably because of poor charge collection in the near-surface region), though the difference is only about 10 eV.

### 9.7.1 Internal fluorescence

'Internal silicon fluorescence' arises by the inverse of the escape peak mechanism: if an incident photon is absorbed in the dead layer and a Si K photon is produced, this may enter the active region and be detected, whereas Auger and photoelectrons are much more likely to be absorbed in the dead layer (fig. 9.10(b)). Theoretically the intensity of the Si fluorescence peak is proportional to the thickness of the dead layer, and decreases rapidly with the incident X-ray energy (Reed and Ware, 1972). The calculated intensity of the peak as a fraction of the intensity of the incident radiation for a typical 0.1 $\mu$m dead layer is generally less than 0.1%. The Si

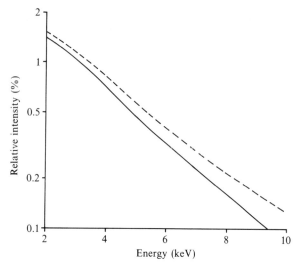

Fig. 9.11 Relative intensity of escape peak compared to parent peak as function of incident X-ray energy, for Si(Li) detector. Solid line – normal X-ray incidence; dashed line – 45° incidence.

peak is thus very small and can generally be neglected. Electrons bombarding the front surface of the detector can also produce a Si peak, but usually this effect is negligible (except for thin-window detectors with ineffective electron trapping or when high energy electrons are used).

## 9.8 Efficiency

The efficiency of a Si(Li) detector, defined as the count-rate per unit probe current, is dependent on the solid angle subtended by the detector at the X-ray source. This may be calculated to a close approximation by dividing the sensitive area of the detector by the square of its distance from the source. The nominal detector area given by the manufacturer may be an overestimate of the actual effective area. This can be determined by measuring the intensity obtained with and without an aperture of exactly known diameter (less than that of the detector) in front of the detector window. The intensity ratio is equal to the ratio of aperture area to detector area.

For purposes of solid angle calculation, the distance should be measured from the actual surface of the detector, which is a few millimetres behind the beryllium window. The manufacturer's figure for the detector–window distance may be used, or it can be determined experimentally by measuring

## 9.8 Efficiency

the intensity under constant conditions with different source–detector distances. The detector–window distance can be obtained from the intercept of a plot of intensity against (source–detector distance)$^{-2}$. With a typical solid angle of 0.004 sterad a probe current of 3 nA gives a total spectrum count-rate of about 10 000 counts s$^{-1}$ for an average specimen, with an accelerating voltage of 20 kV.

In ordinary microprobe analysis a large solid angle is not of great importance, except when analysing materials easily damaged by electron bombardment, necessitating the use of a low current. High collection efficiency is advantageous, however, in the analysis of thin specimens using a fine beam to achieve high spatial resolution, where the X-ray intensities are low. In this case it is desirable to use a detector of large area, even at the expense of some loss of energy resolution, and to place it as close as possible to the specimen, enabling a solid angle in excess of 0.1 sterad to be obtained.

X-rays of up to 20 keV are totally absorbed in a Si(Li) detector of typical thickness (3 mm) and the detection efficiency is close to 100%, apart from attenuation by the window etc. Pulses that are reduced in amplitude owing to trapping and dead layer effects, however, must be considered lost since they appear in the wrong place in the spectrum. Typically a few per cent of the pulses are lost in this way.

Attenuation takes place in the beryllium window, gold surface layer and silicon dead layer. Fig. 9.12 shows the calculated transmission factors for typical thicknesses (8 $\mu$m, 20 nm and 0.1 $\mu$m respectively) as a function of energy. The Si K absorption edge at 1.84 keV causes a significant step in the detector efficiency. The steps due to the Au M edges are considerably less marked than calculations suggest, perhaps because the mass attenuation coefficient (m.a.c.) does not in fact jump suddenly at the edges as usually assumed (Brombach, 1978). It is in any case not strictly valid to calculate absorption by the gold layer using the standard equation, since the thickness is generally non-uniform.

The range of the photoelectrons produced by X-ray absorption in the gold layer is often greater than the thickness of the layer. Some electrons may therefore reach the active region of the detector, giving rise to output pulses. However, these are 'degraded' (of reduced amplitude) and hence lost with respect to the recording of peak intensities (Maenhaut and Raemdonck, 1984), therefore no special allowance for this phenomenon is necessary in calculating absorption.

The rapid increase in the m.a.c.s of beryllium, gold and silicon with decreasing energy causes a steep decline in efficiency below 1 keV. Usually the lowest energy K$\alpha$ peak visible is that of Na at 1.04 keV, though L peaks

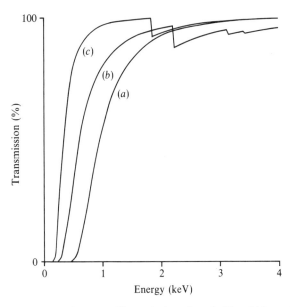

Fig. 9.12 Transmission of (a) beryllium window (8 μm), (b) gold layer (20 nm) and (c) silicon dead layer (0.1 μm) as function of X-ray energy.

of slightly lower energy (e.g. Co Lα at 776 eV) are detectable. The removal of the beryllium window (§9.4.3) greatly improves sensitivity at low energies and allows the detection of the K peaks of light elements down to B ($Z=5$) or even Be ($Z=4$), but absorption by the gold and silicon layers is significant in this region.

### 9.8.1 Varying the collection efficiency

As discussed in the next chapter, e.d. detectors are limited in their ability to handle high count-rates (above a few thousand counts per second typically). Usually the count-rate can be kept within appropriate bounds by suitable choice of probe current. However, this is not always convenient, notably when w.d. and e.d. spectrometers are used simultaneously, e.g. for detecting elements of both high and low concentration (§12.14.1). For w.d. analysis it is often desirable to use a probe current at least ten times higher than for e.d. analysis, hence a means of reducing the intensity of the X-rays reaching the e.d. detector is required.

Some detectors are mounted on a racking mechanism which allows the distance to be varied. A wider range of sensitivity is obtained with a variable aperture in front of the detector. For example, Van Amelsvoort, Smits and Stadhouders (1982) described an adjustable aperture utilising a 'scissors' mechanism operated by the detector racking movement, giving a 60:1 range in sensitivity. A simpler option is to have interchangeable fixed apertures, though this is less flexible. The aperture plate should be made of a material that absorbs X-rays completely throughout the energy range of interest.

The detector response may differ somewhat when an aperture is used. For example, incomplete charge collection effects may be less in the central region of the detector. The reduction in the size of the tails on the low energy sides of the peaks may affect results obtained by spectrum processing procedures described in chapter 12.

### 9.8.2 Efficiency calibration

Some methods used in quantitative e.d. analysis are dependent on knowing the detector efficiency as a function of energy. Efficiency measurement is also of interest in comparing one detector with another and monitoring changes with time. Efficiency at low energies (below 3 keV) is strongly influenced by absorption in the window, gold surface layer and silicon dead layer, the thicknesses of which may be specified approximately by the manufacturer but are impossible to measure directly. Absorption calculations are in any case of doubtful validity because of non-uniformity of thickness.

A useful empirical indication of window thickness etc. can be obtained from measurements of relative peak intensities (e.g. Co $L\alpha$/Co $K\alpha$) in an electron microprobe or s.e.m. However, the absolute value of the ratio depends appreciably on instrumental parameters, including electron incidence angle, X-ray take-off angle and incident electron energy. Measuring this ratio is thus a useful way of monitoring changes with time in a particular instrument, but it is not suitable for absolute efficiency determination.

One way of measuring absolute detection efficiency is to use a radioactive X-ray source of known intensity, but few suitable sources are available for the energy region of interest. Apart from $^{55}$Fe, which emits Mn $K\alpha$ radiation at 5.89 keV, other possibilities are $^{137}$Cs, emitting Ba L radiation at around 4.6 keV, and $^{241}$Am, emitting Np M radiation at around 3.3 keV (Gallagher and Cipolla, 1974). However, at low energies errors due to self absorption in the source are liable to be significant.

An alternative approach is to use X-ray fluorescence to provide suitable

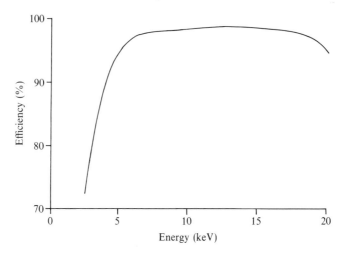

Fig. 9.13 Detection efficiency of typical Si(Li) detector as function of X-ray energy, determined experimentally (Keith and Loomis, 1976).

calibration peaks. For example, Rosner, Gur and Shabason (1975) utilised a fluorescence source emitting Ca and Cl K radiation. With the aim of establishing a standardised method, Stone *et al.* (1981) designed an annular $^{55}$Fe source fitting over the end of the detector and fluorescing a standard glass containing a number of elements with peaks covering a wide energy range. Rather than attempting to derive absolute efficiencies, these authors proposed using a 'window attenuation index' based on measured peak intensities, for the purpose of comparing detectors.

Keith and Loomis (1976) made a thorough study of detector efficiency using a large number of peaks down to 2.6 keV. The basis for the measurements was comparison with an accurately calibrated proportional counter. As shown in fig. 9.13, the maximum efficiency recorded was about 98% (at around 10 keV). At lower energies, the efficiency agrees well with transmission calculations for estimated values of Be, Au and Si thickness.

### 9.8.3 Monitoring contamination layers

The accumulation of oil on the surface of the beryllium window can be monitored by measuring the $L\alpha/K\alpha$ intensity for a pure element, where the $L\alpha$ peak is strongly absorbed by the oil, e.g. zinc (Smith, 1981). Alternatively, a compound containing two suitable elements can be used (Bombelka and Richter, 1988). Given a knowledge of the thickness of the oil layer, a correction for the additional absorption can easily be calculated, for all X-ray energies (Smith, 1981; Cohen, 1987).

Another form of contamination sometimes encountered is ice on the front surface of the detector itself (§9.4.4). The effect of this is very similar to that of oil. Two different methods of measuring ice layer thickness, both employing a $^{241}$Am radioactive source, have been applied by Cohen (1987). In the first, $\alpha$ particles are used, while in the second the M/L intensity ratio in the Np X-ray spectrum emitted by $^{241}$Am is measured. Neither is suitable for *in situ* application to the electron microprobe, however. A more practical approach is to use the method described above, which is sensitive to both ice and oil.

## 9.9 Other detection media

Though the Si(Li) detector occupies a dominant position in e.d. microprobe analysis, and seems likely to continue to do so for the foreseeable future, there are alternative types of solid-state detector which should be mentioned.

### 9.9.1 Germanium

Germanium has similar properties to silicon and has been used for many years for $\gamma$-ray detection. Lithium-drifted germanium has to a large extent been supplanted by high purity germanium, the concentration of impurities in this material being so low that compensation with lithium is not necessary. The factors governing the resolution of germanium detectors are practically the same as for silicon. Slightly better energy resolution is obtainable with germanium because the mean energy per electron–hole pair is smaller. The most important difference is in the efficiency for high energy X-rays, germanium being preferable to silicon above approximately 20 keV. At low energies (below 3 keV) the germanium detector has in the past been considered inferior, but recent developments have greatly improved the performance in this region (Cox, Lowe and Sareen, 1988).

Germanium detectors are rarely used in microprobe analysis, which is concerned almost exclusively with X-ray energies below 20 keV. The detection of high energy K lines of heavy elements has some potential advantages, but with solid samples the spatial resolution is severely worsened at the necessary accelerating voltages (50–100 keV). This does not apply, however, to the analysis of thin specimens in a transmission electron microscope or similar instrument. Advantages of using the high energy K lines of heavy elements include freedom from peak overlaps, small absorption corrections and low continuum intensity. The application of a germanium detector for this purpose has been described by Steel (1986).

### 9.9.2 Mercuric iodide

The liquid nitrogen cooling required for Si(Li) detectors is only a minor inconvenience in a laboratory setting, but for certain specialised applications it is a serious handicap. Alternative methods of cooling have been discussed already (§9.3.1) and are useful in some cases, but a more radical approach is to use a different detection medium which does not require such a low temperature. The most promising candidate is mercuric iodide ($HgI_2$), which has a very low leakage current at room temperature, owing to its large band gap (2.2 eV). In order to realise the potential resolution of $HgI_2$, it is still necessary to cool the FET, however. Moderate cooling of the detector itself is also beneficial. With an FET at 230 K and detector at 270 K (obtained by thermoelectric cooling), Iwanczyk et al. (1986) obtained a resolution of 225 eV at 5.9 keV. The absence of Hg absorption edges in the continuum indicates a thin dead layer, which is favourable for low energy X-ray detection, indeed Iwanczyk et al. (1985) reported the detection of the O K peak at 520 eV.

Electrons and holes are less mobile in $HgI_2$ than in Si and the detector must be thin (e.g. less than 1 mm) in order to avoid undesirable effects resulting from slow charge collection. Consequently the area should be restricted in order to minimise the capacitance. The solid angle for X-ray collection can be increased by using an array of several detectors, which has the incidental advantage of enhancing the high count-rate capability (Warburton and Iwanczyk, 1987).

### 9.9.3 Ion-implanted silicon

Silicon X-ray detectors can be produced by employing ion implantation to produce a sharp p–n junction, with oxide passivation of the surface to reduce leakage current (Kemmer, 1980). These can be used at room temperature. With energy resolution in the region of 1 keV (Burger, Lampert, Henck and Kemmer, 1984), such devices are not useful for electron microprobe analysis in the e.d. mode, but could be competitive with the proportional counter as a detector for w.d. analysis.

# 10

# Electronics for energy-dispersive systems

## 10.1 Preamplifier

X-ray photons absorbed in a Si(Li) detector produce short bursts of current due to electrons and holes moving to the positive and negative electrodes respectively (§9.1). Since the total charge per pulse is very small (e.g. $10^{-16}$ C), a sensitive preamplifier is required and close attention must be paid to minimising noise, which otherwise could seriously degrade the signal. The most suitable available device for the critical first stage of the preamplifier is the junction-type field effect transistor (FET). In its usual form this has three electrodes – 'source', 'drain' and 'gate': the current flowing from source to drain is regulated by the potential on the gate, and an amplified output is obtained. The high gain and high input impedance of these devices are desirable attributes in this application. Cooling of the FET as well as the detector with liquid nitrogen is essential in order to minimise noise: the usual operating temperature of the FET is around 100 K.

A simple FET preamplifier circuit is shown in fig. 10.1(a). The output pulses have a sharp leading edge with a rise time of less than 100 ns (reflecting the time taken for the charge carriers in the detector to reach the electrodes) and an exponential tail. The noise contributed by the coupling capacitor $C_c$ can be avoided by using direct (DC) coupling between detector and FET, as in fig. 10.1(b). However, the current in the detector caused by X-rays and by leakage then drives the FET gate potential increasingly negative and some form of feedback (or charge restoration) is necessary in order to avoid this. In the circuit illustrated in fig. 10.1(b), feedback is provided by the resistor $R_f$. However, this is a source of additional noise, and furthermore the FET gate voltage may still be driven to its limit at high count-rates. Hence other forms of charge restoration are used in practice, as described in the next section.

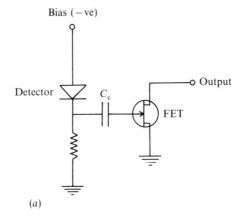

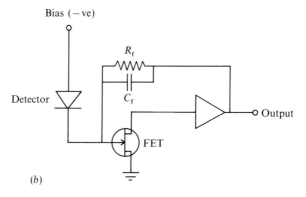

Fig. 10.1 Preamplifier circuits: (a) AC coupled; (b) DC coupled and charge sensitive, with resistive feedback.

An additional feature of the circuit in fig. 10.1(b) is the capacitor $C_f$, which causes the preamplifier to function in 'charge sensitive' mode. Thus the output is proportional to the charge in the pulse rather than the voltage and is therefore independent of differences in detector capacitance.

The FET output pulses are amplified further by circuits outside the cryostat (noise being less critical at this stage), giving pulses suitable for transmission to the main amplifier (§10.3).

### 10.1.1 Charge restoration

The circuits illustrated in fig. 10.1 have some shortcomings, as noted above, hence other systems have been developed. 'Opto-electronic' charge restora-

## 10.1 Preamplifier

tion utilises a light signal, obviating the need for a feedback resistor. In the form described by Kandiah (1966), a photodiode connected to the FET conducts when light from a light-emitting diode (LED) falls upon it (fig. 10.2(a)). The resulting flow of current compensates for the detector current. Charge restoration can be applied either when the DC level at the output reaches a predetermined voltage (as in the original system) or after each pulse (Kandiah, Stirling, Trotman and White, 1968; Kandiah, 1971). The latter approach has the advantage that the DC level at the FET input is kept almost constant, making it easier to achieve a linear response to input pulse height.

Some of the drawbacks of using a separate photodiode are avoided by allowing the light to fall directly on the FET, which achieves the same result (fig. 10.2(b)). In the original system using this technique, feedback was applied continuously (Goulding, Walton and Malone, 1969), whereas

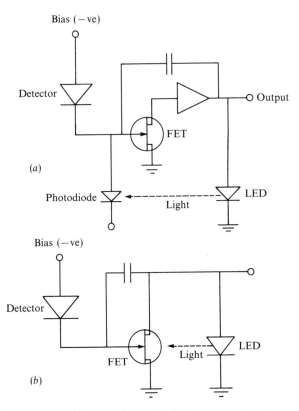

Fig. 10.2 Preamplifiers with opto-electronic charge restoration: (a) using photodiode, (b) with light incident directly on FET.

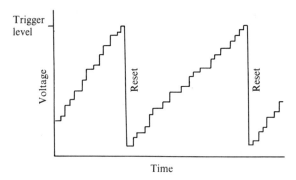

Fig. 10.3 Output signal from preamplifier using pulsed feedback.

*pulsed* opto-electronic charge restoration (Landis, Goulding and Jaklevic, 1970) is used in most current systems. Fig. 10.3 shows the output from a preamplifier operating on this principle. During the 'reset' pulse the main amplifier input is clamped to avoid overload effects. The period for which the system is inactive must be made long enough for transients caused by the reset pulse to decay.

In 'drain feedback', charge restoration is achieved by directly regulating the drain voltage according to the input signal level (Elad, 1972). A pulsed version gives improved performance at high count-rates (Bussolati, Manfredi, Marioli and Krasowski, 1978). With an injection field effect transistor (IJFET), charge restoration can be accomplished by direct injection via an additional electrode. This has the advantage that the recovery time is shorter than with optical feedback, allowing higher throughput rates (Howes and Allsworth, 1986). Also it is no longer necessary to exclude stray light.

## 10.2 Noise

Noise at the preamplifier output causes the baseline upon which the signal pulses are superimposed to fluctuate, thereby introducing random variation in the measured pulse heights. As described in §9.5, such noise plays a major part in determining the energy resolution of the Si(Li) detector (in equations (9.1) and (9.2), $\triangle E_n$ represents the contribution of electronic noise to the peak width). Though some noise originates in the detector itself, a significant amount is contributed by the FET, the characteristics of which are thus very important – FETs are therefore selected individually for low noise by detector manufacturers.

Preamplifier noise consists of three components, distinguished by their frequency distribution. 'Shot noise' has a power spectrum which varies in proportion to frequency ($f$) and is also known as 'parallel noise'. An important source is detector leakage current. 'Current noise' or 'series noise' caused, for example, by fluctuations in the FET channel current, varies as $1/f$, while the third type of noise ('flicker noise') is independent of $f$. The effect of noise on energy resolution can be minimised by frequency filtering, or pulse shaping, in the main amplifier, as discussed below (§10.3.1).

Noise amplitude is a function of detector capacitance, hence the use of a large-area detector (with the object of increasing the X-ray collection efficiency) entails some sacrifice in energy resolution.

External noise sources including vibration and electrical interference can cause loss of energy resolution (peak broadening). Susceptibility to interference may be minimised by avoiding earth loops: the detector should be isolated from the microprobe and earthed only through the cables; also all the electronic units should be earthed at one point only.

## 10.3 Main amplifier

Pulses from the preamplifier are amplified by the main amplifier before passing to the multi-channel p.h.a. (§10.5). The main amplifier may be mounted on the detector housing or remotely in the electronics rack. Apart from amplification, it also has the role of pulse shaping, as described below.

### *10.3.1 Pulse shaping*

The concept of pulse shaping was introduced in §8.3 in connection with proportional counters, for which the requirements, however, are less critical. The desire to optimise the performance of $\gamma$-ray detectors led to detailed theoretical and experimental studies (Fairstein and Hahn, 1965; Goulding, 1966). The same principles apply to Si(Li) X-ray detection systems (Statham, 1981). The main purpose of pulse shaping is to optimise the signal to noise ratio in the main amplifier and thus minimise the widths of the peaks in the e.d. spectrum. A further consideration is the desirability of minimising the total duration of each pulse in order to allow the highest possible pulse throughput rate.

Typical pulse shaping circuits include integrating and differentiating stages. Integration smooths out high frequency noise, while considerably prolonging the rise time of the pulses compared to the steps in the

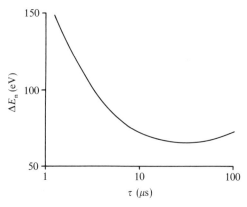

Fig. 10.4 Noise term, $\Delta E_n$, as function of main amplifier time constant, $\tau$.

preamplifier output caused by X-rays. Differentiation, on the other hand, suppresses low frequencies and has the effect of curtailing the time taken for the pulse waveform to decay after peaking, thereby keeping the total duration of the pulse within reasonable bounds.

The simplest form of pulse shaper consists of an *RC* integrator followed by an *RC* differentiator (fig. 8.4), which acts as a bandpass filter, the frequencies passed being of the order of $1/\tau$, where $\tau$ is the time constant ($R \times C$). Fig. 10.4 shows the noise term $\triangle E_n$ as a function of $\tau$, in relation to the three types of noise present in the preamplifier output, as defined in §10.2. For a value $\tau'$ of $\tau$, $\triangle E_n$ is a minimum, giving optimal energy resolution. The maximum pulse throughput rate is proportional to the pulse width and hence $\tau$, so that if $\tau$ is reduced with the aim of increasing the throughput rate, $\triangle E_n$ increases owing to the contribution of series noise. Hence there is an inevitable sacrifice in energy resolution if high throughput rates are required (see §10.3.4). There is also an increase in $\triangle E_n$ for $\tau > \tau'$, but this is mainly of academic interest since there is little incentive to operate in this region.

Alternative forms of pulse shaping offer some improvement in performance by comparison with the simple case just described. A commonly used variant is single differentiation followed by several integrations. The pulse shape obtained is similar to the gaussian function but is somewhat asymmetrical and is known as 'semi-gaussian' (fig. 10.5). Theoretically the ideal shape is a cusp, but this is impracticable: a reasonably close approximation, however, which is easier to realise, is a triangular pulse shape (Tsukada, 1962).

## 10.3 Main amplifier

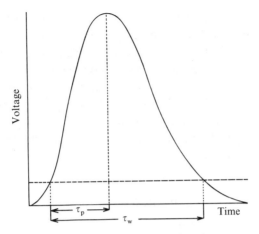

Fig. 10.5 Semi-gaussian pulse shape; $\tau_p$ = peaking time, $\tau_w$ = pulse width.

### 10.3.2 Pole-zero cancellation

Differentiation in the main amplifier produces pulses with an 'undershoot' caused by the long exponential tail. This is undesirable because of its effect on the apparent height of subsequent pulses. The ratio of undershoot to pulse height is equal to the differentiation time constant divided by the input pulse decay time and is typically of the order of 10%. In terms of network theory, the undershoot is caused by the uncancelled 'pole' in the preamplifier and can be corrected by 'pole-zero cancellation' (Nowlin and Blankenship, 1965), which eliminates undershoot and greatly reduces the decay time of the differentiated pulses. Pole-zero cancellation is indispensable for Si(Li) detectors operated at moderate to high count-rates.

### 10.3.3 Baseline restoration

Even in a completely DC-coupled system it is impossible to avoid differentiations with long time constants (introduced for instance by bypass capacitors) which cause undershoots of small amplitude but long decay time. The effect is to depress the mean pulse height and broaden the pulse height distribution to a degree that increases with count-rate. This problem can be overcome by means of 'baseline restoration'. A 'passive' baseline restorer uses only diode circuits, whereas 'active' baseline restoration entails the use of a feedback amplifier to restore the output level to zero.

In a system without baseline restoration, pulse heights are effectively measured relative to the long term mean DC level, whereas the output from a baseline restorer is effectively referred to a relatively short term baseline sample which is subject to noise. Baseline restoration therefore entails loss of energy resolution. With passive restoration the effect is small, but a high level of active restoration causes significant broadening of the peaks.

Karlovac and Gedcke (1973) described an improved form of baseline restoration, in which the restorer is switched off for the duration of each pulse. Such a 'gated' baseline restorer gives better stability at high count-rates without sacrificing resolution.

### 10.3.4 Resolution and count-rate

For a given main amplifier time constant, $\triangle E_n$ tends to increase with increasing count-rate, because the more closely spaced the pulses are, the less time is available for the transient effects of one pulse to decay before the next one arrives. Thus, for a typical system, $\triangle E_n$ increases slowly but steadily throughout the normal range of count-rates (fig. 10.6). With conventional pulse shaping, $\triangle E_n$ is liable to increase sharply at very high count-rates, though this occurs only beyond the range normally used. (The system described in §10.4 does not behave in this way.) Variation in peak width with count-rate is important in quantitative e.d. analysis and should be taken into account in the spectrum processing procedures (§12.7).

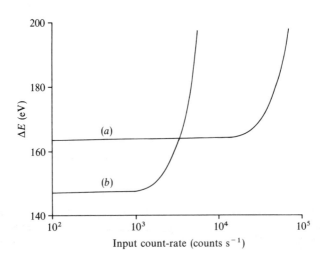

Fig. 10.6 Energy resolution, $\Delta E$ (for Mn K$\alpha$), as function of input count-rate, for (a) short, (b) long pulse shaping time constants.

The dead time of the system (§10.7) is determined by the main amplifier time constant and controls the maximum rate at which pulses can be recorded. If the time constant is reduced, $\triangle E_n$ increases because of the noise characteristics of the detector–FET combination (§10.3.1). Thus, high count-rates are only attainable at the expense of some sacrifice in energy resolution (see fig. 10.6). This should be taken into account when assessing the performance of a particular system. The specification usually refers to the FWHM of the Mn K$\alpha$ peak at a low count-rate (e.g. 1 kHz), whereas for practical purposes the energy resolution at a higher count-rate is more important, especially for quantitative analysis, where the time taken per analysis is governed by the necessity to accumulate a certain total number of counts in order to obtain the required precision.

## 10.4 Harwell pulse processor

The pulse processing system developed at Harwell by Kandiah and coworkers incorporates a number of original features, warranting separate treatment. A simplified block schematic is given in fig. 10.7 and the description of its operation follows Kandiah, Smith and White (1975). This system uses switched time constants, thereby overcoming some of the limitations of the passive pulse shaping circuits described above.

The arrival of a pulse is recognised by a fast discriminator, the output from which initiates a sequence of operations controlled from a central control unit. The equivalent of the main amplifier in a conventional system is the signal processor (fig. 10.8). The input switch S1 is normally closed, but is opened immediately the pulse recognition discriminator is triggered. It is closed only when the input signal has reached full amplitude, thereby ensuring that variations in rise time have no effect. Initially S2 is open and

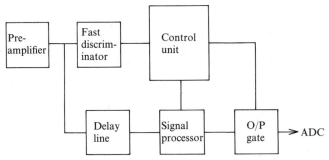

Fig. 10.7 Harwell pulse processor: simplified block schematic.

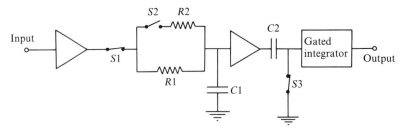

Fig. 10.8 Harwell pulse processor: simplified circuit diagram of signal processor.

the first integrator has a time constant $R1 \times C1$. After the processing of the pulse is complete, $S2$ is closed briefly to allow $C1$ to discharge rapidly ($R2$ being much smaller than $R1$). The differentiator switch $S3$ (normally closed) is opened on the arrival of a pulse and stays open for the duration of the 'on' period of the second (gated) integrator, this being about twice the time constant of the first integrator. It is then closed in order to discharge $C2$. At the same time the output gate is closed to allow the output pulse to reach the analogue-to-digital converter (ADC). The control unit also sends a charge restoration pulse to the preamplifier.

After charge restoration is complete, the system is held in an inactive state for a further time interval in order to allow complete recovery before the next pulse is processed. The total dead time per pulse is thus the sum of the processing time, the restore time and the 'protect time'. The processing time is comparable to the peaking time in a conventional amplifier (§10.3.1) and is related to energy resolution in a similar fashion.

### 10.4.1 Strobed noise peak

A special feature of the Harwell system is the strobed noise peak. In between X-ray pulses, the processing cycle is initiated by 'strobe' pulses from an oscillator (typical frequency – a few hundred hertz). The 'pulse height' thus obtained represents a sample of the baseline in the absence of any signal. A 'noise peak' can thus be displayed, its centroid being located at the zero of the energy scale (fig. 10.9). The width (FWHM) of the noise peak is equal to $\triangle E_n$, the noise term in the expression for energy resolution (§9.5). Automatic zero stabilisation can be implemented by applying a small correction to the DC level of the signal processor output derived from the measured height of each noise pulse. The effect of random fluctuations between pulses averages out and the overall result is to keep the position of the centroid of the noise peak constant.

## 10.5 Multi-channel p.h.a.

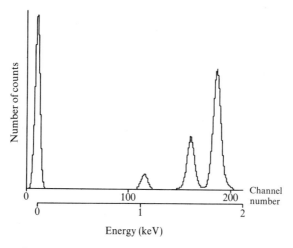

Fig. 10.9 Part of spectrum recorded with Harwell system, showing strobed noise peak at zero energy.

### 10.5 Multi-channel pulse height analysis

For multi-channel pulse height analysis the pulse height is measured by an ADC, which usually operates on the principle of pulse height to time conversion by the 'Wilkinson method'. The input voltage is allowed to charge a capacitor through a diode. When the pulse passes its peak the diode stops conducting and the capacitor holds the peak voltage. The voltage that appears across the diode is sensed by a peak detector circuit which closes the input gate and initiates the discharge of the capacitor from a constant current source. At the same time pulses from an electronic clock are connected to a scaler, and are disconnected when the capacitor voltage falls to zero. The number of pulses recorded by the scaler is proportional to the time taken to discharge the capacitor, and hence to the initial voltage, i.e. the pulse height. A single-channel analyser preceding the ADC prevents it from processing noise and other unwanted signals.

A multi-channel analyser may have its own memory, but more commonly the ADC is connected directly to a computer, part of the memory of which is used for storing the spectrum. Each incoming pulse is assigned a channel address on the basis of the pulse height measurement. The time required for pulse height conversion is equal to the channel address divided by the clock rate, e.g. with a 100 MHz clock the conversion time for channel number 500 is $5\mu s$. Thus the total conversion time is of the order of $10\ \mu s$.

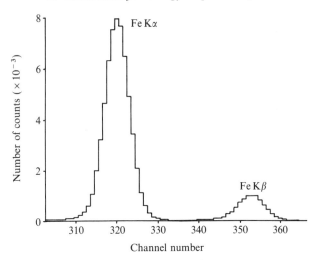

Fig. 10.10 Representation of X-ray spectrum as pulse height distribution histogram.

However, the input pulse width is usually greater than this and therefore determines the dead time per pulse for the system.

The 'conversion gain' determines the number of channels into which the pulse height range covered by the ADC is divided. For most purposes a channel width of 10 or 20 eV is appropriate. There is no significant advantage in using narrower channels, in view of the finite width of the peaks.

The visual display of the pulse height spectrum is presented in histogram form, with channel number along the horizontal axis and the number of counts represented by the vertical height of a bar for each channel (fig. 10.10). A useful option is a logarithmic vertical scale, which enables large and small peaks to be inspected simultaneously. Energy calibration may be checked by reference to known X-ray peaks, two of which (preferably widely separated in energy) are required for adjustment of amplifier gain and zero level. Computer-based systems have many features such as different display modes, facilities for comparing spectra, and markers showing the positions of K, L and M peaks as an aid to identification.

## 10.6 Pulse pile-up

If a second pulse arrives during the rise time of the preceding one, the ADC sees a single pulse of enhanced height (fig. 10.11). When such 'pile-up' occurs neither pulse appears in its proper place in the spectrum, and a spurious pulse is recorded at a higher energy. Pile-up is significant even at moderate count-rates, because of the long time constants used to minimise noise (§10.2). The probability of coincidence between pulses is approximately $n\tau_0$, where $n$ is the count-rate and $\tau_0$ the pulse rise time. Thus for $\tau_0 = 10\,\mu s$ the probability of pile-up occurring at 5000 counts s$^{-1}$ is about 5%.

The most obvious effect of pile-up is the appearance of spurious 'sum peaks' corresponding to the sum of the energies of major peaks in the spectrum (fig. 10.12). Nearly coincident pulses may be eliminated by means of 'pile-up rejection' (Williams, 1968). The usual method of accomplishing this is to have a second amplifier with much shorter time constants in parallel with the main amplifier (fig. 10.13). The output of this fast amplifier is connected to a discriminator, the threshold of which is set just above the noise level. A pile-up inspection circuit detects the presence of pulses

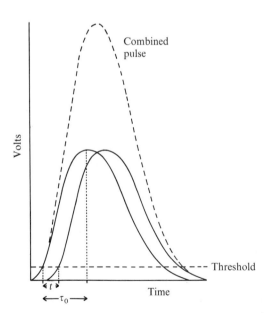

Fig. 10.11 Pulse pile-up: two pulses separated by a short time interval combine to form single larger pulse.

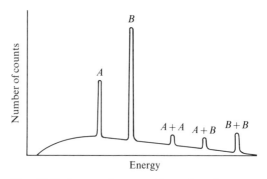

Fig. 10.12 Sum peaks caused by pulse pile-up.

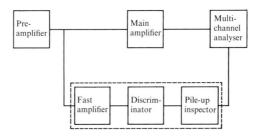

Fig. 10.13 Pile-up rejection system.

separated by less than a given time interval, and applies an 'inhibit' pulse so that the pulse is not recorded. By this means pile-up effects are decreased by a large factor.

The pulse-pair resolution of the fast amplifier is limited by the finite rise time of the preamplifier pulses. Hence pile-up rejection can never be totally effective. However, sum peaks are reduced to a level that is negligible for most purposes.

Unfortunately such high rejection efficiency cannot be maintained down to the lowest X-ray energies because of the poor signal to noise ratio of the fast amplifier (Reed, 1972; Statham, 1977a). The time constants of this amplifier have to be increased to permit the discriminator to distinguish low energy X-ray pulses from noise. This reduces the efficiency of pile-up rejection for higher X-ray energies, but optimum performance for both high and low energies can be obtained by using two rejection channels in parallel, with different time constants (Kandiah *et al.*, 1975). With a windowless detector it is advantageous to use three such channels, one being dedicated to very low energies (below 1 keV).

## 10.7 Dead time

Usually the pulse processing time exceeds the conversion time of the multichannel analyser, and the system effectively has an extendable dead time (see §8.7) determined by the pulse length $\tau_1$. Hence the recorded count rate $n'$ is given by: $n'/n = \exp(-n\tau_1)$, where $n$ is the true count-rate. Thus for $\tau_1 = 50\,\mu s$, $n'/n = 0.61$ at 10000 counts s$^{-1}$, i.e. 39% of the input pulses are lost. For $n > 1/\tau_1$ (20000 counts s$^{-1}$ in this case) $n'$ decreases with increasing $n$ (fig. 10.14). The maximum value of $n'$ is equal to $1/e\tau_1$ (7000 counts s$^{-1}$). As discussed below, the effective dead time is greater than $\tau_1$, owing to pile-up losses.

It is usual for e.d. systems to incorporate automatic dead-time correction, usually by the method of Covell, Sandomire and Eichen (1960), whereby the counting period is extended to compensate for the loss of counts. The counting period is controlled by means of a scaler counting the pulses from a 'clock' oscillator of precisely known frequency, a gate between oscillator and scaler being closed while the ADC processes each X-ray pulse.

The 'Barnhart method' (Russ et al., 1973) uses a somewhat different principle: the number of pulses arriving during the 'dead' periods (as detected by a fast amplifier) is recorded and the time is extended until the same number of extra pulses has been recorded in the spectrum.

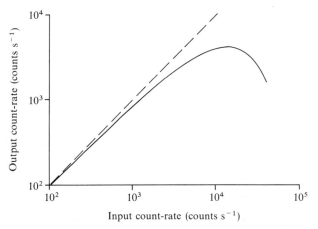

Fig. 10.14 Output count-rate for system with extendable dead time, as function of input count-rate (dashed line – no dead time).

### 10.7.1 Pile-up losses

Counting for a fixed 'live time' makes no allowance for counts lost from peaks due to pile-up. (This loss occurs with or without pile-up rejection, which merely has the effect of preventing pulses affected by pile-up from appearing elsewhere in the spectrum.) The probability of a pulse *not* being followed by another within the rise time $\tau_0$, and therefore appearing in its correct place in the spectrum is $\exp(-n\tau_0)$. The pulse may also be lost as a result of paralysis of the system by the preceding pulse, and the probability of this *not* occurring is $\exp(-n\tau_1)$. The ratio of the apparent peak count-rate $n'_K$ and the true peak count-rate $n_K$ is equal to the product of these probabilities:

$$n'_K/n_K = \exp[-n(\tau_0 + \tau_1)]. \tag{10.1}$$

If the counting period is extended to correct for $\tau_1$ as described above, equation (10.1) becomes:

$$n'_K/n_K = \exp(-n\tau_0). \tag{10.2}$$

Rather than using a calculated correction for pile-up losses it is possible to apply the correction electronically. Various methods have been proposed, one or other of which is used in most commercial systems. Landis *et al.* (1972), for example, modified the Covell system by increasing the period for which the clock pulses are switched off for each recorded pulse from $\tau_1$ to $\tau_1 + \tau_0$. Bartosek, Masek, Adams and Hoste (1972) described another method, whereby whenever a coincidence is detected the clock is stopped until the arrival of the next pulse. Russ *et al.* (1973) proposed a method in which the number of lost pulses is counted separately and the counting period is extended until an equal number of extra pulses has been recorded by the multichannel analyser. This procedure is applied many times per second so that the correction is effectively continuous.

## 10.8 Beam switching

As already noted, the extendable character of the dead time limits the maximum rate at which pulses from a Si(Li) detector can be recorded. However, higher recording rates are possible if the X-ray source is turned off during the processing of each pulse, in which case the dead time becomes non-extendable. Jaklevic, Goulding and Landis (1972) applied this technique to X-ray fluorescence analysis, and Statham, Long, White and Kandiah (1973) used the same idea for electron microprobe analysis, the

## 10.8 Beam switching

electron beam being switched off during the processing of each pulse by means of deflection plates in front of an aperture. The maximum rate at which pulses can be recorded approaches $1/\tau$, where $\tau$ is the processing time. This is around four times higher than with a typical conventional system, given that pile-up in the latter results in the loss of *both* pulses involved, which does not apply in the case of beam switching (see below).

At high count-rates the X-ray signal consists of a train of closely spaced pulses, and the electron beam is on only during the intervals between them. The ratio of the mean current to the full current is $1/(1+n'\tau)$, where $n'$ is the recorded count-rate. If damage to the specimen due to electron bombardment is the limiting factor, and this is assumed to be a function of mean probe current rather than peak current, then higher count-rates are possible with beam switching.

Beam switching automatically provides protection against pile-up, by cutting off the source of X-rays as soon as the arrival of a pulse is detected. This is more effective than conventional pile-up rejection (§10.6) because it depends on recognising single pulses only rather than requiring pairs of pulses to be resolved (Statham, 1977a).

Despite the considerably higher spectrum acquisition rate obtainable with beam switching, this technique is not routinely used, owing to the lack of provision in commercial electron microprobe instruments for fast beam switching.

# 11
# Wavelength-dispersive analysis

## 11.1 Qualitative w.d. analysis

A w.d. spectrometer may be used for qualitative analysis (identification of elements present) by sweeping through the appropriate range of Bragg angles and recording the peaks in the spectrum (fig. 11.1). With older instruments this is carried out by driving the spectrometer continuously, the output being recorded by a ratemeter connected to a chart recorder. In the case of a computer-controlled instrument the spectrometer is moved in small steps, while the digital output is stored in the computer memory and displayed on a v.d.u. screen.

For complete wavelength coverage it is necessary either to change the crystal or to have several spectrometers equipped with different crystals, the latter being more efficient since simultaneous scans covering different wavelength ranges can be carried out. Fig. 11.2 shows the wavelength coverage of the crystals commonly used in w.d. analysis. For longer wavelengths, synthetic multilayer devices are also available (§6.3).

The efficiency of the procedure described above is rather low because for most of the time only background is being recorded. The time required can be reduced by limiting the coverage to regions of the spectrum where peaks of interest may exist. Searching for a large number of elements is still quite time consuming, however. Exclusion of elements thought unlikely to be present may lead to unexpected elements being missed. (In this regard the simultaneous collection of the whole spectrum in e.d. analysis is advantageous.)

X-ray lines may be identified by reference to tables such as those of White and Johnson (1970), in which all known lines are listed in wavelength order. Computers built into electron microprobe instruments commonly incorporate wavelength tables and aids to identification, but may not cover all the minor lines (especially those in the L spectrum).

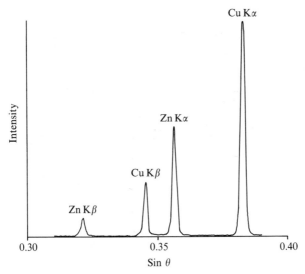

Fig. 11.1 Spectrum of brass recorded with w.d. spectrometer (LiF crystal).

It is desirable to use pulse height analysis (§11.4) for wavelength scans, in order to suppress high order reflections, which add to the complexity of the spectrum. There is usually little ambiguity about the identification of the lines in a w.d. spectrum, owing to the narrowness of the peaks. If there is any doubt, positive identification can invariably be made by seeking other peaks ($\beta$, $\gamma$ etc.) of the element concerned.

## 11.2 Principles of quantitative w.d. analysis

The following discussion refers to 'conventional' analysis, covering atomic numbers from 11 (Na) upwards. The same general principles apply to atomic numbers below 11, but there are various special factors in light element analysis (chapter 18).

Quantitative analysis entails measuring the intensity of a characteristic peak of each element present in the 'unknown' sample and comparing this with measurements on one or more standards under identical instrumental conditions. A significant difference between e.d. and w.d. analysis is that in the former it is not necessary to decide which elements to include before recording the spectrum, whereas in w.d. analysis such a decision, based either on assumption or prior knowledge, is necessary. Decisions must also be made as to which crystals to use (where wavelength coverage overlaps), background offset angles (§11.5), and counting times (§11.7). Choice of

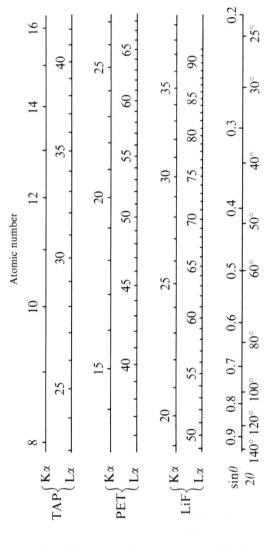

Fig. 11.2 Bragg angles for commonly used crystals, for different atomic numbers.

accelerating voltage is discussed in §11.9. It is advantageous to use as high a probe current as possible from the point of view of minimising the counting time required to achieve a given precision, or minimising the detection limit for a given counting time. However, the choice is sometimes subject to limitations imposed by specimen damage (§11.11).

In w.d. analysis the peak intensities are measured sequentially, hence probe current drift can cause analytical errors. Special attention must therefore be paid to probe current monitoring and stabilisation (§4.10). In principle changes in probe current can be compensated by means of a drift correction, but it is inadvisable to rely on such a correction, other than for small amounts of drift (of the order of 1%).

Measuring a peak intensity involves setting the spectrometer accurately on the top of the peak. This is subject to various hazards, including backlash in the spectrometer mechanism, change in wavelength calibration due to thermal drift etc., and wavelength differences related to chemical bonding. The precautions necessary to avoid errors arising from these factors are discussed in the next section.

The peaks are superimposed on a background which is attributable almost entirely to the continuous X-ray spectrum or bremsstrahlung (§1.6). The intensity measured with the spectrometer set on a peak includes a background contribution which must be subtracted in order to obtain the true peak intensity. It is usual to measure background by offsetting the spectrometer on one or both sides of the peak (§11.5).

An advantage of a computerised system is that conditions for particular types of analysis (list of elements, choice of crystals, spectrometer angles for peaks and backgrounds, p.h.a. settings, counting times, accelerating voltage, probe current) can be stored and recovered quickly when required. Furthermore, if the instrument is sufficiently stable, it is possible to use stored intensities for quantitative analysis, as in e.d. analysis (§12.7.2), thus avoiding the necessity to record data from standards immediately before or after each analysis, though as a rule standard calibrations should be carried out at least once per session. A further advantage of a computer-controlled system is the capability for automated operation, enabling analyses to be carried out, without the presence of the operator, on points in a stored list of stage coordinates.

## 11.3 Peak selection

Fig. 11.3 shows a typical peak profile. The accuracy required in the setting of the Bragg angle $\theta$ is typically about 10 seconds of arc ($5 \times 10^{-5}$ radians).

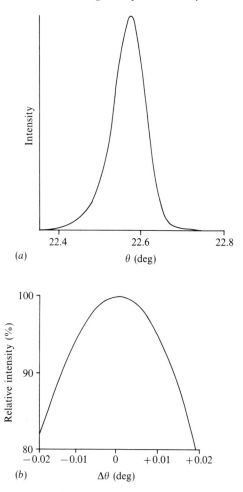

Fig. 11.3 Ca Kα peak recorded with PET crystal: (a) complete peak, (b) central region enlarged.

With a linear spectrometer mechanism (§6.5) the required positional accuracy of the crystal is given by:

$$\triangle x = 2r(\cos\theta)\triangle\theta,$$

where $\triangle\theta$ is the error in $\theta$. Thus, $\triangle x$ is greatest for low values of $\theta$: for example if $r = 150$ mm, $\triangle x = 14\,\mu\text{m}$ at $\theta = 15°$ (for $\triangle\theta = 5 \times 10^{-5}$ rad). In a digitally controlled spectrometer the minimum step size should be smaller than $\triangle x$, implying the need for at least 50 000 steps over the full $\theta$ range.

## 11.3 Peak selection

It is difficult to eliminate backlash completely from the spectrometer mechanism, especially when it becomes worn after prolonged use. It is therefore desirable to approach the final Bragg angle always from the same direction. Another possible source of error is drift in the wavelength calibration, which can be overcome by recalibration at suitable intervals, by reference to a known peak. One cause of such drift is change in the '$d$' spacing of the crystal with temperature (§6.2.3). Of the commonly used crystals, PET is much the most temperature sensitive: at high Bragg angles a difference of only $0.1°$ can cause a change of 1% in the measured peak intensity. Careful attention should therefore be paid to the stability of the ambient temperature.

If crystals are changed in the course of an analysis, it is necessary to redetermine the wavelength calibration, unless the mechanical reproducibility of the crystal position is of a very high order. This requires a 'peak seek' routine (see below) to be carried out. Sometimes, however, the peaks are too small for this to be possible. It is therefore desirable to avoid crystal changes if possible.

Computer control enables automatic 'peak seek' routines to be executed in place of manual 'tuning' for maximum intensity. This entails stepping through a narrow region centred on the presumed position of the peak and recording the number of counts at each point. The simplest assumption is that the position of the peak corresponds to the point giving the highest count. However, unless inconveniently long counting times are used, statistical fluctuations may cause errors in the peak position: it is therefore preferable to apply smoothing to the peak profile, or to use a fitting procedure. Counting time can be economised by utilising a two-stage procedure in which relatively large steps and short counting times are used initially to find the approximate peak position, while in the second stage smaller steps and longer times are used, attention being concentrated on the region close to the centre of the peak.

Small peak shifts due to differences in chemical bonding sometimes occur, most significantly for 'light' elements, where special precautions must be taken (§18.4). For 'normal' elements, such effects can generally be ignored, but if necessary different spectrometer settings can be used for specimen and standard. The effect of chemical shifts can be minimised by avoiding spectrometer conditions that give very sharp peaks (§6.7).

With a vertically mounted spectrometer, the Bragg angle is affected by small changes in the height of the specimen (§6.10.1), and steps must be taken to prevent errors in quantitative w.d. analysis arising from this '$z$ defocussing' effect. The usual procedure is to bring the surface of the specimen to the focal plane of the optical microscope using the $z$ movement of the stage (the focus of the microscope being fixed). The allowable error in the $z$ direction is a few micrometres. The depth of focus of a typical microscope is usually somewhat less than this, hence optical focussing is adequately accurate. With a computer-controlled instrument, the required $z$ settings can be stored together with the $x$ and $y$ coordinates of the points to be analysed, so that a series of points can be analysed completely automatically. It is also important that the position of the beam in the $x$ and $y$ directions should not vary enough to cause defocussing of the spectrometers (§6.10.2).

## 11.4 Pulse height analysis

The function of pulse height analysis is to suppress second (and higher) order reflections from the crystal (§6.2.1) and reduce other unwanted contributions. The mean pulse height of X-rays reflected in the second order is twice that of the first order, hence good discrimination can be obtained in the fairly rare cases where such reflections are closely coincident in Bragg angle. X-rays which arrive at the proportional counter as a result of scattering and which have energies outside the p.h.a. window are also excluded, thereby causing some reduction in background.

The width of the p.h.a. window should normally be set to include almost all of the pulse height distribution (§7.7), thereby avoiding the excessive sensitivity to changes in pulse height which occurs if a narrow window is used. The escape peak (§7.9) should be either wholly included in or excluded from the window. The threshold is set according to the energy of the detected X-rays and therefore has to be changed when the same spectrometer is used for several elements. If the crystal is changed in order to extend the wavelength coverage, it may be desirable to vary the counter voltage, which controls the gain of the counter (§7.4), in order to keep the pulse height approximately constant.

The trouble involved in selecting appropriate p.h.a. conditions can be avoided by operating in the discriminator mode (§8.4) with the threshold set to a value well below the pulse height of the X-rays of interest. This avoids hazards such as pulse height depression at high count-rates (§7.5), broadening of the pulse height distribution due to contamination of the anode wire (§7.8), and gain changes due to variations in counter gas density (§7.3), but entails the sacrifice of the above-mentioned benefits of pulse height analysis.

## 11.5 Background corrections

The purpose of the background correction is to subtract from the measured peak intensity the contribution of background, the principal source of which is the continuous X-ray spectrum. This apparently simple operation requires careful consideration for concentrations below a few per cent, when the background is a significant fraction of the peak.

In fig. 11.4(a) the background is a linear function of wavelength. The background intensity at the centre of the peak can be obtained by calculating the mean of the intensities measured at equal displacements on each side of the peak. It is not unusual for the background to be significantly non-linear, especially at low Bragg angles (fig. 11.4(b)), in which case the above correction method is inaccurate. A correction for curvature may be determined on a specimen containing none of the element concerned, but otherwise of similar composition to the analysed specimen (in order to avoid errors due to differences in absorption). The required correction factor can be derived from the ratio of the intensity measured at the peak position to the mean of the background measurements on each side of the peak.

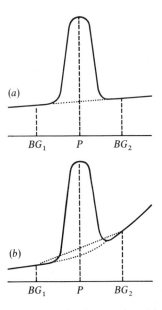

Fig. 11.4 Background measurement – background positions ($BG_1$, $BG_2$) symmetrically located about peak ($P$): linear background (a) gives correct result, curved background (b) gives error.

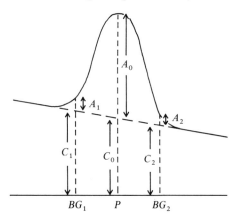

Fig. 11.5 Background measurement – effect of tails of peak at background positions $BG_1$, $BG_2$.

The following argument shows that it is not necessary to measure background so far away from the peak that the latter makes no appreciable contribution to the observed intensity. In the situation illustrated in fig. 11.5, the intensities measured at the background positions $BG_1$ and $BG_2$ include contributions $A_1$ and $A_2$ from the tails of the peak, as well as contributions $C_1$ and $C_2$ from the continuum. The intensity measured at the peak position includes contributions $A_0$ from the peak and $C_0$ from the continuum. The measured peak intensity is equal to $A_0 + C_0$ and the mean background measured at positions $BG_1$ and $BG_2$ is: $(A_1 + C_1 + A_2 + C_2)/2$. Peak minus background is therefore given by:

$$P - B = A_0 + C_0 - (A_1 + C_1 + A_2 + C_2)/2$$
$$= [A_0 - (A_1 + A_2)/2] + [C_0 - (C_1 + C_2)/2].$$

Assuming the continuum to be linear and the background offsets equal, the second bracketed expression above is equal to zero. Further, if $A_1$ and $A_2$ are fixed fractions $a_1$ and $a_2$ of $A_0$, then:

$$P - B = A_0[1 - (a_1 + a_2)/2].$$

Thus, the measured intensity, after subtracting background, is proportional to $A_0$ (the true peak intensity) even though the tails of the peak contribute significantly to the measured background.

According to the above argument, it is permissible to use relatively small background offsets, thus minimising the risk of overlap from adjacent peaks and reducing the effect of curvature. However, the closer the

## 11.5 Background corrections

background positions are to the peak, the greater the risk of error due to changes in the position or shape of the peak, or to inaccurate setting of the spectrometer angle.

Absorption edges should be taken into account when deciding on points at which to measure background. Every peak is accompanied by an absorption edge on the short wavelength side (caused by the same element that produces the peak), beyond which the absorption of the continuous spectrum in the specimen increases and misleadingly low background intensities are observed. A point between peak and edge should therefore be used if possible. Otherwise the background correction should be determined from a measurement on the long wavelength side only. These considerations do not apply to small peaks for which the associated absorption step is negligible.

When the spectrum is so crowded with lines that it is difficult to find places suitable for background determination, a possible solution is to apply 'continuum modelling' as used in e.d. analysis (§12.5), whereby the continuum intensity is extrapolated from a peak-free region, using a suitable formula for the dependence on wavelength (though this is somewhat more difficult for w.d. analysis since the spectrometer efficiency is less predictable). This approach has been applied to rare earths in minerals by Smith and Reed (1981). It has also been used by Merlet and Bodinier (1990) to interpolate between offset and peak positions when a highly accurate background determination is required in order to achieve detection limits in the region of 10 ppm. The spectrometer efficiency factor is derived from measurements on a standard.

Another possibility is to measure the background intensity at the position of the peak using a sample containing none of the element concerned. A correction factor must be applied to allow for the difference in composition between this material and the 'unknown', which affects the continuum intensity (Böcker and Hehenkamp, 1977).

### 11.5.1 Background 'holes'

The procedures for background correction described above depend on the background varying smoothly and monotonically with wavelength. Usually this condition is satisfied, but Self et al. (1990) pointed out the existence of anomalous 'holes' resembling negative lines, which can cause significant errors in trace element analysis. For example, with a LiF crystal the background intensity drops by about 10% at a $2\theta$ value of $36.87°$, which is close to the Au L$\alpha$ line at $36.98°$. Such 'holes' are caused by reflections other than the 'correct' one (200 in the above case), from planes lying at such angles that the reflected rays do not reach the counter. This effect can be minimised by changing the orientation of the crystal lattice relative to the plane of the Rowland circle.

## 11.6 Interferences

If neighbouring peaks are close enough to influence the effective background in the region of the peak of interest, the conventional background correction is inadequate. In the case of a $K\alpha$ peak which is subject to interference from the $K\beta$ peak of a neighbouring element, a correction for the overlap of the peak may be derived by first measuring the ratio between the interfering $K\beta$ line and its parent $K\alpha$ peak, using a suitable standard. The same ratio can then be used for calculating the overlap correction for any other specimen from the measured $K\alpha$ intensity of the interfering element. Ideally the difference in the absorption of $K\alpha$ and $K\beta$ lines should be taken into account, but the effect is usually small. Snetsinger, Bunch and Keil (1968) described the application of this procedure to the determination of V in the presence of Ti, the $K\beta$ line of which lies only 0.009 Å away from V $K\alpha$. Also, the use of calculated factors for correcting overlaps in the spectra of rare earth elements has been described by Roeder (1985).

The background intensity may also be affected by another nearby peak. If this applies on one side of the peak, it can be avoided by measuring background only on the other side. However, it is then necessary to allow for the slope of the continuum, which is taken into account automatically when the usual procedure of averaging the background measured on both sides is used. If necessary, a slope correction can be applied, in similar fashion to the curvature correction described above.

When interference from a neighbouring peak affects both peak and background intensities, a single correction is sufficient, as shown below. In fig. 11.6, the measured peak is '$A$' and the interfering peak '$B$'. The background measured at position $BG_1$ is equal to $A_1 + B_1 + C_1$, and at position $BG_2$ is $A_2 + B_2 + C_2$, while the measured peak intensity is $A_0 + B_0 + C_0$. Hence peak minus background is given by:

$$P - B = A_0 + B_0 + C_0 - (A_1 + B_1 + C_1 + A_2 + B_2 + C_2)/2$$
$$= [A_0 - (A_1 + A_2)/2] + [B_0 - (B_1 + B_2)/2] + [C_0 - (C_1 + C_2)/2].$$

The last term is zero, assuming the continuum is linear and the background offsets are equal. If $A_1$ and $A_2$ are fixed fractions $a_1$ and $a_2$ of $A_0$ and if $B_0$, $B_1$ and $B_2$ are fixed fractions $b_0$, $b_1$ and $b_2$ of $B_3$ (the height of the '$B$' peak), then:

$$P - B = A_0[1 - (a_1 + a_2)/2] + B_3[b_0 - (b_1 + b_2)/2].$$

## 11.7 Counting strategy

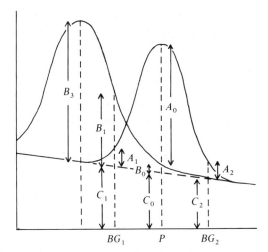

Fig. 11.6 Background measurement – effect of overlap from neighbouring peaks.

Thus, the contribution of '$B$' to the measured intensity of the '$A$' peak can be taken into account by subtracting the second term above, assuming that the peak position and shape do not change and $b_0 - (b_1 + b_2)/2$ is constant. The value of this coefficient can be determined empirically. This procedure also covers the case where more than one peak in the spectrum of element '$B$' causes interference (which can occur with multiple L peaks), provided the relative size of the peaks is constant.

### 11.7 Counting strategy

The s.d. of an intensity measurement in which $n$ counts are accumulated is $n^{1/2}$. There is a probability of 68% that a given determination of $n$ will lie within $\pm 1$ s.d. of the 'true' value, and a 95% probability of it lying within $\pm 2$ s.d. While the *precision* (reproducibility) of the measurement is given by the expression above, the analytical *accuracy* is limited by other considerations, including instrumental factors and errors in matrix corrections. Hence there is usually little advantage in collecting more than $10^5$ counts (for which 2 s.d. = 0.6%).

The higher the count-rate, the shorter the counting time required. Count-rates are determined by the efficiency of X-ray production for the element concerned, the efficiency of the X-ray spectrometer, and the probe current, which may be governed by either specimen damage or instrumental limitations. For typical major element count-rates in the range $10^3$–$10^4$ counts s$^{-1}$, counting times of 10–100 s are needed to meet the above requirement.

For concentrations below 1%, count-rates are typically of the order of 100 counts s$^{-1}$, and the counting time required to accumulate $10^5$ counts may be inconveniently long. A possible remedy is to increase the probe current or accelerating voltage, but this may result in overloading the counting system during the analysis of major elements, in which case the analysis has to be carried out in two stages. In practice lower precision is usually acceptable for minor and trace elements by comparison with major elements.

The counting system can be set to count for a preset time, or else to stop after receiving a preset number of counts. Although the 'preset counts' mode of operation has the advantage of automatically giving a constant relative statistical error, it is inconvenient for widely varying count-rates. 'Preset time' counting is therefore usual.

Peak to background ratios for pure elements are typically in the range 300–1000:1; the background is thus equivalent to a concentration of 0.1–0.3%. In the case of compounds, the dependence of the continuum intensity on mean atomic number should be taken into account. For trace elements ($<0.1\%$) the precision of the background is just as important as that of the peak measurement, and an equal counting time for both is required. At high concentrations the background correction becomes almost trivial, and poor relative precision in the background measurement is acceptable, hence the counting time may be much less than on the peak. For intermediate concentrations, counting time should ideally be divided between peak and background in such a way as to minimise the overall statistical error for a given total time. The optimum ratio of counting times for peak and background is thus equal to the square root of the peak to background ratio. A more detailed treatment of these topics is given by Ancey, Bastenaire and Tixier (1977).

## 11.8 Limits of detection

Assuming counting statistics to be the only relevant consideration, the smallest detectable peak may be defined as 3 s.d. of the background count. The probability of the background count-rate exceeding the 3 s.d. limit due to random fluctuations is 0.2%. The detection limit cannot be reduced indefinitely by accumulating more counts, however, because systematic errors in the background correction eventually become significant. For example, if peak and background measurements are made sequentially, probe current drift may be a limiting factor. Furthermore it is difficult to be sure that the background intensity, extrapolated from measurements made with the spectrometer offset from the peak, is accurate to better than 1%.

Assuming a background correction accuracy of 1% and a peak to background ratio of 1000:1 (which is above average) the ultimate limit of detection is 10 ppm. With a typical background count-rate of 100 counts s$^{-1}$, and assuming the 3 s.d. definition of the limit of detection, a counting time of 1000 s on peak and background would be required to achieve this. In practice shorter counting times are usual: the detection limit is therefore typically a few tens of parts per million, or greater when the peak to background ratio is lower (e.g. for samples of high mean atomic number).

## 11.9 Choice of accelerating voltage

The main considerations in selecting the accelerating voltage are X-ray excitation efficiency and the penetration of the incident electrons in the specimen. Excitation efficiency is dependent on the overvoltage ratio ($U$), defined as $E_0/E_c$, where $E_0$ is the incident electron energy and $E_c$ is the critical excitation energy of the relevant shell (§1.4). For the common elements of atomic number 11–30, $E_c$ for the K shell is in the range 1–10 keV, and an accelerating voltage of 10–25 kV is generally used. Since the characteristic X-ray intensity varies approximately as $(U-1)^{1.67}$ (§13.3), a high $U$ is desirable to give the maximum count-rate. The peak to background ratio also increases with $U$. However, electron penetration increases rapidly with $E_0$, hence spatial resolution deteriorates (§13.9).

Also, the absorption correction increases as the electrons penetrate to a greater depth, and the emerging X-rays have to travel further through the specimen. Since the accuracy of the correction decreases when the correction is large, it is undesirable to use too high an accelerating voltage for quantitative analysis (i.e. above about 25 kV), though the development of improved methods for calculating absorption corrections (§15.6–15.8) has reduced the necessity for observing this limitation. For trace element analysis a higher accelerating voltage is advantageous because of the higher peak to background ratio.

If too low an accelerating voltage is used the X-ray intensities become unduly sensitive to variations in the thickness of the conducting coating (§11.10), where this is required, and carbon contamination (§11.10.1). Unless the coating thickness is accurately controlled, a lower limit of 10 kV is advisable, though a lower voltage may sometimes be desirable for light elements ($Z < 10$) to avoid excessive absorption in the sample.

### 11.9.1 Measurement of accelerating voltage

Matrix corrections (especially the absorption correction) depend on the incident electron energy $E_0$. It is often assumed that $E_0$ is given by the reading of the voltmeter on the high voltage supply for the electron gun. However, this meter may not be accurate and, furthermore, the supply voltage differs from the true accelerating voltage by an amount equal to the gun bias, if it is connected to the grid (fig. 3.2). It is therefore desirable to calibrate the voltmeter in terms of the true accelerating voltage, which determines the electron energy. (After calibration the bias setting should be kept constant.)

This can be accomplished by measuring the intensity of a suitable characteristic X-ray line while the voltage is varied (Sweatman and Long, 1969a). At the point at which the line just appears, the incident electron energy is equal to the critical excitation energy of the element concerned. An accuracy of better than $\pm 200$ V can be obtained. Useful elements for calibration include Zn ($E_K = 9.66$ keV), Se (12.65 keV), Y (17.04 keV), Mo (20.00 keV), Ag (25.52 keV), and Sn (29.19 keV).

Another method of calibration is to determine the high energy cut-off point of the continuous spectrum (the Duane–Hunt limit) using an energy-dispersive detector (see §12.2.1).

## 11.10 Effect of conducting coating

It is necessary for insulating specimens to be coated with a conducting surface layer to provide a path for the probe current and prevent charging. However, such a layer has the effect of reducing the observed X-ray intensity, partly by absorbing some of the energy of the incident electrons and partly through attenuation of the emerging X-rays. These effects may cause small but significant errors in quantitative analysis.

The element most commonly used for coating is carbon, on account of its low atomic number (6), which ensures a minimal effect on X-ray intensities. Also it does not contribute characteristic peaks to the spectrum, at least in the region usually of interest. A theoretically preferable alternative is beryllium (Marshall and Carde, 1984; Marshall, Carde and Kent, 1985), though this presents problems regarding toxicity. (The question of coatings suitable for light element analysis is discussed further in §18.6.) There is sometimes a case for using a heavier element because of its higher thermal conductivity (e.g. aluminium) in order to minimise damage to the specimen due to heating by the beam (§11.11), but this is likely to have a greater effect on X-ray intensities.

## 11.10 Effect of conducting coating

The fractional intensity loss ($\Delta I/I$) caused by a film of thickness $\Delta z$ (cm) and density $\rho$ may be estimated from the expression:

$$\Delta I/I = \{[8.3 \times 10^5/(E_0^2 - E_c^2)] + \mu \mathrm{cosec}\psi\}\rho\Delta z, \qquad (11.1)$$

where $E_0$ and $E_c$ are respectively the incident electron energy and the critical excitation energy (in keV) of the radiation concerned, $\mu$ is the m.a.c. of the coating material, and $\psi$ is the X-ray take-off angle. The first term in the brackets represents the effect of electron energy loss (Reed, 1964) and the second that of X-ray absorption. Kerrick, Eminhizer and Villaume (1973) found good agreement between the above expression and experimental data for carbon films (the density of evaporated carbon being assumed to be 1.3 g cm$^{-3}$).

The small loss of intensity caused by the coating is immaterial provided it is equal for specimens and standards. Reasonably constant thickness may be obtained by coating batches of standards and specimens simultaneously, provided they are mounted at a constant angle and distance from the source. Even then variations in thickness can occur, however, because of non-uniform distribution of evaporant.

It is unnecessary to recoat standards with each new batch of specimens if the thickness can be controlled sufficiently accurately. Methods of monitoring film thickness have been reviewed by Greaves (1970). The simplest is to measure the electrical resistance of a clean glass slide exposed in the evaporation chamber, but this is not entirely satisfactory since the resistance of the film is somewhat dependent on vacuum conditions and the rate of evaporation. Similar objections apply to measuring the optical transmission.

The 'quartz crystal monitor' (Lins and Kukuk, 1960; Lawson, 1967) provides a method for measuring the mass per unit area of evaporated films. This consists of a quartz crystal incorporated in an oscillator circuit, one face of the crystal being exposed in the evaporator. The frequency of oscillation varies with the mass thickness of the film, enabling this to be measured to better than $\pm 5\%$.

With an accelerating voltage of 10 kV the loss of Fe K$\alpha$ intensity caused by a 20 nm carbon film is about 4%, and for Na K$\alpha$, 2% (according to equation 11.1) the difference being due to the different critical excitation energies. The effect of the film decreases with increasing accelerating voltage and is less than 1% for both elements at 20 kV.

Provided the coating thickness on specimens and standards is uniform, an intensity loss of a few per cent is permissible. However, owing to the practical difficulty of achieving uniformity, the thickness should be no more than necessary. If the film is too thin its resistance may be sufficient for a significant voltage to appear at the point of impact of the beam, possibly causing defocussing of the beam and changes in the X-ray intensity. If the film is discontinuous, the bombarded area will charge up to a high voltage and periodically discharge, causing instability in probe current and position, together with low and unstable X-ray intensities. The minimum carbon film thickness required to avoid these effects is about 20 nm.

### 11.10.1 Carbon contamination

The carbon deposit formed on the surface of the specimen at the point of impact of the electron beam (caused by the 'cracking' of hydrocarbons originating from various sources, but mainly from vacuum pump oil) may reduce the observed X-ray intensities, as discussed above. Various measures which can be taken to minimise the rate of deposition of contamination are discussed in §18.7.

## 11.11 Beam damage

Almost the entire energy of the incident electron beam is dissipated in the form of heat at the point of impact, which may be sufficient to cause significant damage to the specimen. Castaing (1951) derived the following expression for the maximum temperature rise $\triangle T$ (K) in a solid sample bombarded with an electron beam of energy $E_0$ (keV), current $i$ ($\mu$A) and diameter $d$ ($\mu$m):

$$\triangle T = 4.8 \, E_0 i / kd, \quad (11.2)$$

where $k$ is the thermal conductivity of the specimen (W cm$^{-1}$ K$^{-1}$). For $E_0 = 20$ keV and $d = 1$ $\mu$m the current required to produce a temperature rise of 100 K is 1 $\mu$A for a typical metal ($k = 1$), 100 nA for a typical crystalline mineral ($k = 0.1$), and 2 nA for an organic compound ($k = 0.002$).

The temperature rise at the point of impact of the beam is reduced by a surface layer of higher thermal conductivity than the specimen (Almasi, Blair, Ogilvie and Schwarz, 1965). According to calculations by Friskney and Haworth (1967), for $E_0 = 30$ keV, $d = 2$ $\mu$m and $i = 0.1$ $\mu$A, the temperature rise of 1430 K for a sample of low thermal conductivity ($k = 0.01$) is reduced to 760 K by a 10 nm aluminium coating.

Reliable quantitative analyses cannot be obtained when heating is sufficient to cause vaporisation at the point of impact of the beam. A pit is then produced, and the assumption of a smooth flat surface for calculating matrix corrections is invalid. Also one or more elements may be selectively removed, altering the local composition. For example hydrated substances readily lose water, and carbon dioxide can easily be driven from carbonates, causing the apparent concentrations of the remaining elements to rise. Other minerals and synthetic materials susceptible to compositional change under electron bombardment, include halides, phosphates, feldspars, and glasses. Most metals have sufficiently high thermal conductivity to be immune from heating effects at normal probe currents.

In ionic compounds it appears that the local electrostatic field attributable to the incident electrons causes migration of ions (Lineweaver, 1963; Fredriksson, 1966; Ribbe and Smith, 1966; Weisweiler and Neff, 1979). The field distribution in insulating specimens has been studied by Cazaux (1986) and its implications discussed by Cazaux and Le Gressus (1991). The temperature rise is also significant, since it affects the rate of diffusion.

The main symptom of beam damage is change in the X-ray intensity with time (assuming that this is not caused by other effects, such as contamination, probe current drift, etc.). Various preventive measures can be applied: the probe current can be reduced, or alternatively the beam may be defocussed so that the energy dissipation is spread over a larger area, but obviously spatial resolution is then sacrificed. The same is true if the specimen is moved continuously under the beam in order to avoid the build-up of damage at any one point.

## 11.12 Standards

Quantitative electron microprobe analysis entails measuring the intensities of the characteristic peaks of the elements of interest in the specimen to be analysed and in one or more standards of known composition. The matrix corrections required to allow for differences in X-ray absorption etc. (§1.10) can be calculated only with a certain accuracy: hence it is advantageous to use standards as close as possible in composition to the specimen. However, for many compositions it is difficult to obtain standards that are both homogeneous and of accurately known composition. This problem can be avoided by using pure elements as standards (where these have suitable physical properties) but errors due to uncertainties in matrix corrections may then be significant. A reasonable compromise is to use a limited number of well characterised simple compounds.

In metallurgy, pure elements are often satisfactory as standards, and in some alloy systems there are intermetallic compounds that can also be used, though they may exhibit excessive compositional variation. Homogeneous binary alloys can be produced in a few systems such as Cu–Au, but most alloys segregate on cooling. Goldstein, Majeske and Yakowitz (1967) described a method of producing alloy standards by rapid chilling from the melt, using a form of 'splat cooling'. A few cubic centimetres of molten alloy is dropped onto a rotating hearth cooled with liquid nitrogen, on which the metal spreads out and solidifies very quickly. The apparatus is filled with argon to prevent oxidation.

In mineralogy pure element standards are reasonably satisfactory for first series transition elements ($Z = 21$–$30$). For elements in the atomic number range 11–20, pure elements are unsuitable in general, but a range of natural and synthetic mineral standards may be used: for example, jadeite ($NaAlSi_2O_6$) for Na, periclase (MgO) for Mg, corundum ($Al_2O_3$) for Al, wollastonite ($CaSiO_3$) for Ca. These have the advantage that matrix corrections are smaller than for pure element standards, also chemical wavelengths shifts are minimised. Some minerals, especially the simpler silicates and oxides, are accurately stoichiometric, making it possible to calculate the concentrations of the elements. The microprobe itself may be used to check such minerals for impurities, and if any are present the major element concentrations can be adjusted accordingly.

If a standard is available in sufficient quantity, bulk chemical analysis can be used to determine its composition, though any errors in the analysis will be transferred to subsequent microprobe analysis. Probably the most accurate compositions are those calculated from the theoretical formulae of simple stoichiometric compounds. Some materials can be tested for non-stoichiometry by physical methods: for example Sweatman and Long (1969a) used $Cr_2O_3$ and CoO tested by electrical measurements, and $TiO_2$ known from its optical properties to be stoichiometric to better than 0.5%.

Synthetic glasses are sometimes used as standards but are rather sensitive to electron bombardment. The composition of a glass standard must be determined after preparation, since volatile elements may be lost from the melt. Tests for homogeneity are also necessary, in case of imperfect mixing in the melt or precipitation during cooling. A procedure for preparing silicate glasses has been described by Smellie (1972).

# 12
# Energy-dispersive analysis

## 12.1 Introduction

The general principles of analysis are the same whether e.d. or w.d. spectrometers are used, but some differences of approach are required because of their different characteristics. This chapter is concerned with procedures applicable specifically to e.d. analysis.

### *12.1.1 Qualitative e.d. analysis*

The e.d. spectrometer is extremely well adapted to qualitative analysis (identification of the elements present in the sample), being both quick and convenient to use for this purpose. Since the whole spectrum is collected simultaneously, it is not necessary to decide where to search for peaks, as is the case with w.d. spectrometers. Commercial e.d. systems are provided with aids such as markers showing the positions of the K, L and M peaks of any specified element, which can be superimposed on the spectrum as a guide to peak identification. Alternatively, by searching a list of peak energies, possible identifications can be obtained. The e.d. spectrometer is also very useful for rapid phase identification based on the peaks present and their approximate relative height. Typically this information takes only a few seconds to acquire.

### *12.1.2 Mapping*

The e.d. spectrometer may be used for producing elemental distribution maps or line scans. For this purpose a 'window' is set up in the region of a peak of interest and the integrated number of counts within the window is used to indicate the concentration of the element concerned. For producing line-scan profiles the output may be fed to an analogue ratemeter. More commonly a digital step scan is used, with the output displayed in histogram form. For two-dimensional 'dot map' images (§5.3) pulses corresponding to X-ray photons with energies within the window are displayed on a screen while the beam scans a raster.

An e.d. detector is insensitive to the position of the X-ray source over a distance of several millimetres, whereas with a w.d. spectrometer the Bragg angle may change sufficiently when the beam is deflected to influence the X-ray intensity appreciably, although there are ways of avoiding this (§6.10.2). The e.d. spectrometer is thus more suitable for large-area scans. Limited count-rate capability (§10.7) is, however, a significant disadvantage of this type of spectrometer, though improvements have occurred recently. By comparison, a w.d. spectrometer has a much higher upper count-rate limit, and the count-rate for low concentrations can be raised by increasing the probe current (subject to specimen damage considerations).

Where peaks are close together, the relatively poor spectral resolution of the e.d. spectrometer is a disadvantage. This can be overcome by carrying out 'stripping' or deconvolution on the spectra recorded at each point, using one of the methods described below. The continuum background can also be removed at the same time. However, a significant amount of computing is required for each point and the time to form a complete image is thus quite long.

### 12.1.3 Quantitative e.d. analysis

As well as qualitative analysis, the e.d. spectrometer is also capable of quantitative analysis. The experimental procedure consists of recording the spectrum for a preset time (sufficient to give the required statistical precision), with the same probe current for specimen and standard, though in e.d. analysis probe current drift affects all elements equally and is thus less serious than in w.d. analysis, where peaks are measured sequentially. Since the whole of the X-ray spectrum is recorded, it is not necessary to decide which elements are of interest before recording the spectrum, as in w.d. analysis. If reliable quantitative results are to be obtained, various instrumental factors discussed previously must be taken into account, including X-ray collimation (§9.4.1), trapping of stray electrons (§9.4.2), extraneous noise (§10.2), pulse pile-up (§10.6), etc. Other relevant considerations, common to both e.d. and w.d. analysis, are discussed in the preceding chapter.

After recording the spectrum (a typical example of which is shown in fig. 12.1) it remains to extract the required peak intensities. In e.d. analysis,

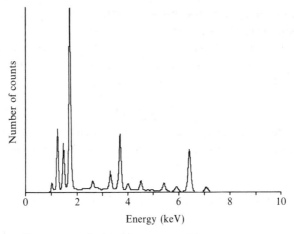

Fig. 12.1 Typical e.d. spectrum (silicate mineral).

overlap between peaks is much more common than in w.d. analysis and background corrections are larger, necessitating a different approach, as described in the following sections.

After determining the true peak intensities, quantitative e.d. analysis follows the same path as when w.d. spectrometers are used, namely comparison with standards of known composition (possibly but not necessarily pure elements) and application of matrix corrections, as described in chapters 13–16. Owing to the constancy of the detection sensitivity of the e.d. spectrometer, frequent reference to standards is not necessary and stored calibration data can be used (§12.7.2). It is even possible to dispense with standards altogether, using appropriate formulae for the efficiency of X-ray production and detection (§12.12), though the accuracy obtainable by such 'standardless' analysis is somewhat inferior.

## 12.2 Energy calibration

The spectrum processing procedures used for quantitative e.d. analysis are sensitive to changes in peak position, which may occur because of drift in the energy calibration of the system. It is therefore desirable that the energy calibration should be monitored regularly. The calibration can be checked by measuring the positions of X-ray peaks of known energy. Assuming the energy scale is linear, both the gain factor (eV/channel) and zero error can be determined from measurements on two peaks, preferably widely separated in energy. For systems with a strobed noise peak (§10.4.1) this can be used instead of a low energy X-ray peak.

The position of a peak may be expressed as the centroid, given by the sum of the product of the number of counts ($n_i$) and the channel number ($i$) divided by the total number of counts. Thus:

$$\bar{i} = \sum n_i i / \sum n_i,$$

where $\bar{i}$ is the centroid expressed as a channel number. In terms of energy, the statistical precision of the centroid of a gaussian peak is given by:

$$\sigma = 0.425 \triangle E N^{-1/2}, \qquad (12.1)$$

where $\triangle E$ is the FWHM of the peak and $N$ is the total number of counts (Raznikov, Dodonov and Lanin, 1977). Thus, for $\triangle E = 100$ eV and $N = 10^5$, $\sigma = 0.14$ eV. Such precision is more than adequate.

### 12.2.1 Duane–Hunt limit

For quantitative analysis (e.d. or w.d.) accurate knowledge of the accelerating voltage is necessary. The easiest way of determining this is to measure the high energy cut-off point of the continuous X-ray spectrum, or the Duane–Hunt limit (§1.6), using an e.d. spectrometer (Solosky and Beaman, 1972). A typical plot is shown in fig. 12.2.

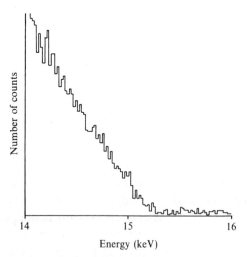

Fig. 12.2 Continuum intensity in vicinity of incident electron energy $E_0$, showing cut-off at Duane–Hunt limit (nominal accelerating voltage = 15 kV, measured $E_0 = 15.2$ keV).

An accuracy of ±100 eV is possible, but a reasonably long counting time is needed, owing to the low intensity of the continuum close to the limit. It is best to use a sample of high atomic number in order to maximise the continuum intensity. The count-rate should not be too high, however, otherwise the presence of pile-up pulses makes it more difficult to locate the limit accurately. It is desirable to check the energy calibration in the region of interest by recording a suitable X-ray peak (using a higher accelerating voltage), e.g. Y K$\alpha$ at 14.96 keV or Rh K$\alpha$ at 20.21 keV.

A first approximation to the cut-off energy may be obtained either visually, or by printing out the spectrum, drawing a line through the continuum, and finding the intercept with the axis. Better accuracy can be achieved with a computed fit. According to Kramers' law (equation (1.1)) the product of intensity and energy is a linear function of energy, hence it is advantageous to plot this quantity.

## 12.3 Peak integration

The simplest approach to the measurement of peak intensities in an e.d. spectrum is to sum the counts in a group of channels centred on the peak position. The wider the energy band included, the greater the total number of counts and the better the statistical precision. However, too wide a band results in increased overlap between adjacent peaks and reduced peak to background ratio. Fig. 12.3 shows the variation of the statistical precision and detection limit as a function of the integrated energy band or 'window' ($\delta E$). A reasonable compromise is to use a band approximately equal to the FWHM of the peak ($\triangle E$) (Short, 1976). For $\delta E = \triangle E$ the integrated area is 76% of the total, but the fact that not all of the peak counts are included is immaterial provided the fraction is constant (which is the case so long as the width and position of the peak are invariant).

### 12.3.1 Peak shift and broadening

Changes in the positions and widths of the peaks in the spectrum are liable to occur as a function of count-rate (§10.3.4). For a gaussian peak it can be shown that if the integrated energy band, $\delta E$, is equal to the FWHM, $\triangle E$, then a peak shift of $0.07\triangle E$ is sufficient to cause a 1% change in the integrated number of counts. The same effect is produced by peak broadening of $0.0085\triangle E$. If errors are to be kept below 1%, shift therefore should not exceed about 10 eV and broadening 1 eV, assuming a typical peak width. (Similar criteria apply to other spectrum processing methods,

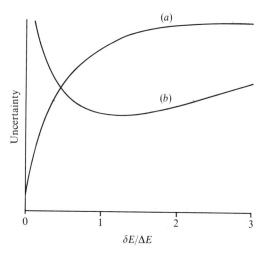

Fig. 12.3 Variation in uncertainty of measurement of large peak (a) and detection limit for small peak (b), as function of ratio of integrated 'window' ($\delta E$) to FWHM ($\Delta E$).

unless corrections for these effects are included.) In systems showing count-rate dependent peak shift or broadening, an appropriate count-rate limit should be observed.

## 12.4 Peak overlap

As a consequence of the relatively poor resolution of e.d. spectrometers, overlap between peaks is much more frequent than with w.d. spectrometers. The K$\alpha$ peaks of Na, Al and Mg, for example, are sufficiently close for the tail of one peak to influence its neighbour, though the effect is small with a detector of good resolution. Much more serious overlap is possible between K$\beta$ and K$\alpha$ peaks, the worst case being V K$\beta$ and Cr K$\alpha$, which are separated by only 15 eV. Overlap also occurs between lines from different shells: e.g. between Pb M$\alpha$ and S K$\alpha$, which are 38 eV apart.

It is reasonable to assume initially that overlap of K lines occurs only between elements differing in atomic number by one, in which case the correction may be expressed thus:

$$I_Z = I'_Z - K_{Z-1}I_{Z-1} - K_{Z+1}I_{Z+1}, \qquad (12.2)$$

where $I'_Z$ and $I_Z$ are the peak intensities for the element of atomic number $Z$ before and after correction for overlap, while $I_{Z-1}$ and $I_{Z+1}$ are the

## 12.4 Peak overlap

intensities for elements $Z-1$ and $Z+1$ (Reed and Ware, 1973). The intensities are obtained by integrating the counts contained within a 'window' centred on the peak, as described in §12.3. The coefficients $K_{Z-1}$ and $K_{Z+1}$ represent the amount of overlap expressed as a fraction of the measured intensity of the overlapping peak (fig. 12.4). Values for the coefficients may be derived from a theoretical peak shape, such as the gaussian function. Alternatively experimental values may be obtained from standards, thereby taking into account the actual form of the peaks.

The intensities $I_{Z-1}$ and $I_{Z+1}$ are assumed to have been corrected for background before substitution in equation (12.2). Also, they should be already corrected for overlap. However, raw intensities can be substituted initially, and after overlap corrections for all elements present have been calculated the procedure may be repeated using the corrected intensities to calculate overlap effects more accurately.

The same principle can be extended to include overlap from elements differing in atomic number by more than one. Table 12.1 gives experimental overlap coefficients for K lines (these are for illustrative purposes only and are not generally applicable, owing to the variation in detector characteristics). The largest coefficients, apart from the case of Zn L overlapping Na K, are for $K\beta$–$K\alpha$ overlaps of elements in the atomic number range 20–25. Small but measurable overlaps occur between elements differing by more than one in atomic number due to the tails on the low energy side of K peaks, caused by incomplete charge collection (§9.6). This effect is most marked for P ($Z=15$), the $K\alpha$ peak of which lies just above the K absorption edge of Si.

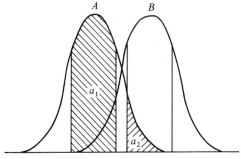

Fig. 12.4 Peak overlap: overlap coefficient for overlap of $A$ on $B = a_2/a_1$.

Table 12.1. Overlap coefficients for Kα lines (expressed as per cent) in e.d. spectrum, from Smith and Gold (1979).

| Over-lapping element (Z) | Overlapped element (Z) | | | | | | | | | | | | | | | | | |
|---|---|---|---|---|---|---|---|---|---|---|---|---|---|---|---|---|---|---|
| | 11 | 12 | 13 | 14 | 15 | 16 | 17 | 19 | 20 | 21 | 22 | 23 | 24 | 25 | 26 | 27 | 28 | 29 |
| 11 | 100 | | | | | | | | | | | | | | | | | |
| 12 | 1.6 | 100 | 0.2 | | | | | | | | | | | | | | | |
| 13 | 0.5 | 1.3 | 100 | 0.1 | | | | | | | | | | | | | | |
| 14 | 0.2 | 0.3 | 0.9 | 100 | 0.7 | | | | | | | | | | | | | |
| 15 | 1.1 | 1.6 | 2.2 | 4.1 | 100 | 0.7 | | | | | | | | | | | | |
| 16 | 0.7 | 0.8 | 1.0 | 1.4 | 2.8 | 100 | 1.2 | | | | | | | | | | | |
| 17 | 0.5 | 0.4 | 0.5 | 0.6 | 0.9 | 2.0 | 100 | 0.1 | | | | | | | | | | |
| 19 | 0.4 | 0.3 | 0.2 | 0.1 | | 0.2 | 0.3 | 100 | 7.2 | | | | | | | | | |
| 20 | 0.4 | 0.3 | 0.1 | 0.1 | 0.1 | 0.2 | 0.3 | 0.6 | 100 | 10.0 | | | | | | | | |
| 21 | 0.3 | 0.2 | 0.1 | 0.1 | | 0.1 | 0.1 | 0.2 | 0.5 | 100 | 12.4 | | | | | | | |
| 22 | 0.3 | 0.2 | | | | | 0.1 | 0.1 | 0.2 | 0.4 | 100 | 13.7 | | | | | | |
| 23 | 0.2 | 0.1 | | | | | | | 0.1 | 0.2 | 0.3 | 100 | 12.7 | | | | | |
| 24 | 0.2 | 0.1 | | | | | | | | 0.1 | 0.2 | 0.3 | 100 | 11.1 | | | | |
| 25 | 0.2 | 0.2 | | | | | | | | | 0.1 | 0.1 | 0.2 | 100 | 8.2 | | | |
| 26 | 1.5 | 0.3 | | 0.1 | | | | | | | | 0.1 | 0.1 | 0.2 | 100 | 5.0 | | |
| 27 | 1.1 | 0.9 | 0.1 | | | | | | | | | | 0.1 | 0.1 | 0.2 | 100 | 2.5 | |
| 28 | 1.2 | 0.6 | 0.4 | | | | | | | | | | | 0.1 | 0.1 | 0.1 | 100 | 1.3 |
| 29 | 24.5 | 0.6 | 0.1 | | | | | | | | | | | | 0.1 | 0.1 | 0.1 | 100 |
| 30 | 282.1 | 2.1 | 1.6 | | | | | | | | | | | | | | | 0.1 |

## 12.5 Continuum modelling

One approach to background corrections in e.d. analysis is to use a mathematical expression representing the shape of the continuum to estimate the background intensity under the peaks. This assumes that the continuum is the only source of background, which is largely true, apart from the effects of pulse pile-up (§10.6), escape peaks (§9.7) and the peak tails mentioned previously.

In modelling the continuum it is necessary to take account of the shape of the continuous X-ray spectrum generated in the sample, the effect of sample composition on this shape (due to absorption, for example) and the variation in the efficiency of the detector with energy. The continuum intensity could be calculated for every channel in the spectrum, but this would be somewhat laborious. If peak integration is used, it is only necessary to obtain a value for the continuum integrated over the 'window' selected for measuring the peak intensity.

The continuum intensity ($I_c$) as a function of X-ray energy ($E$) may be calculated from the Kramers formula (equation (1.1)), with additional factors allowing for absorption in specimen and detector:

$$I_c = k[(E_0 - E)/E] f(\chi) \exp(-\sum \mu_i \rho_i x_i), \tag{12.3}$$

where $k$ is a constant for a given instrument and probe current, $E_0$ is the incident electron energy, $f(\chi)$ is the specimen absorption factor, and $\mu_i$, $\rho_i$ and $x_i$ are respectively the m.a.c., density, and thickness of the absorbers in the X-ray path (detector window, gold surface layer and silicon dead layer).

The function $f(\chi)$ is the factor by which the X-ray intensity is reduced through absorption in the specimen, and $\chi$ is equal to $\mu \mathrm{cosec}\psi$, where $\mu$ is the m.a.c. of the specimen and $\psi$ is the X-ray take-off angle. To a first approximation it may be assumed that the depth distribution of production of the continuum is the same as for characteristic radiation, enabling the same methods for calculating $f(\chi)$ to be applied. For example, the Philibert formula (§15.5) may be used, with $E$ taking the place of $E_c$. Since $f(\chi)$ is dependent on the specimen composition, which is not known until background and other corrections have been applied, this correction must be calculated iteratively.

The term in equation (12.3) representing absorption by the detector window etc. requires knowledge of $\mu$, $\rho$ and $x$ for each absorber. Although manufacturers give approximate figures, the exact thicknesses are usually unknown. However, values of $\rho_i x_i$ can be determined empirically by finding the best fit to the continuum between about 0.5 and 3.5 keV, where absorption is significant (Ware and Reed, 1973). This approach assumes the thicknesses to be uniform. In reality there are likely to be local variations in thickness, in which case the exponential absorption factor in equation (12.3) is not strictly valid, though reasonable results are obtained with it in practice.

Uncertainties in equation (12.3) may be taken into account by introducing an empirical energy dependent correction term $F(E)$:

$$I = k[F(E,\chi,\chi_i) + F(E)], \qquad (12.4)$$

where:

$$F(E,\chi,\chi_i) = [(E_0 - E)/E]f(\chi)\exp(-\sum \mu_i \rho_i x_i),$$

as proposed by Ware and Reed (1973). Between 3 and 10 keV, $F(E)$ can be given a constant value $F_0$. In order to determine $k$ and $F_0$, background measurements at two preferably widely separated energies are required. One suitable energy is 2.95 keV, corresponding to the Ar K$\alpha$ peak, which is absent other than very exceptionally. (Ag L$\alpha$ and Th M$\alpha$ are also close to this energy.) The escape peak of Ti is of similar energy and should be 'stripped' (§12.9) before the background intensity is calculated. Usually it is easy to find a region free of peaks at the high energy end of the spectrum, for example around 10 keV. Below 3 keV the uncertainties in equation (12.4) are more serious and $F(E)$ is not constant. Empirical values of $F(E)$ in this region can be obtained using standards of known composition. Ware and Reed (1973) found that $F(E)$ values could be assumed to be independent of sample composition and accelerating voltage to a good approximation. Since $F(E)$ is partly dependent on instrumental characteristics, it must be determined empirically for every system.

An example of continuum modelling applied to a spectrum containing Na, Mg, Al and Si peaks is shown in fig. 12.5. In this case the tails of the peaks overlap, so that the true background cannot be observed directly between 1 and 2 keV.

Computer programs for quantitative e.d. analysis using peak integration and background subtraction based on continuum modelling have been described by Myklebust, Fiori and Heinrich (1979) and Ware (1981).

### 12.5.1 Modifications to Kramers' law

Kramers' law underestimates the continuum intensity at low energies (Reed, 1975), though in the correction procedure described above, this is counteracted by $f(\chi)$ being overestimated (see §12.5.3). The 'constant' is

## 12.5 Continuum modelling

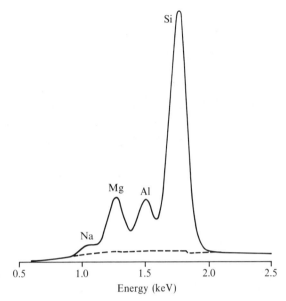

Fig. 12.5 Fitted continuum (dashed line) for specimen containing Na, Mg, Al and Si.

also found to be a function of atomic number (e.g. Hehenkamp and Böcker, 1974). It is thus desirable that more rigorous expressions for both production and absorption of continuum X-rays should be used.

A better representation of the shape of the continuum can be obtained by modifying the energy dependent term to: $(E_0 - E)/E^{1+a}$, a value of 0.21 for the coefficient $a$ being appropriate for Cu, according to theoretical calculations (Reed, 1975). Values for other elements calculated by Statham (1976b) range from 0.15 ($Z = 79$) to 0.26 ($Z = 13$). Production of continuum (unlike characteristic) radiation is not isotropic, though the anisotropy is not marked, owing to the effect of electron scattering. This effect can be taken into account by varying $a$ slightly as a function of X-ray take-off angle, as shown by Statham.

Several different variations on the Kramers expression have been proposed. Lifshin, Ciccarelli and Bolon (1975), for example, fitted a quadratic expression to experimental spectra, and this has been used by Fiori, Myklebust, Heinrich and Yakowitz (1976) for background corrections in e.d. analysis. Small, Leigh, Newbury and Myklebust (1987) derived the following expression for the shape of the continuum: $[(E_0 - E)/E]^m$, with $m = 1.05 + 0.00599 E_0$ ($E_0$ in keV), by fitting to a large set of experimental data.

### 12.5.2 Carbon reference spectrum

The uncertainties associated with estimating the absorption in the detector window etc. can be avoided by recording the continuum from a suitable reference material and applying a correction for the differences in composition between this and the sample. Smith, Gold and Tomlinson (1975) chose pure carbon (diamond) as the continuum standard, on the grounds of the absence of peaks in the spectrum (fig. 12.6). The following expression, proposed by Smith and Gold (1979), can be used to scale the continuum intensity to allow for the difference in composition between carbon and a sample of mean atomic number $\bar{Z}$:

$$I_c = b\bar{Z}^n[(E_0 - E)/E]^x, \qquad (12.5)$$

where $n$ and $x$ are functions of $E$ and $\bar{Z}$ which are determined empirically. This procedure allows for the variation in the shape of the continuum as well as the overall dependence of continuum intensity on atomic number. Smith and Gold included a correction for electron backscattering, which modifies the continuum shape, owing to the $Z$ dependence of the energy distribution of backscattered electrons (§14.7).

### 12.5.3 Absorption of the continuum

In modelling the shape of the continuum it is necessary to include a term for the absorption of the continuum X-rays in emerging from the specimen. The factor $f(\chi)$ included in the expressions given previously is analogous to that used for correcting characteristic intensities for absorption (§15.1). However, the cross-section for continuum production, as given by Kirkpatrick and Wiedmann (1945) for example, has an energy dependence

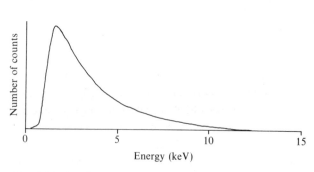

Fig. 12.6 Continuum recorded from carbon ($E_0 = 15\,\text{keV}$).

which differs significantly from that for characteristic X-ray production. Continuum production is thus concentrated much more toward the ends of the electron trajectories. The effect on the depth distribution is moderated by electron scattering, but nevertheless the mean production depth is somewhat greater for continuum X-rays.

Reed (1975) proposed changing the constant in equation (15.7) to $4.0 \times 10^5$ in order to modify the Philibert correction (§15.5) for application to continuum radiation. This gives results which agree quite well with Monte-Carlo calculations (Statham, 1976b). August and Wernisch (1991a) used the 'multiple reflection' model (§14.4) to derive the depth distribution of the continuum and proposed a simple modification to the function $\phi(\rho z)$ used to represent the depth distribution of characteristic X-rays (§15.1) to adapt it to the continuum.

## 12.6 Digital filtering

A completely different approach to background corrections is to apply purely mathematical methods which avoid the need for a detailed knowledge of the shape of the continuum. One possibility is to use Fourier analysis, whereby the spatial frequency distribution of the spectrum is filtered to remove the low frequency components which represent background, while leaving the high frequency content which represents the peaks (Russ, 1972; Hay, 1985). Though reasonably effective, the approach described below is usually preferred for application to quantitative e.d. analysis.

In the widely-used alternative approach a 'filter function' of limited width is applied to the recorded spectrum so as to produce a modified form of that spectrum, in which background is eliminated (Mariscotti, 1967). The filtered peaks are altered in shape but similarly filtered pure element peaks can be fitted to the spectrum, as described in the next section, thereby enabling the relative peak intensities to be obtained.

The filtering process can be described by the expression:

$$y'_i = \sum_{k=-s}^{k=+s} h_k y_{i+k}. \qquad (12.6)$$

The number of counts in channel $i$ of the filtered spectrum ($y'_i$) is thus the sum of the products of the filter function coefficients ($h_k$) and the number of counts in each channel in the original spectrum ($y_{i+k}$) between channels $i-s$ and $i+s$. The operation described by equation (12.6) is carried out for each channel in the spectrum.

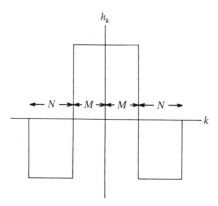

Fig. 12.7 'Top hat' filter function.

The most appropriate form of filter is the 'top hat' function (Robertson, Prestwich and Kennett, 1972), which consists of a central region ($k = -M$ to $+M$) in which $h_k$ has a constant positive value, with negative lobes of width $N$ on each side. The total width in channels is $2(M+N)+1$. The values of $h_k$ are chosen so that the total area is zero. An example of such a function is shown in fig. 12.7. This type of filter has the advantage that the calculations required are economical of computer time owing to their simplicity.

The effect of filtering a gaussian peak is shown in fig. 12.8. The filtered peak is similar to the inverse of the second differential of the original peak, but is smoothed as a result of the summation of the counts in a number of channels. Owing to the symmetry and zero total area of the 'top hat' filter function, the output ($y'_i$) is zero in the absence of peaks, provided that the background is linear (although it may have a slope). Subject to the above condition, background is eliminated without the need for detailed knowledge of its shape, as required in the continuum modelling approach. This method is thus particularly useful in cases were modelling is difficult, e.g. for particles and thin films.

Intuitively the width of the filter function should approximate to that of the peaks (Schamber, 1977). The optimum filter for detecting small peaks superimposed on background is relatively wide, but this is undesirable in the case of a small peak adjacent to a large one, hence a compromise choice must be made. For a FWHM of 7.5 channels appropriate values of $M$ and $N$ are 3 and 4 respectively (Statham, 1977b). Peaks widths vary with energy, of course, but it is inconvenient to vary the filter width, so in practice a constant value is used.

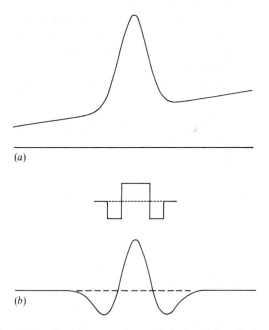

Fig. 12.8 Effect of 'top hat' filter: peak profile before (a) and after (b) filtering.

## 12.7 Least squares fitting

A more refined method of determining X-ray intensities than simple summation of channel contents over a fixed energy range (§12.3) is to fit a function representing the peak profile, using the well known 'least squares' method, which entails minimising the sum of the squares of the errors. This approach can be applied where several, possibly overlapping, peaks are present, in which case the need for explicit overlap corrections is obviated by fitting profiles to all the peaks. Background must, however, be removed first either by using a continuum model or by filtering (see above). In the latter case the reference peaks used for fitting must also be filtered.

For illustrative purposes we may consider the fitting of an unfiltered isolated background-free peak using the gaussian function:

$$y = a \exp\{-[(x-b)/c]^2\}, \tag{12.7}$$

where $y$ is the number of counts contained in channel number $x$. For convenience, equation (12.7) may be rewritten in the form:

$$\ln y = a_0 + a_1 x_i + a_2 x_i^2, \tag{12.8}$$

where $a_0 = \ln a - (b/c)^2$, $a_1 = 2b/c^2$, and $a_2 = -1/c^2$. If the actual number of counts in channel number $x$ is $Y_x$, then the optimum values for $a_0$, $a_1$ and $a_2$ in equation (12.8) and hence $a$, $b$ and $c$ in equation (12.7) can be determined by finding the minimum of $\sum (\ln Y_x - \ln y_x)^2$. Setting the differentials of this function to zero leads to the following simultaneous equations:

$$\begin{aligned}
a_0 n + a_1 \sum x + a_2 \sum x^2 &= \sum (\ln Y), \\
a_0 \sum x + a_1 \sum x^2 + a_2 \sum x^3 &= \sum (x \ln Y), \\
a_0 \sum x^2 + a_1 \sum x^3 + a_2 \sum x^4 &= \sum (x^2 \ln Y),
\end{aligned} \qquad (12.9)$$

which can be solved for $a_0$, $a_1$ and $a_2$ by standard methods. The gaussian coefficients $a$, $b$ and $c$ may be obtained from the relations given above. The peak area is equal to $1.772ac$.

### 12.7.1 Fitting multiple peaks

The preceding discussion refers only to fitting a single peak, whereas real spectra generally contain several peaks, requiring a corresponding number of separate profile functions to be fitted. If the positions and widths of the peaks can be considered known, the problem is reasonably simple, because the equations are linear. In the general case where the peak positions and widths are not predetermined, the equations are non-linear, making the problem considerably more difficult (see §12.7.3).

So far gaussian peak profiles have been assumed: it is preferable, however, to use experimental profiles, thereby automatically taking into account any non-gaussian features of the peaks. Gehrke and Davies (1975), for example, fitted spectra from pure oxides to the spectra of compounds composed of mixtures of oxides: this procedure is not rigorous as regards the background, however, and it is better to use the filtering process described above to remove background prior to least squares fitting.

### 12.7.2 Fitting filtered spectra

The combination of digital filtering with least squares fitting of peak profiles is known as the 'filter-fit' method (Schamber, 1973, 1977; Statham, 1977b). There are distinct advantages in deriving the reference profiles from experimental spectra, thereby avoiding the complications involved in modelling the detector response precisely. Since most of a pure element spectrum consists of background which gives zero output on filtering, it is

## 12.7 Least squares fitting

only necessary to store the filtered spectrum in the region of the peaks, thereby achieving a useful saving in memory space. If pure elements are unavailable, profiles may be derived from the spectra of compounds, provided there are no other lines in the region of interest.

Usually the characteristics of Si(Li) detectors are quite stable, hence stored profiles can be used and there is no need for frequent recalibration. Obviously profiles can only be used for fitting spectra in which the energy resolution is the same, hence if the pulse processing time constants are changed, a different set of profiles must be used. To obtain absolute elemental concentrations, it is necessary to record the spectra of reference standards and specimens under identical conditions, including constant accelerating voltage and probe current. However, differences in current can be taken into account by recording the spectrum of a pure element (e.g. Co) with each batch of specimens and relating all peak intensities to that element. This is particularly useful for instruments not equipped with probe current monitoring facilities (such as most s.e.m.s).

Sources of error in the filter-fit method are discussed in §12.8.

### 12.7.3 Non-linear fitting

Serious errors can arise when quite small amounts of peak shift and broadening occur, if it is assumed that peak positions and widths are fixed. On the other hand, if these are treated as unknowns, analytical solutions for the resulting non-linear equations cannot be obtained. They can, however, be reduced to approximate 'linearised' versions which are valid within certain limits (Schamber, 1981) enabling exact solutions to be obtained by iteration, though care is required in order to avoid non-convergence or convergence to a false minimum (Statham, 1978). Peak position and width parameters should preferably be constrained, for example by assuming that the energy scale is linear and peak widths obey a relationship of the type given by equation (9.2) (Nullens, Van Espen and Adams, 1979).

An alternative approach is to use a 'trial and error' method, whereby synthesised spectra are compared to the recorded spectrum, and that giving the best fit is found. The 'sequential simplex' method (Spendley, Hext and Himsworth, 1962) can be used for this purpose, as described by Fiori, Myklebust and Gorlen (1981). In this procedure the spectrum is reconstructed using estimated peak heights and is compared to the original. Revised heights are then derived using a routine which ensures convergence. This process is repeated until there is no further improvement in the fit.

## 12.8 Filter-fit method – sources of error

In the filter-fit method as commonly used for quantitative e.d. analysis, elemental profiles filtered with a 'top hat' filter function are fitted by the least squares method to the filtered spectrum of the specimen (§12.7.2). The filtering process (§12.6) eliminates background provided it is linear, but errors are possible if it is curved. The background in e.d. spectra shows significant curvature, especially in the low energy region (fig. 12.5). The effect of background curvature has been studied by Statham (1977b) and is found to cause only small errors in practice. Kitazawa, Shuman and Somlyo (1983) showed that small spurious concentrations of Na, which tend to occur owing to background curvature, can be corrected by including the continuum itself (using a reference spectrum recorded from a carbon specimen) in the fitting procedure as if it were a peak.

Another possible problem is the presence of steps in the continuum caused by absorption edges (fig. 12.9). Usually the effect is negligible because the step height is related to the concentration of the element and is never more than a small fraction of the peak height. The worst case is when the peak of another element is close to the step: then an error in the height of this peak equivalent to about one third of the size of the step can occur (Statham, 1977b).

The reference profiles normally include all the lines for the particular shell of the element concerned, in order to take full account of inter-element overlaps. However, the implicit assumption that the relative intensities of the lines are constant is not strictly valid because of differential absorption. Schamber, Wodke and McCarthy (1980) calculated the errors in selected examples and found that in the worst case (Zn K edge between Pt L$\alpha$ and L$\beta$ lines) the error could amount to 4% of the concentration (of Pt). A possible remedy is to fit the L$\beta$ peak separately, using only L$\alpha$ for calculating the concentration.

For some elements, lines from more than one shell fall within the observed spectrum (e.g. Zn K$\alpha$ at 8.64 keV and L$\alpha$ at 1.01 keV). In such cases it is appropriate to treat each shell separately, deriving the concentration from the fitted intensity of one and discarding the other.

In the usual form of the filter-fit procedure there is no allowance for variations in the positions and widths of the peaks, and errors ensue if such variations occur. For example, Schamber et al. (1980) found that a shift of 6 eV causes an error of 1% in the case of a peak with a FWHM of 160 eV. The worst case is a small peak adjacent to a large one, where the effect varies

## 12.8 Filter-fit method – sources of error

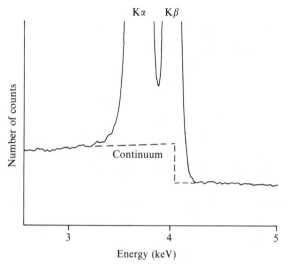

Fig. 12.9 Step in continuum (CaSiO$_3$ specimen) caused by Ca K absorption edge (dashed line represents continuum under peaks).

both in size and sign according to the separation of the peaks. According to Statham (1977b), a shift of 1/30 of the FWHM causes a maximum error in the small peak corresponding to 2.5% of the large one, with a separation of 2/3 of the FWHM.

### 12.8.1 Derivative references

Kitazawa *et al.* (1983) proposed a method for reducing the sensitivity of the filter-fit procedure described above to changes in peak position and width. This depends on the fact that a spectrum in which the peaks are shifted and broadened can be expressed, to a good approximation, as a linear combination of the original spectrum and its first and second derivatives. These effects can thus be nullified by including the derivatives in the fitting procedure.

The effectiveness of this approach was confirmed in tests by McMillan, Baughman and Schamber (1985). However, the use of derivative references has an adverse effect on the statistical precision of the peak area determination and gives unsatisfactory results for peaks containing less than about 5000 counts (Bostrom and Nockolds, 1989).

## 12.9 Escape peak stripping

The Kα lines of elements above 18 in atomic number have significant escape peaks 1.74 keV below each main peak, their intensity being 0.1–1% of the parent peak (§9.7). In the event of interference between an escape peak and the Kα peak of a lighter element, a small concentration of that element may thus be recorded when none is in fact present.

Escape peak interference may be treated as a form of peak overlap and can be corrected by subtracting from the measured intensity of the peak suffering interference the appropriate fraction of the intensity of the interfering peak. Since the escape peak usually does not exactly coincide with the Kα peak, the effective intensity ratio is less than that given in fig. 9.11, and must be derived either empirically or by calculation, taking into account the peak shape and the width of the integrated energy band.

Alternatively the escape peak may be 'stripped' by subtracting from the count recorded in channel $i$ the escape peak intensity derived from channel $i+a$, where $a$ is the number of channels corresponding to 1.74 keV. This has the advantage that it automatically takes account of the effect of the escape peak mechanism as it applies to the continuum (though this is small). The intensity ratio has to be calculated for the energy of every channel, using an expression such as equation (9.4).

## 12.10 Analysis at high count-rates

A spectrum accumulation rate of around 5000 counts s$^{-1}$ is commonly used for quantitative e.d. analysis, this being governed by the dead time of the system (§10.7). However, improved preamplifier performance enables higher count-rates to be obtained, using shorter pulse shaping time constants, without excessive loss of energy resolution. Quantitative analysis is thus possible at count-rates of 20 000 counts s$^{-1}$ or higher (Reed, 1990b). The advantages of this are that either similar results can be obtained in a shorter time than previously, or higher precision and lower detection limits can be achieved in the same time.

At 'normal' count-rates, 'sum peaks' caused by unresolved coincidences in the pulse processing system (§10.6) are generally negligible, but this is not so at high count-rates. For example, a significant spurious concentration of Cr occurs in the analysis of samples containing Ca and Si, owing to the Si+Ca sum peak. In principle sum peaks can be 'stripped' from the spectrum using a calculation based on the characteristics of the pulse recognition circuitry used in the pile-up rejection system (Statham, 1977a), but a simpler option if the 'filter-fit' procedure is used for spectrum processing is to include stored sum peaks (which can be recorded at a very high count-rate) amongst the reference profiles (Reed, 1990b). It is not necessary to include all possible sum peaks – only those which cause interference need be considered, and the number of these is quite limited (at least for K lines).

Another phenomenon which becomes significant at high count-rates is the loss of counts due to pulse pile-up, which occurs when the interval between pulses is less than the resolving time of the pile-up rejection system (§10.6), in which case the pulse formed by the combination of the two unresolved pulses appears in the 'wrong' place in the spectrum: hence as far as peak intensity measurements are concerned, such pulses are effectively lost. The fraction lost is dependent on the total spectrum count-rate and may amount to several per cent at high count-rates. Errors can occur in quantitative analysis, therefore, when specimen and standard spectra are recorded at different rates. However, a simple calculated correction is effective (Gui-Nian and Turner, 1989; Reed, 1990b).

## 12.11 Peak to background ratio method

A method of quantification based on measuring the ratios of the heights of the peaks in the spectrum to the continuum intensity, as proposed by Hall (1968), is widely used for thin film analysis. This offers the possibility of taking into account the usually unknown thickness of the film, which is a problem peculiar to this type of analysis. A similar approach can be applied to solid samples (Cobet and Traub, 1971; Small et al., 1978; Statham and Pawley, 1978; Wendt and Schmidt, 1978). In this case the apparent concentration $C'$ (analogous to that defined by equation (1.4) for conventional analysis) is given by:

$$C' = [(P/B)/(P/B)_o]C_o,$$

where $P$ and $B$ are the peak and background intensities respectively and the subscript 0 refers to the standard. In order to obtain the true concentration, $C$, it is necessary to apply corrections as in equation (1.5), but in this case the correction factors include terms representing the dependence of the continuum on composition.

The main purpose of this approach is to improve the reliability of the results for samples of irregular geometry, such as particles or specimens with rough surfaces, the effect of such irregularity on absorption being similar for characteristic and continuum radiation of the same energy. It suffers from the disadvantage that the statistical precision of the peak to background ratio is governed by the number of background counts accumulated, which is usually much less than the number of peak counts.

A necessary condition for the successful application of this method is that the measured background should be a true reflection of the continuum intensity and not influenced by scattered radiation or the various spectral artefacts discussed in chapters 9 and 10. It is obviously impossible to measure directly the background underneath a peak, and in e.d. spectra the peaks are often too close together to allow values adjacent to the peak on each side to be obtained. It is therefore necessary either to remove the peaks by a spectrum stripping process and integrate the remaining counts in the region of each peak (Statham, 1978) or to use the continuum modelling approach described in §12.5 to extrapolate the continuum intensity from peak-free regions (Small et al., 1978).

## 12.12 Standardless e.d. analysis

The need for standards can be obviated by using calculated pure element intensities (see §13.6). This concept is particularly well suited to e.d. analysis, on account of the predictable and reproducible efficiency of the Si(Li) detector.

An alternative possibility which avoids the need for knowledge of the spectrometer efficiency is to use empirical data for the variation of intensity with atomic number (Blum and Brandt, 1973; Barbi, Skinner and Blinder, 1976). This approach is not strictly 'standardless', in that the initial calibration is based on standards, though intensities can be derived by interpolation in the case of elements for which no standard is available.

A standardless version of the peak to background ratio method described in the previous section has been proposed by Heckel and Jugelt (1984). This entails calculating the continuum intensity as a function of composition in addition to the peak intensities.

## 12.13 Precision and accuracy

The precision (reproducibility) of quantitative e.d. analysis is governed by counting statistics in essentially the same way as for w.d. analysis (§11.7). The s.d. of a peak measurement is thus equal to the square root of the number of counts. This is a minimum value: additional errors may be introduced by the spectrum processing procedure, also not all of the counts in the peak are necessarily 'useful'. For example, if peak integration is used

## 12.13 Precision and accuracy

as described in §12.3, only the counts within the 'window' are utilised. Also, when background corrections are significant, the statistical error in the peak intensity is the sum of the errors in peak and background and is thus greater than for the peak alone. The same applies in the case of overlaps from other peaks.

The precision obtainable in e.d. analysis can be estimated as follows. For a count-rate of 5000 counts s$^{-1}$ and a counting time of 100 s, the spectrum contains a total of 500 000 counts. Typically about half are in the continuum, hence for a pure element the characteristic peaks contain about 250 000 counts (though this will vary considerably). The K$\beta$ peak may contain 10% of these and about 30% of those in the K$\alpha$ peak may be 'wasted', as noted above, leaving a 'useful' total of 150 000. The precision expressed as $\pm 2$ s.d. is $\pm 0.5$%. For a concentration of 10% this becomes $\pm 1.6$%. As mentioned previously, the errors are greater when background and overlap are significant. Also the spectrum processing procedure may increase the errors. The precision can be improved, of course, by collecting more counts.

The accuracy of quantitative e.d. analysis is governed to a large degree by the same considerations that apply to w.d. analysis, such as the accuracy of matrix corrections and of the assumed compositions of standards, which limit the attainable accuracy to about $\pm 1$% (relative). As regards instrumental factors, e.d. analysis has some advantages. For example, the *ratios* of concentrations in a given analysis are immune to probe current drift, owing to the parallel collection of the spectrum.

For low concentrations errors in background corrections become important and this applies at higher concentrations (below about 1%) in e.d. analysis than in w.d. analysis. Also peak overlaps are relatively common and although available methods of deconvolution are effective, the results are susceptible to errors that are either inherent in the methods used or result from instrumental factors such as peak shift. Errors of this type are discussed in more detail in §12.8.

### 12.13.1 Detection limits

The discussion of statistical precision at the beginning of the preceding section can easily be extended to include the estimation of detection limits. The useful integrated peak area for a pure element was previously assumed to contain 150 000 counts. For a typical peak to background ratio of 100, the background is therefore 1500. Hence the detection limit defined as 3 s.d. is 116 counts, corresponding to a concentration of 0.08%, which is a

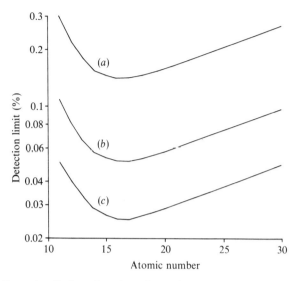

Fig. 12.10 Detection limit as function of atomic number, derived from measured intensities: (a) $\bar{Z}=79$, (b) $\bar{Z}=14$, (c) as (b), but four times as many counts accumulated.

reasonably representative figure. Detection limits are considerably higher for specimens of high atomic number, owing to the greater continuum intensity.

More detailed information on detection limits can be obtained from experimental peak and background intensities for different elements. Such data have been used to derive the curves plotted in fig. 12.10. The measured background intensities have been normalised to mean atomic numbers of 14 (curve (a)) and 79 (curve (b)). The variation with peak energy reflects the combined effects of varying efficiency of characteristic and continuum X-ray generation and detection efficiency. Curves (a) and (b) are based on a spectrum accumulation time which gives 100 000 counts in the Co K$\alpha$ peak. Lower detection limits can, of course, be obtained by counting for a longer time (curve (c)). The data plotted in fig. 12.10 are for isolated peaks: if overlaps are present, detection limits are higher.

## 12.14 Energy- versus wavelength-dispersive analysis

The main characteristics of e.d. and w.d. spectrometers are compared in table 12.2. For qualitative analysis the high collection efficiency and parallel data recording of the e.d. spectrometer give it an overwhelming advantage. However, for quantitative analysis the situation is less clear cut.

## 12.14 Energy- versus wavelength-dispersive analysis

Table 12.2. *Comparison between e.d. and w.d. spectrometers.*

|  | e.d. | w.d. |
|---|---|---|
| X-ray collection efficiency | high | low |
| Data recording mode | parallel | serial |
| Count-rate capability | low | high |
| Spectral resolution | low | high |
| Peak to background ratio | low | high |

The high collection efficiency remains an advantage for specimens that are easily damaged by the electron beam and require the use of a low current, but the advantage disappears when a higher current can be used to compensate for the lower efficiency of w.d. spectrometers. The higher peak to background ratio obtained in w.d. analysis is a definite advantage for low concentrations. Also, the superior resolution of the w.d. spectrometer makes it better able to deal with spectra containing closely-spaced lines (though procedures for deconvolving overlapping peaks in e.d. spectra are quite effective).

The limited count-rate capability of the e.d. spectrometer generally determines the time taken per analysis. Provided a higher current can be used, much higher count-rates (and therefore shorter counting times) for major elements are feasible in w.d. analysis, but the fact that measurements are made sequentially must be taken into account. Also time is wasted in moving between the required spectrometer positions. If the number of elements included in the analysis is small (and especially if there is only one per spectrometer), w.d. analysis may be quicker, for a given statistical precision. As the number of elements increases, the balance of advantage moves in favour of e.d. analysis. It should also be noted that the time taken for setting up and calibration is considerably less for e.d. analysis.

### *12.14.1 Combined e.d. and w.d. analysis*

The above discussion shows that e.d. and w.d. analyses each have their advantages, therefore it is clearly desirable to have both types of spectrometer. Furthermore, in some applications it is beneficial to combine both forms of analysis, using e.d. data for major elements and w.d. data for minor and trace elements (Schwander and Gloor, 1980; Ware, 1991). For combined e.d./w.d. analysis it is desirable to reduce the sensitivity of the e.d. spectrometer, for example by means of an aperture (§9.8.1), so that a probe current suitable for w.d. analysis can be used.

# 13

# X-ray generation and stopping power

## 13.1 Introduction

As discussed in §1.9, the characteristic X-ray intensity emitted by an element in a compound is proportional to the mass concentration of that element, given that the mass of the sample penetrated by the incident electrons is constant. Electron penetration is a function of not only the incident electron energy (which is constant for a given analysis) but also the 'stopping power' of the sample, which depends somewhat on atomic number. Hence concentrations calculated on the basis of the above assumption require correcting. The correction is applied by means of the 'stopping power factor', $F_s$ (§1.10). In the following sections, equations for the generated characteristic X-ray intensity are developed, leading to expressions for $F_s$. Intensity equations also have other applications, including analysis without standards, where pure element intensities are predicted rather than measured (§13.6). Electron penetration also governs the range of electrons in solid targets (§13.7) and is thus relevant to the question of the spatial resolution of microprobe analysis, which is discussed in §13.9.

## 13.2 Ionisation cross-section

For characteristic X-ray intensity calculations the probability of inner-shell ionisation as a function of the kinetic energy of the incident electrons needs to be known. This may be expressed in terms of the ionisation cross-section ($Q$) of the shell concerned. By definition the number of ionisations ($dn$) per increment of electron path length ($dx$) is equal to $Q$ multiplied by the number of atoms per unit volume. Hence for a pure element:

$$dn = Q(N\rho/A)dx, \tag{13.1}$$

where $N$ is Avogadro's number ($6.02 \times 10^{23}$), while $\rho$ is the density and $A$ the atomic weight of the element concerned. The total number of ionisations per incident electron may be calculated by integrating equation (13.1).

The Bethe (1930) formula for $Q$ may be written in the form:

$$QE_c^2 = a \ln(bU)/U, \qquad (13.2)$$

where $E_c$ is the critical excitation energy, $U$ is the overvoltage ratio ($E/E_c$), and $a$ and $b$ are constants. Mott and Massey (1949) deduced from theoretical calculations by Burhop (1940) that $b = 2.42$ for the K shell, but $Q$ then fails to fall to zero at $U = 1$, as required. Worthington and Tomlin (1956) proposed the expression:

$$b = [0.41 + 0.59\exp(1 - U)]^{-1},$$

which agrees with Mott and Massey for high $U$, but gives $Q = 0$ at $U = 1$. Green and Cosslett (1961) suggested that for $U < 20$ the simplifying assumption $b = 1$ may be used, and proposed the following expression for the K shell:

$$QE_c^2 = 7.92 \times 10^{-20} \ln U/U, \qquad (13.3)$$

in which $E_c$ is in keV and $Q$ in cm². Fig. 13.1 shows $QE_c^2$ as a function of $U$, calculated from this equation. The shape of the curve is essentially the same for other shells. Expressions for $Q$ have been discussed in greater detail by Powell (1976, 1990).

It follows from equation (13.3) that for a typical $E_c$ value of a few keV, $Q$ is of the order of $10^{-20}$ cm². The K shell thus presents a very small 'target' of

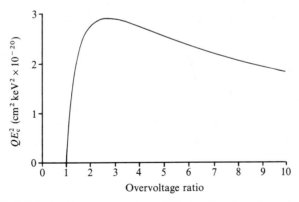

Fig. 13.1 K-shell ionisation cross-section: $QE_c^2$ as a function of overvoltage ratio ($U$).

about 0.001 nm diameter, compared with an overall atomic diameter of about 0.1 nm. An incident electron is therefore much more likely to interact with outer electrons than ionise the K shell. Most electrons lose all their energy in such interactions, and only a small proportion cause K-shell ionisation before coming to rest.

Since $QE_c^2$ is approximately constant (for a given $U$), $Q$ decreases rapidly with increasing atomic number. Thus the cross-section of the K shell of Cu ($Z=29$) is less than 1/50 of that of Na ($Z=11$), owing to the increasingly tight binding of the K electrons with increasing nuclear charge. As far as characteristic X-ray intensities are concerned, however, this trend is counteracted to a considerable degree by the rapid increase in the fluorescence yield with $Z$ (fig. 1.4).

## 13.3 Characteristic X-ray intensity

Incident electrons lose energy through inelastic scattering, which includes various kinds of interaction with target atoms. The excitation of single atomic electrons gives rise to losses ranging from a few eV up to several keV in the case of inner-shell ionisation (though this occurs comparatively infrequently, as noted above). Large energy losses associated with the production of bremsstrahlung (continuum X-rays) are also infrequent. The number of inelastic collisions suffered by an incident electron with an initial energy of 10 keV or more is therefore large, and it is permissible to adopt the 'continuous slowing down approximation', in which energy is assumed to be lost continuously, so that there is a unique relationship between path length $x$ and energy $E$.

Equation (13.1) may now be integrated to give the total number of ionisations produced along the trajectory of an electron with initial energy $E_0$:

$$n = -\int_{E_c}^{E_0} Q(N\rho/A)(dx/dE)dE, \tag{13.4}$$

in which the minus sign takes account of $dx/dE$ being negative. Defining the 'stopping power' ($S$) of the bombarded material as $-dE/d(\rho x)$, equation (13.4) becomes:

$$n = (N/A)\int_{E_c}^{E_0} (Q/S)dE. \tag{13.5}$$

## 13.3 Characteristic X-ray intensity

The approximate 'Thomson–Whiddington law' gives the energy $E$ of electrons of initial energy $E_0$ after passing through a solid layer of thickness $x$: $E_0^2 - E^2 = c\rho x$, (Whiddington, 1912), in which the constant $c$ is about $3 \times 10^5$ keV$^2$ cm$^2$ g$^{-1}$. Green and Cosslett (1961) used the differential form:

$$S = c/2E, \qquad (13.6)$$

with equation (13.3) for $Q$, to obtain the following expression from equation (13.5):

$$n = 9.54 \times 10^4 \, (R/Ac) \int_{E_c}^{E_0} (\ln U/U)(E/E_c^2) dE$$

$$= 9.54 \times 10^4 (R/Ac) \int_1^{U_0} \ln U \, dU$$

$$= 9.54 \times 10^4 (R/Ac)(U_0 \ln U_0 - U_0 + 1),$$

where $U_0 = E_0/E_c$ and $R$ is a factor taking into account the loss of ionisation resulting from electron backscattering (see chapter 14).

Multiplying $n$ by the fluorescence yield $\omega$ gives the characteristic X-ray intensity in photons per incident electron:

$$I = 9.54 \times 10^4 (\omega R/Ac)(U_0 \ln U_0 - U_0 + 1). \qquad (13.7)$$

Substituting $0.365(U_0 - 1)^{1.67}$ for $(U_0 \ln U_0 - U_0 + 1)$ gives an error of less than 10% for $1.5 < U_0 < 16$, thus:

$$I = 3.48 \times 10^4 (\omega R/Ac)(U_0 - 1)^{1.67}. \qquad (13.8)$$

Experimental intensity measurements generally indicate a slightly smaller exponent than the 1.67 in equation (13.8), which can be explained by the increase in $c$ with electron energy. Green and Cosslett (1968) found that an exponent of 1.63 fitted K$\alpha$ intensity measurements best for elements of atomic number 6–47.

Equation (13.8) gives the total intensity per incident electron radiated into a solid angle of $4\pi$. A more useful quantity is the intensity in photons per second per millisteradian solid angle, per nanoamp probe current, which may be obtained by changing the constant in equation (13.8) to $1.73 \times 10^{10}$. The K intensity calculated for typical conditions is plotted in fig. 13.2. Though useful for estimating the intensity of characteristic radiation for some purposes, the approach used here is inadequate for calculating atomic number corrections, for which a more accurate stopping power expression is required, as discussed below.

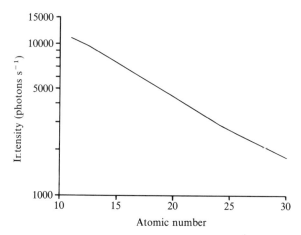

Fig. 13.2 Calculated K radiation intensity, $I_K$, as a function of atomic number for solid angle of 1 msterad (probe current 1 nA, accelerating voltage 25 kV).

## 13.4 Bethe's stopping power expression

Bethe (1930) obtained the following expression for $S$ (in keV cm² g⁻¹) from a wave mechanical treatment of the hydrogen atom:

$$S = 7.85 \times 10^4 (1/E)(B/A), \tag{13.9}$$

where the electron energy $E$ is in keV and $B$ is a dimensionless quantity called the 'stopping number', given by:

$$B = Z \ln(kE/J), \tag{13.10}$$

in which $k$ is a constant and $J$ is the 'mean excitation energy' (or 'mean ionisation potential') of the atom. The value $k = 2$ used in the original Bethe expression applies to heavy ions, whereas for electrons the appropriate value is $\sqrt{(e/2)}$, or 1.166 (Bethe and Ashkin, 1953). The stopping power correction is derived from the Bethe expression (§13.5).

### 13.4.1 Mean excitation energy

Bloch (1933) deduced from the Thomas–Fermi statistical model of the atom that the mean excitation energy $J$ should be proportional to $Z$ ('Bloch's law'). This dependence on $Z$ reflects the increasing binding energy of the electrons. Bloch gave 13.5 eV as the value of the constant of proportionality, but the value more commonly used is 11.5 eV, as

## 13.4 Bethe's stopping power expression

determined by Wilson (1941) from measurements of the penetration of 4 MeV protons in aluminium.

More important than the exact numerical value of the Bloch constant is the question of whether it is truly constant or varies with Z, and here the evidence is conflicting. For example Caldwell (1955) found support for Bloch's law in experimental proton stopping data, while Bakker and Segrè (1951) concluded that there are significant variations in $J/Z$ below $Z=26$. Berger and Seltzer (1964) favoured an empirical expression attributed to Sternheimer:

$$J/Z = 9.76 + 58.8\, Z^{-1.19}, \tag{13.11}$$

applicable to atomic numbers above 12. For low-Z elements Berger and Seltzer used individual values not lying on a smooth curve.

Other empirical expressions have been proposed specifically for the purpose of calculating stopping power corrections in electron microprobe analysis (e.g. equation (13.17)).

### 13.4.2 Shell effects

Bethe's stopping power theory assumes that the incident electron energy is much greater than that of the bound electrons with which inelastic scattering interactions take place. This is valid for outer electrons with low binding energy, but not necessarily for the inner shells. For example, the K-shell binding energy of a heavy element may greatly exceed the incident electron energy, and it is then inappropriate to use a value for $J$ that includes a contribution from the K electrons. In any case there must obviously be a limit to the range of applicability of the simple Bethe expression, since $S$ becomes negative when $E < J/1.166$.

According to Livingston and Bethe (1937) the stopping number $B$ may be split into the contribution of the K shell ($B_K$) and that of all the other electrons combined ($B'$), where $B_K = 1.81\ln(kE/E_K)$ and $B' = (Z-1.81)\ln(kE/J')$. At high energies $B = B' + B_K$, hence:

$$\ln J = [1 - (1.81/Z)]\ln J' + (1.81/Z)\ln E_K,$$

where $E_K$ is the K-shell excitation energy and $J'$ is the effective mean excitation energy of the atom excluding K electrons. For example, for copper ($Z=29$) $J = 334$ eV (assuming $J/Z = 11.5Z$) and $J' = 270$ eV. The effect of substituting $J'$ for $J$ at $E = 10$ keV is to reduce $S$ by 4%. For heavy elements, the reduced contribution of L electrons must also be considered.

Taking account of shell effects rigorously adds considerably to the complexity of stopping power calculations. Various simpler alternative ways of avoiding the anomalous behaviour of the Bethe expression at low energies have therefore been suggested. For example, Rao-Sahib and Wittry (1974) proposed using a parabolic expression below the inflection in the Bethe function which occurs at an energy of 6.34 J. Love, Cox and Scott (1978a) also employed a parabolic expression, but extended its use to cover all energies, thereby replacing the Bethe formula altogether (see §13.5.3). Another alternative form is given in §13.5.4.

Joy and Luo (1989), however, pointed out that these expressions predict a steep increase in $S$ at very low energies which is not in accord with reality. Instead, they proposed substituting $J$ in the Bethe expression by $J'$, given by:

$$J' = J/(1 + mJ/E),$$

where $m$ is a constant with a value of approximately 0.85. This modification simulates the reduction in the effective value of $J$ at low energies due to shell effects in a way which, though not theoretically rigorous, gives more realistic behaviour than the parabolic expressions mentioned above. Substituting $J'$ in equation (13.10), and assuming $b = 1.166$, we have:

$$B = Z \ln[(1.166E/J) + 1], \tag{13.12}$$

which can be substituted in equation (13.9) (Duncumb, 1991). The effect on the behaviour of $S$ at low energies is shown in fig. 13.3.

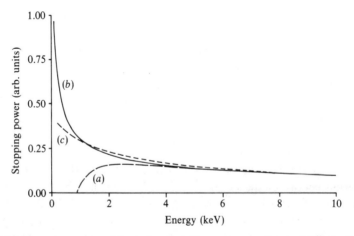

Fig. 13.3 Stopping power of Au as function of energy according to (a) Bethe (1930), (b) Rao-Sahib and Wittry (1974) and (c) Joy and Luo (1989).

## 13.5 Stopping power corrections

So far only pure elements have been considered. Equation (13.4) may be modified for compounds by replacing $N\rho/A$ (the number of atoms per unit volume in the pure element) by $CN\rho/A$ (the number of atoms of the element concerned per unit volume in the compound), where $C$ is the mass concentration of that element. The ratio of the characteristic X-ray intensity $(I)$ from the specimen to that emitted by a standard $(I_0)$ containing a concentration $C_0$ of the element concerned is therefore given by:

$$I/I_0 = [C \int_{E_c}^{E_0} (Q/S) \mathrm{d}E]/[C_0 \int_{E_c}^{E_0} (Q/S_0) \mathrm{d}E]. \tag{13.13}$$

Since $Q$ is not related to composition, it has the same value in both parts of the above expression. Hence, if it is assumed that $S$ and $S_0$ both have the same dependence on $E$, the integration is superfluous and equation (13.13) becomes:

$$I/I_0 = CS_0/C_0 S. \tag{13.14}$$

The correction is thus a function of the difference in stopping power between specimen and standard. Given this approximation, the factor $F_s$ as defined in §1.10 can be equated to $S$.

Castaing (1960) proposed using the approximation: $S = Z/A$, which reflects the number of electrons per unit mass, this being less for heavy elements with their more neutron-rich nuclei, but Bethe's law (§13.4) is more accurate. The constant and the $1/E$ term in equation (13.9) disappear in the ratio $S/S_0$ and are therefore redundant. This ratio varies only slowly with $E$ and hence can be calculated for a suitable mean energy, $\bar{E}$. Most commonly this is derived from the expression:

$$\bar{E} = (E_0 + E_c)/2, \tag{13.15}$$

as proposed by Thomas (1963), though there is a case for the alternative form: $\bar{E} = (2E_0 + E_c)/3$, on the grounds that high energy electrons contribute most to the total X-ray intensity.

The correct method for calculating the stopping power of a compound may be deduced by considering a layer of thickness $\triangle x$ containing two elements with mass concentrations of $C_1$ and $C_2$. The layer can be treated as separate layers of pure element 1 of thickness $\triangle x_1$ and of pure element 2 with thickness $\triangle x_2$. The mass thicknesses of these are related to the mass thickness of the compound layer thus: $\rho_1 \triangle x_1 = C_1 \rho \triangle x$, $\rho_2 \triangle x_2 = C_2 \rho \triangle x$.

The energy loss in each pure element layer is given by: $\Delta E_1 = S_1 \rho_1 \Delta x_1$, $\Delta E_2 = S_2 \rho_2 \Delta x_2$. The total energy loss for the compound layer is thus given by $\Delta E = \Delta E_1 + \Delta E_2 = (C_1 S_1 + C_2 S_2) \rho \Delta x$. Hence for a compound: $\bar{S} = \sum C_i S_i$.

The stopping power factor, $F_s$ (§1.10), is equivalent to $S$ (subject to the approximation that the integration in equation (13.13) can be eliminated, as described above). It differs from the other factors in equation (1.6) in that its numerical value does not have absolute significance. However, this is unimportant given that the correction factor takes the form of a ratio. If desired, $F_s$ can be expressed as the ratio of $S$ for the specimen to that for the pure element, so that its meaning is more obvious. In this case $F_s$ for the standard, where this is not the pure element, must be expressed similarly.

### 13.5.1 Duncumb–Reed method

The atomic number correction procedure of Duncumb and Reed (1968) has been widely used. This incorporates a separate factor for backscattering, as discussed in §14.9. The stopping power is obtained from the simplified form of Bethe's expression:

$$S = (Z/A)\ln(1.166 E/J). \tag{13.16}$$

In calculating $\bar{E}$ from equation (13.15), the value of $E_c$ used is that for element A, the analysed element. For compounds, $\bar{S} = \sum C_i S_i$ (see above).

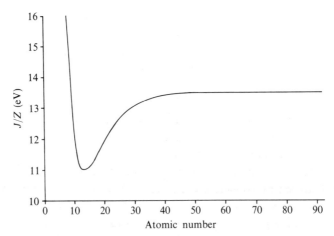

Fig. 13.4 Mean excitation energy: $J/Z$ as a function of $Z$, according to Duncumb and Da Casa (1967).

## 13.5 Stopping power corrections

The mean excitation energy $J$ (in eV) is obtained from the expression of Duncumb and Da Casa (1967):

$$J/Z = 14[1 - \exp(-0.1Z)] + 75.5/Z^{Z/7.5} - Z(100+Z). \quad (13.17)$$

This relationship was obtained by fitting experimental electron probe data for binary compounds with small absorption and fluorescence corrections. The curve of $J/Z$ versus $Z$ (figure 13.4) rises rapidly in the low Z region. For $Z<6$ it is necessary to make the arbitrary assumption that $J/Z$ is constant in order to avoid unrealistic values. Experimental evidence in favour of the Duncumb–Da Casa $J$ values, based on analyses of compounds of known composition, was obtained by Duncumb, Shields-Mason and Da Casa (1969) and Sweatman and Long (1969a,b). Various other expressions for $J/Z$ have also been used.

### 13.5.2 Philibert–Tixier method

Philibert and Tixier (1968a,b) showed that $\int_{E_c}^{E_0}(Q/S)dE$ can be expressed in terms of the logarithmic integral function, thereby avoiding the approximation involved in dropping the integration. Though theoretically advantageous, this method gives results which do not differ significantly from those obtained by the simpler approach described above. Also it suffers from the drawback that the logarithmic integral can only be obtained from tables or by numerical integration.

### 13.5.3 Love–Cox–Scott method

The procedure for calculating stopping power corrections proposed by Love, Cox and Scott (1978a) is also based on evaluating $\int_{E_c}^{E_0}(Q/S)dE$, but uses the following alternative expression for $S$ which behaves better than Bethe's equation at low energies (§13.4.2), while agreeing for $E \gg J$:

$$S = (Z/AJ)/[1.18 \times 10^{-5}(E/J)^{1/2} + 1.47 \times 10^{-6}(E/J)].$$

(Note: this gives $S$ in keV cm$^2$ g$^{-1}$ with $E$ and $J$ in keV.) Using equation (13.3) for $Q$, the following expression for $F_s$ is obtained by integration:

$$F_s = (Z/A)/\{1 + 16.05(J/E_c)^{1/2}[(U_0^{1/2}-1)/(U_0-1)]^{1.07}\}. \quad (13.18)$$

For compounds, the mean value of $Z/A$ (calculated using mass concentrations) is used and the mean value of $J$ is obtained from the relationship:

$$\ln \bar{J} = \sum (C_i Z_i/A_i) \ln J_i / \sum C_i Z_i/A_i, \text{ with } J_i = 13.5 Z_i.$$

This averaging procedure is equivalent to using the mass concentration average of $S$, when this is derived from the Bethe expression.

### 13.5.4 Pouchou–Pichoir method

An alternative formulation which also avoids the problem of the behaviour of Bethe's expression at low energy is used in the 'PAP' correction method (Pouchou and Pichoir, 1987). The stopping power is represented by the expression: $S = (1/\bar{J})[\sum (C_i Z_i/A_i)][1/f(V)]$, where $\bar{J}$ is defined as in the previous section, with $J_i = Z_i[10.04 + 8.25\exp(-Z_i/11.22)]$. The function $f(V)$, where $V = E/\bar{J}$, is given by:

$$f(V) = \sum_{k=1}^{k=3} D_k V^{p_k},$$

with the coefficients taking the following values:

$$D_1 = 6.6 \times 10^{-6}, \ D_2 = 1.12 \times 10^{-5}(1.35 - 0.45\bar{J}^2), \ D_3 = 2.2 \times 10^{-6}/\bar{J},$$
$$p_1 = 0.78, \ p_2 = 0.1, \ p_3 = -(0.5 - 0.25\bar{J}).$$

Pouchou and Pichoir combined the above expression for $S$ with the following modified form of equation (13.2): $Q = a\ln U/E_c^2 U^m$, where $a$ is a constant and $m$ takes the following values for different shells: $0.86 + 0.12\exp(-Z_A/5)$ (K), 0.82 (L), and 0.78 (M). The resulting expression for the intensity $I$, obtained by integrating $Q/S$ is:

$$I = (U_0/V_0)(\sum C_i Z_i/A_i)^{-1} \sum \{D_k[(V_0/U_0)^{p_k}] \quad (13.19)$$
$$[T_k U_0^{T_k} \ln U_0 - U_0^{T_k} + 1]/T_k^2\},$$

where $T_k = 1 + p_k - m$. In the 'PAP' correction procedure (§15.8) the intensity (corrected for backscattering) is incorporated into the $\phi(\rho z)$ function.

## 13.6 Standardless analysis

In conventional quantitative analysis the concentration of a given element is derived from the ratio of the intensity measured on the specimen to that measured on a standard. Sometimes a standard may be unavailable, in which case the intensity can be calculated, using an expression such as equation (13.8) (Russ, 1974). This gives the *generated* intensity, which must be multiplied by the absorption and spectrometer efficiency factors in order to obtain the observed intensity. This approach is particularly suitable for use with e.d. spectrometers, for which the efficiency is predictable (§12.12), but can also be applied to w.d. spectrometers (Wernisch, 1985).

Expressions for the generated intensity such as equation (13.8) give the total emission for the shell concerned. For the purpose of analysis, however, a particular line is measured (K$\alpha$ in the case of the K shell). It is therefore necessary to include a factor representing the fraction of the total intensity for the shell which is contained in the appropriate line. For the K$\alpha$ line this factor varies smoothly and only slowly with atomic number. However, in the case of the L$\alpha$ line, Coster–Kronig transitions (§A.7) have a significant effect, causing the intensity factor to vary with $Z$ in a non-smooth fashion which should be taken into account in calculations for standardless analysis (Lábár, 1987).

## 13.7 Electron range

The range $x_r$ of an electron of initial energy $E_0$ along its trajectory may be determined from the integration:

$$x_r = \int_{E_0}^{0} (dx/dE)dE = \int_{0}^{E_0} (1/\rho S)dE.$$

Substituting $S = c/2E$ (Thomson–Whiddington law – §13.3), we have:

$$\rho x_r = (2/c\rho) \int_{0}^{E_0} E\, dE = E_0^2/c. \tag{13.20}$$

In reality, the 'constant' $c$ increases somewhat with $E_0$, so that $x_r$ increases with $E_0$ more slowly than suggested by equation (13.20).

Bethe's expression for $S$ can be used, but owing to the breakdown of the formula at low energy ($E < J$), the integration must be terminated at a finite energy $E_1$. Worthington and Tomlin (1956) chose $E_1 = 1.03\, J$, giving:

$$\rho x_r = 9.37 \times 10^{-6}(A/Z)J^2 Ei(t_0) \tag{13.21}$$

where $t_0 = 2\ln(1.166 E_0/J)$ and $Ei(t)$ is the exponential integral function defined as:

$$\int_{0}^{t} [\exp(x)/x]$$

Figure 13.5 shows the 'Bethe range' calculated from equation (13.21) for various elements.

A simplified version of equation (13.21) can be obtained by taking the first term only of the series expansion of the exponential integral function (Archard and Mulvey, 1963):

$$\rho x_r = 1.44 \times 10^{-11}(A/Z)[E_0^2/2\ln(0.101 E_0/Z)],$$

assuming $k=2$ and $J=11.5Z$ in equation (13.10).

It should be noted that the Thomson–Whiddington law is concerned with the *most probable* energy loss occurring in a distance d$x$, whereas Bethe's law gives the *mean* energy loss. Owing to the skewness of the energy distribution, the former is smaller than the latter. Also, the range given by equation (13.20) is the maximum depth of penetration, whereas the Bethe range (equation (13.21)) refers to the path length.

Love et al. (1978a) derived the following expression for range from the stopping power as given in §13.5.3:

$$\rho x_r = (A/Z)(7.73 \times 10^{-6} J^{1/2} E_0^{3/2} + 7.35 \times 10^{-7} E_0^2). \tag{13.22}$$

(This gives $\rho x_r$ in g cm$^{-2}$, with $E_0$ and $J$ in keV.)

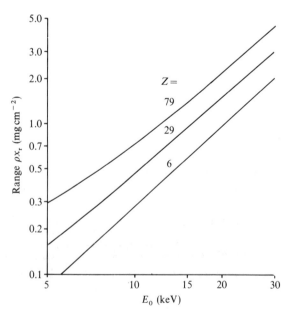

Fig. 13.5 Bethe range ($\rho x_r$), after Bishop and Poole (1973).

## 13.8 Lenard coefficient

The transmission of electrons through thin films may be represented approximately by an exponential function (Lenard, 1895):

$$i = i_0 \exp(-\sigma\rho x), \qquad (13.23)$$

where $i$ and $i_0$ are the transmitted and incident currents respectively, $x$ is the film thickness and $\sigma$ is the 'Lenard coefficient', which is analogous to the mass attenuation coefficient for X-rays. Cosslett and Thomas (1964a) obtained the following empirical expression for $\sigma$ (in $cm^2 g^{-1}$):

$$\sigma = 1.4 \times 10^5 \, E_0^{-1.5}, \qquad (13.24)$$

where $E_0$ is in keV. The Lenard coefficient is used in the Philibert absorption correction (§15.5) and in the fluorescence correction (§16.2.1) to represent the depth distribution of X-ray production.

## 13.9 Spatial resolution

In a scanning image spatial resolution may be defined as the minimum distance apart at which two objects are distinguishable, whereas in quantitative analysis a more rigorous definition is appropriate, namely how large a particle must be to obtain the required analytical accuracy. In either case the effective resolution depends on the spatial distribution of X-ray production in the analysed region. The X-ray source is roughly spherical (fig. 13.6(a)), with emission concentrated around the centre and falling off with distance from the point of impact of the electrons. Spatial resolution is related to the lateral X-ray distribution function (fig. 13.6(b)).

Bishop (1965) computed the spatial distribution of Cu $K\alpha$ emission in pure copper for $E_0 = 29$ keV by the Monte-Carlo method (§14.6). Numerical integration of Bishop's data gives a lateral distribution curve from which the effective resolution with respect to scanning images is estimated to be 0.7 $\mu$m (Reed, 1966). However, this neglects the finite probe diameter. Duncumb (1960) found that with correct choice of operating conditions to give the maximum X-ray intensity for a given resolution, the overall resolution is 1.6 times that resulting from electron scattering, hence the practical resolution limit in this case is 1.1 $\mu$m.

The depth distribution function $\phi(\rho z)$ is approximately independent of atomic number for a given incident electron energy (§15.1), and the same applies to the lateral distribution. However, it is somewhat influenced by the critical excitation energy $E_c$, since once the energy of the incident

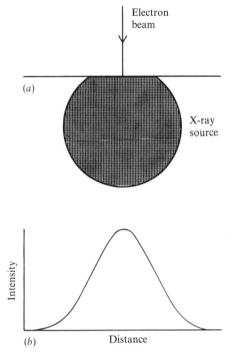

Fig. 13.6 Lateral X-ray intensity distribution.

electrons falls below $E_c$ they no longer produce characteristic X-rays. The dependence on $E_0$ and $E_c$ may be taken into account by a factor $(E_0^{1.5} - E_c^{1.5})$ derived from equation (13.24). Using the data for copper referred to above, the following general expression for resolution $d$ (in $\mu$m) may be derived (Reed, 1966):

$$d = 0.077(E_0^{1.5} - E_c^{1.5})/\rho, \quad (13.25)$$

in which $E_0$ and $E_c$ are expressed in keV. Fig. 13.7 is a nomogram for estimating $d$, based on equation (13.25). In principle $d$ can be made arbitrarily small by reducing $E_0$. However, the X-ray intensity decreases rapidly as $E_0$ approaches $E_c$. Also the difference in $E_c$ for different elements must be taken into account.

By analogy with experimental depth distribution curves, it may be assumed that the outer parts of the lateral distribution function are approximately exponential. It can then be shown that the spatial resolution for quantitative analysis (defined as the particle size required to contain 99% of the X-ray production) is about three times $d$ as calculated from

## 13.9 Spatial resolution

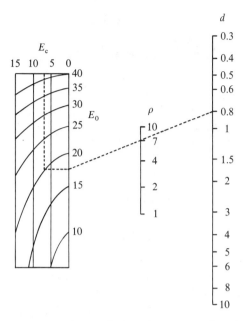

Fig. 13.7 Nomogram for spatial resolution, $d$ ($\mu$m), as a function of density ($\rho$), incident electron energy ($E_0$) and critical excitation energy ($E_c$), both in keV (Reed, 1966).

equation (13.25) (Reed, 1966). For example, for iron ($E_c = 7.1$ keV, $\rho = 7$), the estimated value of $d$ at 20 keV is 0.8 $\mu$m, hence the resolution for quantitative analysis is about 2.4 $\mu$m.

The effect of the density on spatial resolution is important. For example with a biological sample ($\rho \approx 1$), the resolution at 20 keV as defined above exceeds 10 $\mu$m. With such samples high spatial resolution requires the use of a low accelerating voltage, or alternatively a thin specimen in which the electrons do not spread out laterally.

# 14
# Electron backscattering

## 14.1 Elastic scattering

The backscattering of electrons from solid targets is caused mainly by elastic scattering, in which incident electrons are deflected with negligible loss of energy. Some electrons escape after experiencing single scattering events in which the deflection exceeds 90°. More commonly, however, backscattered electrons suffer multiple deflections before emerging (fig. 14.1). Deflections associated with inelastic scattering are small.

### *14.1.1 Single scattering*

Elastic scattering is caused mainly by interactions with the nuclei of target atoms. The theory of the deflection of $\alpha$ particles by the positive charge on the nucleus developed by Rutherford (1911) can also be applied to electrons, though in this case the force is attractive (fig. 14.2). The angular deflection, $\theta$, is given by:

$$\cot(\theta/2) = 2p/b, \tag{14.1}$$

where $b = 1.44 \times 10^{-10} Z/E$ (in which $E$ is the electron energy in keV) and $p$ is the 'impact parameter', that is the minimum distance (in cm) between the undeflected electron path and the nucleus.

The probability of deflection by an angle greater than $\theta$ can be represented as a cross-section, $\sigma(\theta)$, given (in cm²) by:

$$\sigma(\theta) = 1.62 \times 10^{-20} Z^2/E^2 \tan^2(\theta/2). \tag{14.2}$$

For some purposes it is more appropriate to use the differential form which gives the probability of scattering through angle $\theta$, per unit solid angle ($\Omega$):

$$d\sigma/d\Omega = 1.29 \times 10^{-21} Z^2/E^2 \sin^4(\theta/2). \tag{14.3}$$

## 14.1 Elastic scattering

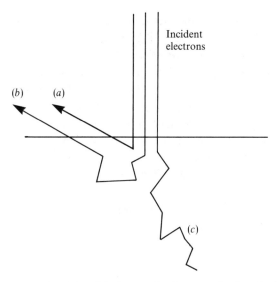

Fig. 14.1 Electron trajectories: (a) electron backscattered after single large-angle deflection, (b) ditto after multiple deflections, (c) electron coming to rest in target.

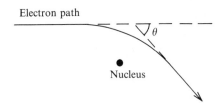

Fig. 14.2 Rutherford scattering – deflection of electron by atomic nucleus.

The formulae above are applicable to large deflections, where the incident electron approaches the nucleus closely. For small angles, where the orbital electron cloud is not completely penetrated, it is necessary to take account of screening of the nuclear charge by inner-orbital electrons (Wentzel, 1927). To allow for screening, equation (14.3) is modified to:

$$d\sigma/d\Omega = 5.16 \times 10^{-21} Z^2/E^2 (1 - \cos\theta + 2\alpha)^2, \quad (14.4)$$

where $\alpha$ is the 'screening parameter'. Various expressions have been proposed for this parameter, a typical example being that used by Bishop (1966c): $\alpha = 3.4 \times 10^{-3} Z^{2/3}/E$. The effect of screening is greatest for high atomic numbers and low electron energies.

A more rigorous quantum mechanical treatment of elastic scattering gives results which sometimes differ significantly from the Rutherford model. The 'Mott cross-section' derived from this approach cannot be expressed in simple analytical form, however. Tabulated values obtained by numerical methods are given by Reimer and Lödding (1984).

To calculate the probability of an electron being deflected through an angle greater than $\theta$ in a single scattering event while passing through a thin layer of thickness $z$ and density $\rho$, the cross-section must be multiplied by the number of atoms per unit area, $N\rho z/A$ (where $N$ is Avogadro's number). Thus, for Cu ($Z=29$), $\sigma(90°)$ is $3.4 \times 10^{-20}$ cm² (for $E=20$ keV) according to equation (14.2), and the probability of an electron being deflected by more than 90° in passing through a layer 0.1 μm thick is 0.03 (assuming $\rho = 8.9$).

### 14.1.2 Multiple scattering

In a solid target an electron with an initial energy of around 20 keV suffers a considerable number of elastic scattering events (typically several hundred) before coming to rest. To calculate these individually is laborious and a considerable economy can be achieved by a collective treatment, whereby the total effect of a number of deflections occurring over a finite path length is represented by a single deflection. In the theory of Goudsmit and Saunderson (1940), for example, the deflection resulting from such multiple scattering is expressed as a Legendre series, which can be calculated using the Rutherford single scattering cross-section (see above). This approach can be applied to thin films, or short sections of the electron trajectories in solid targets, where the number of elastic collisions is of the order of 10.

After more collisions the direction of travel becomes random and the motion of the electrons may be regarded as a form of diffusion. This occurs at the 'diffusion depth' ($z_d$). Owing to the increasing frequency of elastic scattering with increasing atomic number, and the increasing mean deflection angle, diffusion is established nearer the surface in heavy elements than in light ones. Experimental values of $z_d$ are in reasonable agreement with the expression:

$$z_d = 12 \, x_r/(Z+8), \tag{14.5}$$

where $x_r$ is the Bethe range (§13.7), though this gives somewhat low values for $z_d$ at high $Z$ (Cosslett, 1964).

## 14.2 Electron backscattering coefficient

The electron backscattering coefficient $\eta$ is defined as the fraction of incident electrons that do not remain within the sample. It is difficult to derive $\eta$ directly from fundamental scattering theory, hence experimental measurements are the main source of information on this parameter.

Bishop (1966a) measured $\eta$ for various elements using a large chamber in which the number of electrons reaching the specimen after 'rebackscattering' from the walls was small and could be corrected easily. A positive bias voltage was applied to the specimens in order to prevent the escape of secondary electrons. These results (and many other experimental studies) show that $\eta$ increases approximately linearly with $Z$ up to about 30, beyond which the curve flattens out somewhat, reaching slightly above 0.5 for the highest atomic numbers (fig. 5.4). There is also a relatively minor variation with $E_0$: $\eta$ decreases with $E_0$ for low $Z$ and increases for high $Z$ (fig. 14.3).

It is generally assumed that $\eta$ is a smoothly varying function of $Z$, but Heinrich (1968) observed irregular variations in the atomic number range 24–29, correlated with $Z/A$. This is probably related to the dependence of the stopping power on $Z/A$ (§13.3).

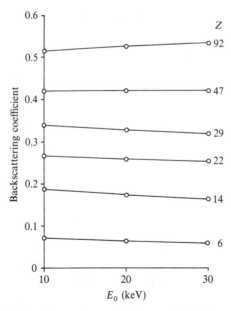

Fig. 14.3 Variation in backscattering coefficient $\eta$ with incident electron energy $E_0$ for various elements (after Bishop (1966a)).

### 14.2.1 Empirical expressions for $\eta$

Various expressions for $\eta$ based on experimental data have been proposed. If the dependence on incident electron energy is neglected, $\eta$ may be expressed as a function of $Z$ only, as in the following polynomial (Love and Scott, 1978b):

$$\eta = -5.23791 \times 10^3 + 1.5048371 \times 10^{-2} Z - 1.67373 \times 10^{-4} Z^2 + 7.16 \times 10^{-7} Z^3, \quad (14.6)$$

which is fitted to experimental data for $E_0 = 20$ keV.

The relatively weak dependence of $\eta$ on $E_0$ seen in fig. 14.3 can also be represented by suitable empirical formulae. For example, Love and Scott (1978b) use the following expression:

$$\eta = \eta_{20}[1 + a\ln(E_0/20)], \quad (14.7)$$

in which: $a = -0.11128 + 3.0289 \times 10^{-3} Z - 1.5498 \times 10^{-5} Z^2$ and $\eta_{20}$ is obtained from equation (14.6). Other expressions taking account of the dependence on $E_0$ have been proposed, for example by Hunger and Küchler (1979) and Neubert and Rogaschewski (1980).

## 14.3 Simplified backscattering models

It is not a straightforward matter to calculate $\eta$ for a solid target, owing to the geometrical complexity of the electron trajectories. The highly simplified approaches described below are only approximate, but they give some insight into the factors governing the variation of $\eta$ with $Z$.

Everhart (1960) proposed a model in which the incident electron travels into the target in a straight line, its energy as a function of depth being derived from the Thomson–Whiddington law (§13.3). The probability of scattering through a sufficiently high angle to reemerge is calculated at each increment of depth from the Rutherford single scattering equation (§14.1.1) and integrated over the range of the electron. The resulting formula is:

$$\eta = (a - 1 + 0.5^a)/(a + 1),$$

in which $a$ is, to a good approximation, a linear function of $Z$ and is independent of the initial electron energy. The theoretical value of $a$ gives low values for $\eta$, hence Everhart proposed the empirical expression: $a = 0.045Z$. Archard (1961) showed that a similar relationship can be derived theoretically if electrons experiencing two deflections are taken into account. The above equation correctly predicts the observed dependence of

## 14.5 Transport equation

$\eta$ on $Z$ in the low to medium atomic number range, but fails for high atomic numbers. McAfee (1976) derived the energy distribution of the backscattered electrons from the Everhart model and obtained reasonable agreement with experiment for atomic numbers up to 40.

In the models of Bethe, Rose and Smith (1938) and Archard (1961), the electrons are assumed to travel straight into the target until they reach the 'diffusion depth' (§14.1.2), whereupon they diffuse uniformly in all directions. The following formula for the backscattering coefficient was obtained by Archard:

$$\eta = (7Z - 80)/(14Z - 80).$$

The model is clearly invalid at low atomic numbers, since the above expression becomes negative for $Z < 12$. Archard therefore proposed using it in combination with the Everhart model described above for low atomic numbers. A more refined diffusion model has been used by Kanaya and Ono (1978) to calculate $\eta$ and other parameters.

### 14.4 Multiple reflection model

Cosslett and Thomas (1965) considered an imaginary thin layer at depth $z$ in a solid target and calculated the total effect of electrons 'reflected' repeatedly from the imaginary surfaces above and below the layer. This model enables $\eta$ and other parameters to be calculated using experimental data from thin films. Other useful results can also be obtained, including the depth distribution of X-ray production (Lantto, 1979; August and Wernisch, 1991b).

### 14.5 Transport equation

If the continuous slowing down approximation (§13.3) is assumed, the distribution function representing the position and velocity of electrons can be expressed as a Boltzmann-type transport equation. This is only amenable to analytical solution in certain special cases of limited practical use, but can be solved by numerical methods, with some approximations (Bethe *et al.*, 1938). Reasonable agreement with experiment was obtained by Brown and Ogilvie (1966) and Brown, Wittry and Kyser (1969) for transport equation calculations of X-ray production in solid targets. However, a significant drawback of this approach is that it neglects large angle scattering.

## 14.6 Monte-Carlo method

The best prospect for calculating $\eta$ and related parameters accurately is to simulate individual electron trajectories as realistically as possible. The 'Monte-Carlo' technique is commonly used for that purpose. This entails computing the trajectories in three dimensions, scattering angles being determined from a probability function, using a random number to replicate the role of chance. To obtain reasonable precision for $\eta$ etc., a large number of trajectories (e.g. 1000 or more) must be calculated. Approximations are usually made in the model in order to keep the computing time within reasonable limits, though this constraint has become less important, with advances in computer performance. An example of electron trajectories computed by the Monte-Carlo method is shown in fig. 14.4.

The backscattering coefficient, $\eta$, can be determined simply by counting the number of electron trajectories that recross the surface. It is also possible to determine the energy distribution of the b.s.e.s, which is required for the calculation of the loss of X-ray production (§14.8). Further, by using a suitable expression for the ionisation cross-section (§13.2), the generation of X-rays can be simulated, enabling the depth distribution function $\phi(\rho z)$ to be determined, from which the absorption correction factor can be derived (§15.1). Monte-Carlo calculations of $\eta$ can be used for testing the validity of the Monte-Carlo model by comparing the results with experimental data, since $\eta$ can be measured relatively easily.

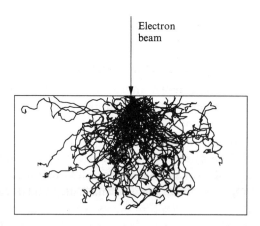

Fig. 14.4 Electron trajectory simulation by Monte-Carlo method: pure Cu, $E_0 = 20$ keV, dimensions of rectangle $= 2 \times 1$ $\mu$m (by courtesy of P. Duncumb).

### 14.6.1 Monte-Carlo models

Ideally the physical processes involved in the scattering and slowing down of incident electrons should be replicated precisely, but this would require excessively complex and time consuming calculations, therefore all actual models are simplified to some degree. In early versions the simplifications were more drastic, in view of the limitations of the computers then available. Subsequently greater rigour has been introduced, as described below.

Green (1963a) divided the electron paths into $0.1\,\mu$m steps and used experimental data for thin films for the angular distribution of the electrons scattered at each step. With a similar procedure, Bishop (1965) obtained reasonable agreement between calculated and observed $\eta$ values for Cu. Agreement between theoretical and experimental b.s.e. energy distributions was less good, owing to the approximations in the model.

Bishop (1966b,c; 1967) developed a more flexible procedure using theoretical scattering formulae. Trajectories were divided into 25 equal steps and the energy loss was calculated from Bethe's law (§13.4). The scattering at each step was derived from the multiple scattering theory of Goudsmit and Saunderson (1940), using the screened Rutherford cross-section (§14.1.1) for single scattering. Shimizu, Murata and Shinoda (1966b) used a similar approach, but with the step length decreasing with decreasing energy, which is somewhat preferable. Shinoda, Murata and Shimizu (1968) refined this procedure by modifying the screening parameter to fit experimental thin film scattering data.

Curgenven and Duncumb (1971) developed a model suitable for small computers, based on a simplified form of single scattering model. The electron paths are divided into a fixed number of steps (e.g. 50), and the Rutherford formula (equation (14.1)) is applied at each step to derive the scattering angle. The impact parameter is selected randomly between zero and an upper limit $p_0$, which is adjusted empirically to compensate for the effect of the approximations in the model. Some refinements to this model are described by Love, Cox and Scott (1977).

In principle greater realism can be obtained by treating elastic scattering events individually, as in the procedure of Murata, Matsukawa and Shimizu (1971), where the step length is equated with the mean free path for elastic scattering. Also, the statistical spread in the energy loss, which is neglected in the continuous slowing down approximation (§13.3), can be taken into account by applying a spreading function to the energy loss derived from the Bethe equation (Shimizu et al., 1975).

Shimizu et al. (1976) treated the different energy loss processes separately, while Reimer and Krefting (1975) applied a 'hybrid' approach, the continuous slowing down approximation being used to represent the collective effect of small energy losses only, with large losses treated separately. These latter authors also used Mott scattering cross-sections (§14.1.1) for large deflections and added the further refinement of allowing for the effect of fast secondary electrons 'knocked on' by the primary electrons.

### 14.7 Energy distribution of backscattered electrons

Knowledge of $\eta$ alone is insufficient for calculating the loss of X-ray intensity caused by electron backscattering, since the potential of an electron for X-ray production is a function of its energy, and b.s.e.s lose a varying amount of energy before emerging. Calculating the loss of X-ray intensity thus requires knowledge of their energy distribution.

Fig. 14.5 shows the differential energy distribution function $d\eta/dW$, where $W = E/E_0$, for various atomic numbers. For heavy elements the distribution is strongly peaked at high energies, because of high angle elastic scattering near the surface. By comparison, electrons backscattered by a light element mostly originate from a greater depth and have lost more of their initial energy. In the range of energies relevant to microprobe

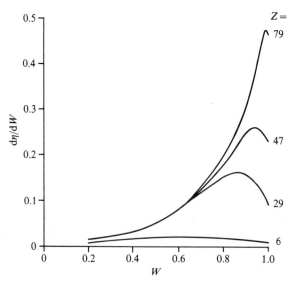

Fig. 14.5 Energy distribution of backscattered electrons as function of $W (= E/E_0)$ (after Bishop (1966)).

analysis, the energy distribution is practically independent of $E_0$, though for heavy elements the peak near $W=1$ decreases somewhat with decreasing energy, especially below 10 keV.

The distribution curves in fig. 14.5 were measured with an energy analyser recording electrons backscattered at 45° to the surface of the sample. Kanter (1957) found the angular distribution to be quite closely proportional to the cosine of the angle relative to the incident beam (Lambert's law). This was confirmed by Bishop (1966a), who also compared energy distributions at different angles and found that the peak of the energy distribution moves to lower energy with increasing angle.

## 14.8 X-ray intensity loss

The fractional X-ray intensity loss due to electron backscattering ($\eta_x$) is equal to the intensity lost divided by the total intensity that would be obtained with no backscattering. It is closely related to the electron backscattering coefficient $\eta$, but is numerically smaller because the energy lost by b.s.e.s before escaping decreases their potential for producing X-rays. The factor $R$, by which the X-ray intensity is reduced owing to electron backscattering, is equal to $1-\eta_x$.

In order to derive $\eta_x$ it is necessary to calculate the X-ray intensity that the b.s.e.s would produce if they remained within the sample. This entails using the integral in §13.3 to give the intensity for b.s.e.s with energy between $W$ and $W+\mathrm{d}W$ and then integrating with respect to $W$ $(=E/E_0)$ between $W_0$ $(=E_c/E_0)$ and 1 $(E=E_0)$. The resulting intensity is then divided by the total intensity which would be obtained if there were no backscattering, in order to obtain $\eta_x$. Thus:

$$\eta_x = \int_{W_0}^{1} (\mathrm{d}\eta/\mathrm{d}W) \int_{E_c}^{WE_0} (Q/S)\mathrm{d}E\,\mathrm{d}W \bigg/ \int_{E_c}^{E_0} (Q/S)\mathrm{d}E$$

$$= \int_{W_0}^{1} \eta(W)(Q/S)\mathrm{d}W \bigg/ \int_{E_c}^{E_0} (Q/S)\mathrm{d}E, \qquad (14.8)$$

where $\eta(W)$ is the fraction of incident electrons backscattered with energy exceeding $WE_0$. Assuming Bethe's formula for $S$ (equation (13.9)) and equation (13.3) for $Q$, equation (14.8) becomes:

$$\eta_x = \int_{W_0}^{1} \eta(W)\ln W\,\mathrm{d}W \bigg/ \int_{W_0}^{1} \ln W\,\mathrm{d}W. \qquad (14.9)$$

According to equation (14.7), $\eta_x$ is a function of $W_0$ and $Z$ only and is independent of $E_0$, though this is not strictly true because $\eta(W)$ is somewhat dependent on $E_0$.

### 14.8.1 Experimental determination of $\eta_x$

Derian and Castaing (1966) measured $\eta_x$ by placing a thin film of one element above a polished solid sample of another, with a small hole in the film for the beam to pass through. The X-ray intensity generated in the thin film by electrons backscattered from the solid sample was measured, enabling $\eta_x$ to be determined (after applying various corrections). Results for Cu and Au agree well with the values calculated from experimental electron backscattering data (Bishop, 1968).

## 14.9 Backscattering corrections

The backscattering correction factor $F_b$ (§1.10) is given by:

$$F_b = 1/(1 - \eta_x) = 1/R.$$

The factor $R$ is, to a reasonable approximation, a function of $Z$ and $W_0$ only (§14.8). Various methods for calculating $R$ are described in the following sections; the results which they give are generally similar.

Owing to the decrease in $R$ with $Z$ (which is related to the increase in $\eta$), the effect of backscattering is to lower the uncorrected concentration when the mean atomic number of the specimen is higher than that of the standard (and vice-versa). This opposes the effect of the difference in stopping power (§13.5), but the latter predominates in the overall 'atomic number correction'.

### 14.9.1 Duncumb–Reed R values

Duncumb and Reed (1968) used the integration described in §14.8 to obtain $R$ values from the experimental electron backscattering data of Bishop (1966a), which are given in table 14.1. For values of $W_0$ and $Z$ between those tabulated, linear interpolation may be used.

### 14.9.2 Springer polynomial

Springer (1967a) proposed the following polynomial:

$$R = a_0 + a_1 W_0 + a_2 W_0^2 + a_3 W_0^3 + a_4 W_0^4,$$

Table 14.1. *The backscattering factor R according to Duncumb and Reed (1968).*

| | $W_0$ | | | | | | | | | | |
|---|---|---|---|---|---|---|---|---|---|---|---|
| Z | 0.01 | 0.1 | 0.2 | 0.3 | 0.4 | 0.5 | 0.6 | 0.7 | 0.8 | 0.9 | 1.0 |
| 0 | 1.000 | 1.000 | 1.000 | 1.000 | 1.000 | 1.000 | 1.000 | 1.000 | 1.000 | 1.000 | 1.000 |
| 10 | 0.934 | 0.944 | 0.953 | 0.961 | 0.968 | 0.975 | 0.981 | 0.988 | 0.993 | 0.997 | 1.000 |
| 20 | 0.856 | 0.873 | 0.888 | 0.903 | 0.917 | 0.933 | 0.948 | 0.963 | 0.977 | 0.990 | 1.000 |
| 30 | 0.786 | 0.808 | 0.828 | 0.847 | 0.867 | 0.888 | 0.911 | 0.935 | 0.959 | 0.981 | 1.000 |
| 40 | 0.735 | 0.760 | 0.782 | 0.804 | 0.827 | 0.851 | 0.878 | 0.907 | 0.938 | 0.970 | 1.000 |
| 50 | 0.693 | 0.718 | 0.741 | 0.764 | 0.789 | 0.817 | 0.847 | 0.881 | 0.919 | 0.959 | 1.000 |
| 60 | 0.662 | 0.688 | 0.713 | 0.737 | 0.764 | 0.793 | 0.825 | 0.862 | 0.904 | 0.950 | 1.000 |
| 70 | 0.635 | 0.663 | 0.687 | 0.713 | 0.740 | 0.770 | 0.805 | 0.844 | 0.889 | 0.941 | 1.000 |
| 80 | 0.611 | 0.639 | 0.665 | 0.691 | 0.718 | 0.750 | 0.785 | 0.826 | 0.874 | 0.932 | 1.000 |
| 90 | 0.592 | 0.613 | 0.639 | 0.665 | 0.695 | 0.730 | 0.767 | 0.811 | 0.862 | 0.924 | 1.000 |
| 99 | 0.578 | 0.606 | 0.634 | 0.661 | 0.691 | 0.725 | 0.763 | 0.806 | 0.858 | 0.921 | 1.000 |

Table 14.2. *Coefficients for use in the Springer polynomial for R (see text).*

| i | $b_{0i}$ | $b_{1i}$ | $b_{2i}$ | $b_{3i}$ | $b_{4i}$ |
|---|---|---|---|---|---|
| 0 | $1.0088 \times 10^2$ | $-7.6070 \times 10^{-1}$ | $-3.5702 \times 10^{-3}$ | $1.6329 \times 10^{-4}$ | $-9.6521 \times 10^{-7}$ |
| 1 | $-6.1134 \times 10^{-1}$ | $6.0271 \times 10^{-1}$ | $1.6222 \times 10^{-2}$ | $-4.5936 \times 10^{-4}$ | $2.5267 \times 10^{-6}$ |
| 2 | $-9.1447 \times 10^{-1}$ | 2.9326 | $-1.7636 \times 10^{-1}$ | $2.8558 \times 10^{-3}$ | $-1.3294 \times 10^{-5}$ |
| 3 | $-7.0753 \times 10^{-1}$ | $-4.6855$ | $2.9116 \times 10^{-1}$ | $-4.6797 \times 10^{-3}$ | $2.1597 \times 10^{-5}$ |
| 4 | 1.3735 | 1.9015 | $-1.2703 \times 10^{-1}$ | $2.1144 \times 10^{-3}$ | $-9.8423 \times 10^{-6}$ |

with coefficients calculated from the expression:

$$a_i = b_{0i} + b_{1i}Z + b_{2i}Z^2 + b_{3i}Z^3 + b_{4i}Z^4.$$

Coefficients determined by fitting to R values of Bishop (1966c) are given in table 14.2. Results obtained by this method are similar to those of Duncumb and Reed (see above). Fig. 14.6 shows R as a function of Z for different values of $W_0$ obtained from this polynomial.

### 14.9.3 Love–Scott method

Love and Scott (1978a) use the following expression:

$$R = 1 - \eta[I(U_0) + \eta G(U_0)]^{1.67}, \tag{14.10}$$

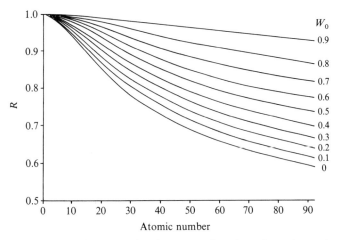

Fig. 14.6 Backscattering correction factor $R$ as function of atomic number and $W_0$ ($=E_c/E_0$) (after Springer (1967a)).

in which $\eta$ is obtained from equations (14.6) and (14.7). The following expressions for $I(U_0)$ and $G(U_0)$ were obtained from Monte-Carlo calculations (Scott and Love, 1991):

$$I(U_0) = 0.3\{-U^{-1} + \exp[1.5(1-U_0^{-0.25})]\},$$
$$G(U_0) = (0.368 - 0.075\ln U_0)\exp[-(1-2.3U_0^{-4})].$$

### 14.9.4 Pouchou–Pichoir method

Pouchou and Pichoir (1987) use the following expression in the 'PAP' correction procedure (§15.8):

$$R = 1 - \eta \bar{W}[1 - G(U_0)],$$

in which the various parameters are derived from the following expressions:

$$\eta = 1.75 \times 10^{-3}\bar{Z}_p + 0.37[1 - \exp(-0.015\bar{Z}_p^{1.3})],$$
$$\bar{Z}_p = (\sum C_i Z_i^{0.5})^2, \quad \bar{W} = 0.595 + (\eta/3.7) + \eta^{4.55},$$
$$G(U_0) = \{U_0 - 1 - [1 - U_0^{-(1+a)}]/(1+a)\}/(2+a)J(U_0),$$
$$J(U_0) = 1 + U_0(\ln U_0 - 1), \quad a = (2\bar{W} - 1)/(1 - \bar{W}).$$

This approach is based on the work of Coulon and Zeller (1973).

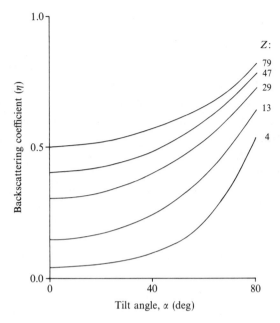

Fig. 14.7 Variation in backscattering coefficient with specimen tilt angle (after Drescher, Reimer and Seidel (1970)).

## 14.10 Non-normal electron incidence

It is observed that $\eta$ increases with $\alpha$, the angle between the beam and the perpendicular to the specimen surface, as shown by Kulenkampff and Spyra (1954) and others (fig. 14.7). This behaviour is caused by the reduction in the effective depth of penetration of the incident electrons and is most marked for low atomic numbers.

There are various expressions for $\eta(\alpha)$, the backscattering coefficient for electron incidence angle $\alpha$. The following were obtained by fitting to experimental data:

$$\eta(\alpha) = 0.891(\eta_0/0.891)^{\cos\alpha}, \qquad (14.11)$$

where $\eta_0$ is the value of $\eta$ for normal electron incidence (Darlington, 1975), and:

$$\eta(\alpha) = \eta_0(1.032/\eta_0)^a,$$

where $a = 0.895(1 - \cos^m \alpha)$ and $m = 1.36 - \eta_0$ (Neubert and Rogaschewski, 1980). Gaber (1987) derived the following expression from Monte-Carlo calculations:

$$\eta(\alpha) = 2.123 \, \exp[g + (K/Z)] - 1.195,$$

where $g = -0.123\cos\alpha - 0.0244$ and $K = -5.126\cos\alpha - 0.520$.

The energy distribution of b.s.e.s is also affected by electron incidence angle and this should be taken into account when calculating $R$ for non-normal incidence. It is found that the change in mean b.s.e. energy with angle is related to $\eta$ in the same way as the change with $Z$: there is thus a unique relationship between the mean energy and $\eta$ for all angles and atomic numbers. Hence, if $R$ is expressed in terms of $\eta$ (as in equation (14.10)) it is merely necessary to modify $\eta$ (using equation (14.11), for example) to allow for non-normal electron incidence (Love et al., 1978a).

## 14.11 Backscattered electron imaging and analysis

The strong dependence of $\eta$ on $Z$ (fig. 5.4) provides the basis for distinguishing between areas of different composition in b.s.e. images (§5.5.1). Measurements of $\eta$ for films of different thickness show that most b.s.e.s originate from a quite shallow depth, as evidenced by the fact that $\eta$ typically reaches its solid-sample value for a thickness of only 30% of the electron penetration distance (Cosslett and Thomas, 1965). This explains why better spatial resolution can be obtained in b.s.e. images than in X-ray images (§5.5).

By measuring the b.s.e. signal, a form of quantitative analysis is also possible. However, it is necessary to take account of the variation in detection sensitivity with electron energy (given that the energy distribution of backscattered electrons varies with $Z$) by measuring the signal for pure elements of different atomic number and constructing a calibration curve which can be used to determine the mean atomic number of 'unknown' samples (Ball and McCartney, 1981; Robinson, Cutmore and Burdon, 1984). The relative concentrations of two elements in a binary compound can be determined in this way, provided their identity is known.

Some assumption about how to average $\eta$ must be made in the case of compounds. According to Danguy and Quivy (1956) the effective mean atomic number can be calculated from the relationship:

$$\bar{Z} = (n_A Z_A^2 + n_B Z_B^2)/(n_A Z_A + n_B Z_B),$$

where $n_A$ and $n_B$ are the atomic concentrations of elements 'A' and 'B'.

An alternative possibility is to calculate the mean value of $\eta$ using the relationship proposed by Castaing (1960):

## 14.11 B.s.e. imaging and analysis

$$\bar{\eta} = C_A \eta_A + C_B \eta_B,$$

where $C_A$ and $C_B$ are the mass concentrations of 'A' and 'B'. Support for this approach is provided by experimental evidence (e.g. Bishop, 1966a) and Monte-Carlo calculations (e.g. Herrmann and Reimer, 1984).

Specimen current measurements can be used for the same purpose (e.g. Philibert and Weinryb, 1963; Heinrich, 1964), the ratio of the specimen current (the current flowing from the specimen to ground) to the beam current (as measured with a Faraday cup) being equal to $1 - \eta$. The effect of secondary electrons can be eliminated by applying a bias voltage to the specimen.

# 15

# Absorption corrections

## 15.1 Introduction

The attenuation of X-rays passing through a given thickness of absorber is easily calculated by means of equation (1.2), assuming the mass attenuation coefficient, $\mu$, (§1.7) is known. In electron microprobe analysis X-rays are produced over a range of depths from the surface of the sample down to the maximum depth penetrated by incident electrons before their energy falls below $E_c$. It is therefore necessary to carry out an integration in order to derive the effective absorption factor, for which purpose the shape of the depth distribution of X-ray production must be known.

The depth distribution function $\phi(\rho x)$ represents the X-ray intensity per unit mass depth ($\rho z$) relative to that produced in an isolated thin layer. The distance travelled in the specimen by X-rays generated at depth $z$ is $z\mathrm{cosec}\psi$, where $\psi$ is the X-ray take-off angle (fig. 15.1), and the factor by which the intensity is reduced is $\exp[-\mu(\mathrm{cosec}\psi)\rho z]$. The absorption factor is thus given by:

$$f(\chi) = \int_0^\infty \phi(\rho z)\exp(-\chi\rho z)\mathrm{d}(\rho z) / \int_0^\infty \phi(\rho z)\mathrm{d}(\rho z), \qquad (15.1)$$

where $\chi = \mu\mathrm{cosec}\psi$. The denominator in equation (15.1) represents the area of the $\phi(\rho z)$ curve and serves the purpose of normalisation, so that $f(\chi) \to 1$ as $\chi \to 0$. The absorption correction factor $F_a$ (§1.10) is equal to $1/f(\chi)$.

Fig. 15.2 shows the typical features of $\phi(\rho z)$. At the surface, X-ray emission is enhanced by scattered electrons travelling upwards from below, hence $\phi_0$, the value of $\phi(\rho z)$ for $\rho z = 0$, is greater than 1. The initial rise with depth is caused by deflection of the incident electrons increasing the mean path length per unit depth (until the directions of the electron trajectories are eventually randomised). After passing through a maximum, $\phi(\rho z)$

## 15.1 Introduction

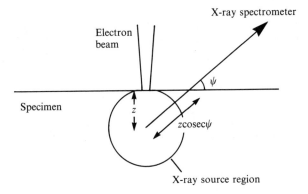

Fig. 15.1 Geometry of X-ray absorption.

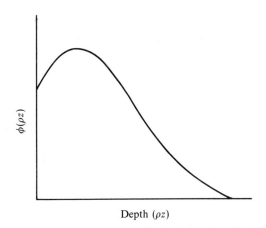

Fig. 15.2 X-ray depth distribution function $\phi(\rho z)$.

decreases due to the scattering and deceleration of the electrons, finally falling to zero.

Experimental studies (see below) show that the shape of $\phi(\rho z)$ is always qualitatively similar, in accord with theoretical scaling laws applying to electron scattering (Spencer, 1955). As $E_0$ increases, $\phi(\rho z)$ expands rapidly along the $\rho z$-axis owing to increasing electron penetration (§13.7). The absorption correction therefore depends strongly on $E_0$. There is also a secondary dependence on $E_c$, which determines where along their trajectories the electrons cease to be able to produce characteristic X-rays. The shape of $\phi(\rho z)$ changes with atomic number as a result of increasing high angle scattering. The effect of each of these factors is illustrated in fig. 15.3.

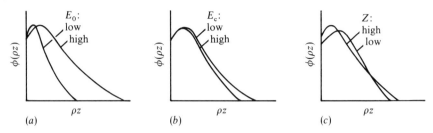

Fig. 15.3 Effect on $\phi(\rho z)$ of variation in (a) $E_0$, (b) $E_c$ and (c) $Z$.

In calculating absorption corrections, $f(\chi)$ has to be evaluated for specimen and standard. The calculation of $\chi$ requires knowledge of the m.a.c.s of the constituent elements for the radiation concerned (§15.12), and their mass concentrations. Since the composition of the analysed specimen is not known initially, uncorrected concentrations are used for the first calculation and the corrected concentrations obtained are substituted to derive a closer approximation to the 'true' correction factors. This iterative procedure is continued until convergence occurs (§17.3).

Much effort has been devoted to developing general methods of calculating $\phi(\rho z)$ (§§15.4–15.8). Most attention has been paid to the case of normal electron incidence. For non-normal incidence, modifications are required, as discussed in §15.10. As well as being used for calculating absorption corrections, the $\phi(\rho z)$ function is also applicable to the analysis of samples containing layers of different composition (§15.13).

## 15.2 Experimental determination of $\phi(\rho z)$ and $f(\chi)$

The most direct way to determine $\phi(\rho z)$ is the 'tracer method' developed by Castaing and Descamps (1955) and subsequently used by many others (Castaing and Hénoc, 1966; Shimizu et al., 1966a,b: Shinoda, 1966; Brown, 1969; Vignes and Dez, 1968; Dürr, Hofer, Schultz and Wittmaack, 1971; Brown and Parobek, 1972, 1973, 1974, 1976; Weisweiler, 1975a; Sewell, Love and Scott, 1985a). In this method a 'sandwich' specimen is produced by vacuum evaporation of successive layers, so that a thin layer of one element is embedded at various depths in a matrix of another (fig. 15.4). Measurements of the characteristic X-ray emission from the tracer element enable $\phi(\rho z)$ to be determined, after due allowance for absorption and fluorescence effects. Typical $\phi(\rho z)$ curves obtained by the tracer method are shown in fig. 15.5.

In the 'wedge method' (Schmitz, Ryder and Pitsch, 1969), $\phi(\rho z)$ is

## 15.2 Experimental determination of $\phi(\rho z)$ and $f(\chi)$

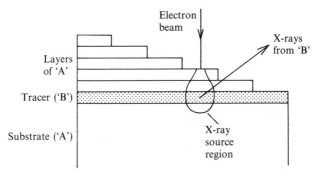

Fig. 15.4 Determination of $\phi(\rho z)$ by tracer method.

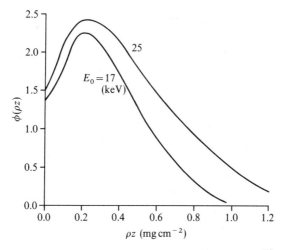

Fig. 15.5 $\phi(\rho z)$ curves determined by the tracer method: V tracer, Ti matrix (after Vignes and Dez, 1968).

derived from X-ray intensity measurements at points along a wedge of one element mounted in contact with a second (fig. 15.6). The thickness is deduced from the distance along the wedge. The intensity is proportional to $\int_0^{\rho z} \phi(\rho z) d(\rho z)$, which may be differentiated to obtain $\phi(\rho z)$. The slope has a negligible effect when the wedge angle is only a few degrees.

The 'variable take-off angle method' (Green, 1963b, 1964; Shimizu et al., 1966b, 1967; Small et al., 1991) involves measuring the characteristic X-ray intensity from a pure element at different X-ray take-off angles. Since it is

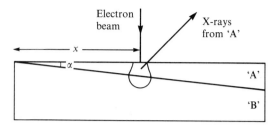

Fig. 15.6 Wedge method for determining $\phi(\rho z)$; thickness of layer of 'A' obtained by measuring $x$, given $\alpha$.

impossible to vary the take-off angle in a conventional electron microprobe, special apparatus is required. This approach was first used in studies of the depth of X-ray production in X-ray tubes (Ham, 1910; Kirkpatrick and Hare, 1934). Absolute determination of $f(\chi)$ requires knowledge of the intensity with zero absorption, which cannot be measured directly. Green solved this problem by plotting the intensity on a logarithmic scale against $\chi$, giving a nearly straight line easily extrapolated to $\chi = 0$.

The 'specimen tilt' method, in which $\chi$ is changed by varying the angle of the specimen (Castaing, 1951; Kirianenko, Maurice, Calais and Adda, 1963), enables conventional instruments with fixed spectrometers to be used. This has the drawback that the concurrent change in the angle of electron incidence also affects $\phi(\rho z)$. A constant (but non-normal) incidence angle can be maintained while varying $\psi$, by tilting the specimen and rotating it about the electron beam axis (Gennai, Murata and Shimizu, 1971).

## 15.3 Graphical correction methods

Early attempts to satisfy the need for practical methods of estimating absorption corrections involved the presentation of experimental data in the form of curves of $f(\chi)$ against $\chi$ (fig. 15.7). These enable absorption factors to be obtained for any $\chi$, but are restricted as to the element and accelerating voltage. Green (1963b) incorporated the effect of $E_c$ by plotting curves for different values of $E_0 - E_c$; separate sets of curves are required, however, for different atomic numbers. With the advent of the on-line computer as a normal adjunct to the electron microprobe, graphical methods fell into disuse in favour of analytical expressions, as described in the following sections.

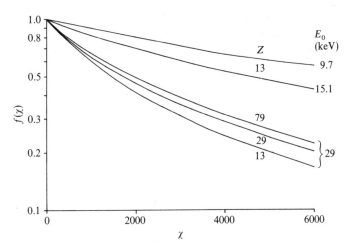

Fig. 15.7 Empirical $f(\chi)$ curves for different $Z$ and $E_0$ (after Castaing (1960)).

## 15.4 'Phi-rho-z' models

As explained previously, knowledge of the depth distribution of X-ray production, represented by the function $\phi(\rho z)$, or 'phi-rho-z', is required in order to derive the absorption factor $f(\chi)$. It is advantageous if $\phi(\rho z)$ can be expressed in integrable form, leading to an analytical expression for $f(\chi)$. Information on the shape of $\phi(\rho z)$ and its dependence on the various relevant parameters can be derived both from experiment (§15.2) and from theoretical approaches described in §§14.4–14.6, of which the Monte-Carlo method is most commonly used. The $\phi(\rho z)$ models described in the following sections contain expressions which are fitted to available data from these sources and experimental intensity measurements on specimens of known composition.

The factors controlling $\phi(\rho z)$ are closely related to the behaviour of electrons in relation to penetration and scattering, as treated in chapters 13 and 14 respectively, though it is traditional to keep the calculation of the corrections separate. The oldest $\phi(\rho z)$ model is that of Philibert, described in the following section, which forms part of what is commonly known as the 'ZAF method'. The term 'phi-rho-z method' is commonly used for calculation procedures based on the more sophisticated models described in §§15.6–15.8, though the distinction is somewhat artificial. The main advantage of this latter group of models is that they give better results in cases of severe absorption and are thus preferable for light element analysis (see chapter 18). In most cases, the results of calculations using different models do not differ greatly.

## 15.5 Philibert method

Philibert (1963) assumed a constant ionisation cross-section so that the X-ray intensity is proportional to the number of electrons and their mean path length in each layer. The number of electrons $n(\rho z)$ at depth $z$ is given by Lenard's law (§13.8): $n(\rho z) = n_0 \exp(-\sigma \rho z)$. Scattering is taken into account by the function $R(\rho z)$, representing the mean electron path length per unit depth (as distinct from $R$ used in chapter 14 to represent the backscattering factor). Below the 'diffusion depth' $z_d$ (14.1.2), $R(\rho z)$ has a constant value ($R_\infty$). In this region there are as many electrons travelling upwards as downwards and since their directions are random the mean path length is increased by a factor of 2, hence $R_\infty = 4$. It is further assumed that $R(\rho z)$ increases with depth from the surface value $R_0$ according to the expression:

$$R(\rho z) = R_\infty - (R_\infty - R_0)\exp(-k\rho z),$$

where $k$ is related to $z_d$. Hence:

$$\phi(\rho z) = [R_\infty - (R_\infty - R_0)\exp(-k\rho z)]n(\rho z). \quad (15.2)$$

Substituting the Lenard expression for $n(\rho z)$ and putting $R_0 = \phi_0$:

$$\phi(\rho z) = 4\exp(-\sigma z) - (4 - \phi_0)\exp[-\sigma(1 + 1/h)\rho z], \quad (15.3)$$

where $h = \sigma/k$. The diffusion depth $z_d$ corresponds to the point at which $R(\rho z)$ effectively reaches its limiting value $R_\infty$, and $k$ may be derived from the relationship: $\exp(-k\rho z_d) = 0.1$. From the expression of Bothe (1927) for the distribution of angular deflections in multiple scattering it may be deduced that $k$ is proportional to $Z^2/AE_0^2$ and, since $\sigma$ is approximately proportional to $E_0^{-2}$, it follows that $h$ is proportional to $A/Z^2$. Since $F_a = 1/f(\chi)$, and:

$$f(\chi) = \int_0^\infty \phi(\rho z)\exp(-\chi \rho z)\mathrm{d}(\rho z) \bigg/ \int_0^\infty \phi(\rho z)\mathrm{d}(\rho z),$$

it follows from equation (15.3) that:

$$F_a = [1 + (\chi/\sigma)][1 + h/(1+h)(\chi/\sigma)]/\{1 + h(\phi_0/4)(\chi/\sigma)/[1 + h(\phi_0/4)]\}. \quad (15.4)$$

## 15.5 Philibert method

Table 15.1. *Values of the adjusted Lenard coefficient σ for substitution in equation (15.5), after Philibert (1963).*

| $E_0$(keV) | σ | $E_0$ | σ | $E_0$ | σ |
|---|---|---|---|---|---|
| 10 | 9600 | 17 | 4850 | 24 | 2725 |
| 11 | 8700 | 18 | 4450 | 25 | 2550 |
| 12 | 7850 | 19 | 4075 | 26 | 2375 |
| 13 | 7100 | 20 | 3725 | 27 | 2200 |
| 14 | 6450 | 21 | 3450 | 28 | 2075 |
| 15 | 5900 | 22 | 3200 | 29 | 1950 |
| 16 | 5350 | 23 | 2950 | 30 | 1820 |

In view of the paucity of data for $\phi_0$ available at the time, and to simplify the formula, Philibert introduced the approximation $\phi_0 = 0$, giving:

$$F_a = [1 + (\chi/\sigma)]/\{1 + [h/(1+h)](\chi/\sigma)\}. \tag{15.5}$$

For compounds the mean value of $h$ is calculated from: $\bar{h} = \sum C_i h_i$ (using mass concentrations). Philibert assumed $h = 1.2A/Z^2$ and used empirical σ values given in table 15.1.

As shown in fig. 15.8, the approximation $\phi_0 = 0$ significantly changes the shape of $\phi(\rho z)$ near the surface, but X-rays from this region travel only a short distance in the specimen and usually are not significantly absorbed (except in the case of long wavelengths). The models described below in §§15.6–15.8 do not use this approximation.

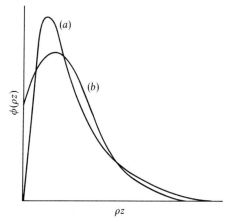

Fig. 15.8 $\phi(\rho z)$ according to simplified Philibert model (a), compared to true shape (b).

### 15.5.1 The effect of critical excitation potential

Castaing and Descamps (1955) found that for pure elements $f(\chi)$ increases (i.e. absorption decreases) with atomic number for a fixed incident electron energy $E_0$. In deriving equation (15.5), Philibert made the implicit assumption that the observed variation is entirely dependent on electron scattering. However, this neglects the effect of the critical excitation energy $E_c$, which determines the point along the electron trajectory at which characteristic X-ray production ceases, and is partly responsible for the variation in the depth of X-ray production with $Z$ in pure elements (Green, 1963b, 1964).

Duncumb and Shields (1966) proposed that the dependence on $E_c$ should be taken into account by using a modified form of $\sigma$:

$$\sigma = 2.39 \times 10^5/(E_0^{1.5} - E_c^{1.5}), \tag{15.6}$$

based on the observation by Cosslett and Thomas (1964a) that electron penetration varies as $E_0^{1.5}$. The constant was obtained by fitting $f(\chi)$ to the curves of Castaing and Descamps (1955) and Green (1963b, 1964).

Heinrich (1967) proposed a modified form of equation (15.6):

$$\sigma = 4.5 \times 10^5/(E_0^{1.65} - E_c^{1.65}), \tag{15.7}$$

in which $E_0$ and $E_c$ are in keV. This version was tested by Duncumb et al. (1969) on a wide range of microprobe data for compounds of known composition and found preferable to equation (15.6).

### 15.5.2 Variants of Philibert method

Several alternative expressions for $\sigma$ and $h$ have been put forward. For example Love et al. (1975) proposed the use of the following variants, based on fitting the simplified Philibert formula to a large set of experimental data for binary compounds:

$$\sigma = 6.8 \times 10^5/(E_0^{1.86} - E_c^{1.86}), \quad h = 0.85 A/Z^2.$$

Ruste and Zeller (1977), on the other hand, made $\sigma$ somewhat dependent on atomic number and $h$ on electron energy.

Attempts have been made to use the full Philibert formula (equation (15.4)), thereby avoiding the approximation $\phi_0 = 0$ (e.g. Reuter, 1972). However, in recent times attention has been concentrated on different forms of $\phi(\rho z)$ model, as described in the following sections.

Several authors (Ziebold and Ogilvie, 1964; Belk, 1966; Borovskii and

Rydnik, 1968) have advocated a simplification whereby $\phi(\rho z)$ is represented as a single exponential function, giving:

$$F_a = 1/[1 + (\chi/\sigma)]. \tag{15.8}$$

This is obviously very easy to calculate and gives reasonable results when absorption is moderate.

### 15.5.3 Other exponential models

Various other expressions for $\phi(\rho z)$ have been proposed using exponential functions, which have the advantage of easy integrability. For example, the 'XPP' model of Pouchou and Pichoir (1988, 1991) utilises the following expression:

$$\phi(\rho z) = A\exp(-a\rho z) + (B\rho z + \phi_0 - A)\exp(-b\rho z),$$

which when integrated gives:

$$F(\chi) = A/(a+\chi) + (\phi_0 - A)/(b+\chi) + B/(b+\chi)^2.$$

Equation (15.11) is used for calculating $\phi_0$, and from this and the total generated intensity ($F$), the slope of $\phi(\rho z)$ at the surface and the mean depth of X-ray production, the parameters $a$, $b$, $A$ and $B$ are derived.

## 15.6 Rectilinear models

Bishop (1974) proposed a 'square model' for $\phi(\rho z)$, in which X-ray production is assumed to be constant from the surface down to a distance equal to twice the mean depth of production, $\bar{z}$. Though a gross approximation, this model has the merit of extreme simplicity. Also, compared to the simplified Philibert model described above, it has the advantage of not falling to zero at the surface.

The expression for $f(\chi)$ from this model is:

$$f(\chi) = [1 - \exp(-2\chi\rho\bar{z})]/2\chi\rho\bar{z}.$$

Bishop proposed using the expression for $\rho\bar{z}$ implicit in the Philibert equation (§15.4): $\rho\bar{z} = (2h+1)/\sigma(h+1)$ and further suggested applying a correction factor to $f(\chi)$ derived from Monte-Carlo calculations (§14.6) or experimental data, in order to compensate for the difference between the true shape of $\phi(\rho z)$ and the 'square' shape assumed.

Sewell et al. (1985a) gave the following expression for $\rho\bar{z}$ derived from experimental data:

$$\rho\bar{z} = \rho x_r \ln U_0 / [(2.4 + 0.07Z) \ln U_0 + 1.04 + 0.48\eta],$$

where $\rho x_r$ is the electron range, as given by equation (13.22), and $\eta$ is the electron backscattering coefficient (§14.2). For compounds: $\rho\bar{z} = \sum C_i \rho\bar{z}_i$. This simple procedure is satisfactory for moderate absorption. An alternative method of calculating $\rho\bar{z}$ has been suggested by Markowicz, Storms and Van Grieken (1986).

A closer approximation to reality is achieved in the 'quadrilateral model' (Love, Sewell and Scott, 1984; Sewell, Love and Scott, 1985b), in which $\phi(\rho z)$ is defined by $\phi_0$ and the following parameters:

$\phi_m$ — the maximum value of $\phi(\rho z)$,
$z_m$ — the depth at which $\phi(\rho z)$ reaches its maximum,
$z_r$ — the depth at which $\phi(\rho z)$ falls to zero.

Between these points $\phi(\rho z)$ is assumed to follow straight lines (fig. 15.9). The equation for $f(\chi)$ obtained from the quadrilateral model is:

$$f(\chi) = 2(P+Q)/R,$$

where:

$$P = 1 - h + \chi(r - m) + h\exp(-\chi r) - \exp(-\chi m),$$
$$Q = [\exp(-\chi m) - 1](1 - h)r/m,$$
$$R = (r - m)(m + hr)\chi^2, \text{ and } h = \phi_m/\phi_0,$$

in which $m = \rho z_m$ and $r = \rho z_r$. Sewell et al. (1985) derived the following expressions for the various parameters in the above equations, based on Monte-Carlo and experimental data:

$$h = a_1 - a_2 \exp(-a_3 U_0^x),$$

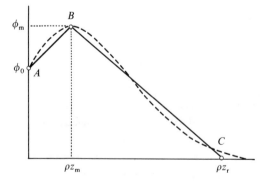

Fig. 15.9 Quadrilateral approximation for $\phi(\rho z)$.

where:

$$a_1 = 2.2 + 1.88 \times 10^{-3} Z,$$
$$a_2 = (a_1 - 1)\exp(a_3),$$
$$a_3 = 0.01 + 7.19 \times 10^{-3} Z,$$
$$x = 1.29 - 1.25\eta,$$

and

$$m = [0.29 + (0.662 + 0.443 U_0^{0.2})Z^{-0.5}]\rho\bar{z},$$

where $\rho\bar{z}$ is obtained from the expression given previously and $r$ is derived from the following relationship:

$$\rho\bar{z} = (m^2 + hr^2 + hmr)/3(m + hr).$$

This method gives significantly better results than the square model in cases of heavy absorption (Sewell, Love and Scott, 1985c).

## 15.7 Gaussian models

The approximately gaussian shape of experimental $\phi(\rho z)$ curves noted by Wittry (1958) led to the development of the 'offset gaussian' model, illustrated in fig. 15.10(a) (Kyser, 1972; Tanuma and Nagashima, 1983, 1984). Packwood and Brown (1981) proposed an alternative form with the gaussian curve centred at $\rho z = 0$ and modified in the near-surface region by an exponential 'transient' function (fig. 15.10(b)). The equation for $\phi(\rho z)$ in this case is:

$$\phi(\rho z) = \gamma \exp[-\alpha^2(\rho z)^2]\{1 - [(\gamma - \phi_0)/\gamma]\exp(-\beta\rho z)\}.$$

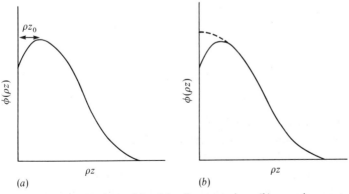

Fig. 15.10 Gaussian $\phi(\rho z)$ models: (a) offset gaussian, (b) gaussian centred at surface, with additional 'transient' function.

Brown and Packwood (1982) used numerical integration to obtain $f(\chi)$ for this function, but Bastin, van Loo and Heijligers (1984) derived the following analytical form allowing faster computation:

$$\int_0^\infty \phi(\rho z)\exp(-\chi\rho z)\mathrm{d}(\rho z) = \alpha^{-1}[\gamma P(y)\chi/2\alpha - (\gamma-\phi_0)P(y)(\beta+\chi)/2\alpha], \quad (15.9)$$

where $P(y)$ is the complementary error function, which can be expressed as a polynomial.

The above expression represents the emitted intensity after absorption and includes the stopping power factor. The generated intensity is given by

$$\int_0^\infty \phi(\rho z)\mathrm{d}(\rho z) = \alpha^{-1}[\gamma - (\gamma-\phi_0)P(y)\beta/2\alpha]. \quad (15.10)$$

The absorption factor $f(\chi)$ can be obtained by dividing equation (15.9) by equation (15.10).

Various expressions for $\alpha$, $\beta$, $\gamma$ and $\phi_0$, based on theoretical considerations, fitting to experimental $\phi(\rho z)$ curves, and fitting to analytical data, have been proposed. Bastin et al. (1984) gave the following expression for $\alpha$:

$$\alpha = 1.75 \times 10^5 E_0^{-1.25}(U_0-1)^{-0.55}[\ln(1.166 E_0/J)/E_c]^{0.5},$$

with

$$J = 9.29Z(1+1.287Z^{-2/3}).$$

The following expressions for the other parameters are as given by Bastin, Heijligers and van Loo (1986):

$$\beta = \alpha Z^n/A, \text{ with } n = Z/(0.4765+0.5473Z),$$
$$\gamma = 1+(U_0-1)/[0.3384+0.4742(U_0-1)] \text{ for } U_0 < 3$$

or

$$\gamma = [5\pi(U_0+1)/U_0\ln(U_0+1)][\ln(U_0+1)-5+5(U_0+1)^{-0.2}] \text{ for }$$
$U_0 > 3$.

Equation (15.11) is used for $\phi_0$.

A modified form of this correction procedure has been described by Bastin and Heijligers (1991). This has some features in common with the 'PAP' method described in the next section, including the expression for the generated X-ray intensity which is used to derive the area under the $\phi(\rho z)$ curve.

## 15.8 Parabolic model

In the model used to derive the correction procedure known as the 'PAP method', Pouchou and Pichoir (1986, 1987) represent $\phi(\rho z)$ by two parabolic functions:

$$\phi_1(\rho z) = A_1(\rho z - m)^2 + B_1, \text{ for } z = 0 \text{ to } z_c,$$

and

$$\phi_2(\rho z) = A_2(\rho z - r)^2, \text{ for } z = z_c \text{ to } z_r,$$

as shown in fig. 15.11, with m and r having the same meaning as in §15.6. Where the two parabolas meet (at $z = z_c$), the following conditions must be satisfied:

$$\phi_1(\rho z) = \phi_2(\rho z) \text{ and } d\phi_1(\rho z)/d(\rho z) = d\phi_2(\rho z)/d(\rho z).$$

A further condition is that $d\phi_2(\rho z)/d(\rho z) = 0$ at $z = z_r$.

It follows that:

$$\rho z_c = 1.5\{[F - (\phi_0 r/3)]/\phi_0 - d^{1/2}/[\phi_0(r-m)]\},$$

where:

$$d = (r-m)\{F - (\phi_0 r/3)\}\{(r-m)F - \phi_0 r[m + (r/3)]\}.$$

Under certain extreme conditions this does not give a satisfactory result, in which case a modified procedure is adopted (Pouchou and Pichoir, 1991). The coefficients of the parabolas can be expressed thus:

$$A_1 = \phi_0/m\{c - r[(c/m) - 1]\}, \quad A_2 = A_1(c-m)/(c-r), \quad B_1 = \phi_0 - A_1 m^2,$$

where $c = \rho z_c$.

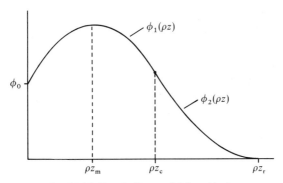

Fig. 15.11 Parabolic model for $\phi(\rho z)$.

The integral $F(\chi) = \int_0^\infty \phi(\rho z)\exp(-\chi\rho z)\mathrm{d}(\rho z)$ is obtained in two parts, $F_1(\chi)$ and $F_2(\chi)$, corresponding to the two parabolic functions, calculated from the following expressions:

$$F_1(\chi) = (A_1/\chi)(\{(c-m)[r-c-(2/\chi)] - 2/\chi^2\}\exp(-\chi c) \\ - (c-m)r + m[c-(2/\chi)] + 2/\chi^2),$$

$$F_2(\chi) = (A_2/\chi)(\{(r-c)[r-c-(2/\chi)] + 2/\chi^2\}\exp(-\chi c) \\ - (2/\chi^2)\exp(-\chi r)).$$

The shape parameters are derived as follows:

(1) $r = \{Q_0 + (1-Q_0)\exp[-(U_0-1)\bar{Z}/40]\}\{1 + U_0^{-a}\}s$, where:

$a = \bar{Z}^{0.45}$,

$s = (\sum C_i Z_i/A_i)^{-1} \sum \{\bar{J}^{(1-p_k)} D_k [E_0^{(1+p_k)} - E_j^{(1+p_k)}]/(1+p_k)\}$,

$Q_0 = 1 - 0.535\exp[-(21/\bar{Z}_n)^{1.2}] - 2.5 \times 10^{-4}(\bar{Z}_n/20)^{3.5}$,

$\bar{Z} = \sum C_i Z_i$ and $\ln\bar{Z}_n = \sum C_i \ln Z_i$.

The summation with respect to $k$ is for $k = 1$–$3$; values of $D_k$ and $p_k$ are given in §13.5.4.

(2) $m = G_1 G_2 G_3 r$, where:

$G_1 = 0.11 + 0.41\exp[-(\bar{Z}/12.75)^{0.75}]$,

$G_2 = 1 - \exp[-(U_0-1)^{0.35}/1.19]$,

$G_3 = 1 - \exp[-(U_0-0.5)\bar{Z}^{0.4}/4]$.

(3) $\phi_0$ is obtained from equation (15.11).

(4) $F$ is obtained from the following relationship: $I = Q(E_0)F$, where $I$ is the generated X-ray intensity and $Q(E_0)$ is the ionisation cross-section (for $E = E_0$). This arises from the definition of $\phi(\rho z)$ as the intensity at depth $z$ relative to that from an isolated thin film, as given by $Q$. The method described in §13.5.4 is used to calculate $I$, with the backscatter factor derived as in §14.9.4. The stopping power and backscattering corrections are thus integrated with the absorption correction, giving a combined 'ZA' factor.

The absorption factor is given by:

$$f(\chi) = [F_1(\chi) + F_2(\chi)]/F.$$

## 15.9 Surface ionisation

The value of $\phi(\rho z)$ at the surface, $\phi_0$, sometimes known as the 'surface ionisation', plays an important role in the models described above. It is especially significant in light element analysis, where absorption is commonly severe and only X-rays produced near the surface contribute significantly to measured intensity. It is also relevant to the analysis of thin films on substrates.

As explained previously (§15.1), $\phi_0$ is greater than unity, because of the contribution of scattered electrons travelling upwards from below the surface. Hence $\phi_0$ is related to $\eta$, the electron backscattering coefficient (§14.2), and increases with atomic number as $\eta$ increases. The additional X-ray production in the surface layer is, however, also a function of the energy and direction of the backscattered electrons. Hence $\phi_0$ is not simply proportional to $\eta$.

It is relatively easy to determine $\phi_0$ experimentally by measuring the X-ray intensity from a surface layer of known thickness, and a considerable amount of such data is available. This has led to the development of empirical expressions such as that of Reuter (1972):

$$\phi_0 = 1 + 2.8[1 - 0.9 U_0^{-1}]\eta.$$

A somewhat different form, which behaves more correctly as $U_0$ approaches 1, has been proposed by Pouchou and Pichoir (1987) for use in the PAP method described in the previous section:

$$\phi_0 = 1 + 3.3[1 - U_0^{-(2-2.3\eta)}]\eta^{1.2}. \tag{15.11}$$

Values of $\phi_0$ derived from this expression are plotted in fig. 15.12.

An alternative relationship between $\phi_0$ and $\eta$ has been derived by Love, Cox and Scott (1978b), on the basis of Monte-Carlo calculations:

$$\phi_0 = 1 + [\eta/(1+\eta)][I(U_0) + G(U_0)\ln(1+\eta)],$$

where:

$$I(U_0) = 3.43378 - 10.78720 U_0^{-1} + 10.97628 U_0^{-2} - 3.62286 U_0^{-3},$$

and:

$$G(U_0) = -0.59299 + 21.55329 U_0^{-1} - 30.55248 U_0^{-2} + 9.59218 U_0^{-3}.$$

An analytical expression for $\phi_0$ based on experimental b.s.e. data has also been put forward by August and Wernisch (1990).

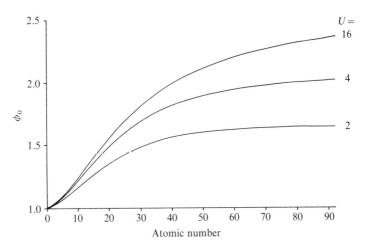

Fig. 15.12 Surface ionisation function $\phi_0$ derived from expression of Pouchou and Pichoir (1986).

## 15.10 Non-normal electron incidence

In electron microprobes the electron beam is usually incident on the specimen perpendicularly to the surface, though some instruments have non-normal incidence (§2.4, 4.3) and tilted specimens are common in s.e.m.s. Since $\phi(\rho z)$ is affected by the electron incidence angle, absorption correction procedures must be modified for non-normal incidence.

Reducing the angle between beam and surface decreases the mean depth of X-ray production and reduces absorption. The simplest assumption is that the depth is reduced by a factor $\cos\alpha$, where $\alpha$ is the angle between the beam and the normal to the surface (see fig. 15.13(b)), so that it is merely necessary to modify $\chi$ thus:

$$\chi = \mu\cos\alpha\,\text{cosec}\,\psi. \tag{15.12}$$

From measurements of $f(\chi)$ using the variable take-off angle method (§15.2), however, Green (1964) deduced that this approximation leads to under-estimation of absorption. Bishop (1965) proposed the following alternative expression:

$$\chi = \mu(1 - 0.5\sin^2\alpha)\,\text{cosec}\,\psi,$$

though a later study cast doubt on its validity (Bishop, 1968). Several subsequent investigations have shown that no simple trigonometric factor is entirely satisfactory, because the shape of $\phi(\rho z)$ changes significantly with $\alpha$ (fig. 15.14).

## 15.10 Non-normal electron incidence

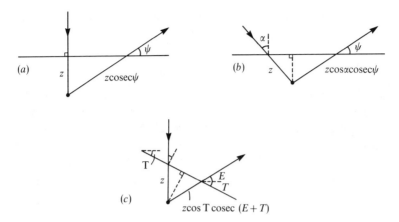

Fig. 15.13 X-ray absorption paths for different geometries: (a) normal electron incidence, (b) inclined electron incidence, (c) tilted specimen.

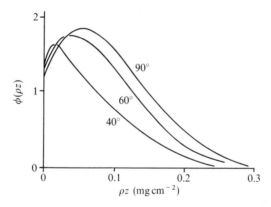

Fig. 15.14 Changing shape of $\phi(\rho z)$ with varying incidence angle, as determined by tracer method (after Sewell, Love and Scott (1987)).

Sewell et al. (1987), on the basis of tracer measurements and Monte-Carlo calculations, derived modified expressions for the parameters used in the quadrilateral model (§15.6). Tests on experimental analytical data gave substantially better results than the $\cos\alpha$ approximation described above. Modifications to the 'XPP' model (§15.5.3) to allow for non-normal electron incidence have been described by Pouchou, Pichoir and Boivin (1990).

It should be noted that non-normal electron incidence has a significant effect on backscattering corrections, as described in §14.10. There is no effect on the penetration factor and the effect on fluorescence corrections is small.

### 15.10.1 Tilted specimens

Tilting the specimen decreases absorption both by reducing the depth of X-ray production, owing to the non-normal electron incidence, and by increasing the X-ray take-off angle (fig. 15.13(c)). With the specimen tilted at an angle $T$ to the horizontal, and the X-ray path at an angle of elevation $E$ to the horizontal, the take-off angle is $E+T$, and the electron incidence angle $\alpha$ is equal to $T$. If it is assumed that the effect of non-normal electron incidence can be taken into account by means of a cosine term, as in equation (15.12), then the appropriate expression for $\chi$ is $\mu\cos T\cosec(E+T)$. When the direction of tilt is at an angle $A$ (the azimuth angle) to the direction of the spectrometer (fig. 15.15), $\cosec(E+T)$ is replaced by: $1/(\sin T\cos A\cos E+\sin E\cos T)$ (Moll, Baumgarten and Donnelly, 1980).

## 15.11 Surface roughness

In the discussion of absorption so far, ideal geometry (as illustrated in fig. 15.1) has been assumed. In particular, the surface of the specimen is taken

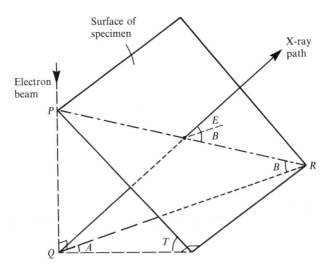

Fig. 15.15 Geometry of tilted specimen where tilt is not directed towards X-ray spectrometer.

to be perfectly planar. Topographic variations on a horizontal scale of the order of 1 $\mu$m can affect the path length traversed by the emerging X-rays, in which case the absorption factors as calculated by the methods described will be incorrect. Since, in general, such topography takes the form of irregular variations in height, there is no way in which it can be taken into account in calculating the correction, and if such effects are significant, errors in the results can be expected.

An indication of the possible size of such errors can be obtained by using equation (1.2), from which it follows that, for a moderate m.a.c. value of 100 cm$^2$ g$^{-1}$ and a medium density of 5 g cm$^{-2}$, the intensity is reduced by 1% in travelling a distance of 0.2 $\mu$m in the absorbing material. For $\mu = 10\,000$ cm$^2$ g$^{-1}$ (a typical value for heavily absorbed radiation) the corresponding distance is only 2 nm. These calculations show the importance of surface smoothness for quantitative analysis, especially where long wavelength X-rays are concerned, as in the case of elements of low atomic number.

## 15.12 Mass attenuation coefficients

In order to calculate absorption corrections, the m.a.c.s (§1.7) of all elements present for each of the characteristic lines used for analysis are required. Experimental data are available only for certain elements and wavelengths, and it is not yet possible to calculate m.a.c.s from theory with complete confidence. The coefficients therefore have to be derived from interpolation formulae fitted to experimental data.

Kramers (1923) predicted theoretically the following relationship: $\mu = kZ^4\lambda^3$, where $\lambda$ is the wavelength of the absorbed radiation and $Z$ the atomic number of the absorber. However, experimental evidence indicates that the exponents are neither constant nor exact integers, and furthermore $\mu$ may not be a smooth function of $Z$. Hence it is preferable to use the more general empirical expression:

$$\mu = C\lambda^n. \tag{15.13}$$

At absorption edges the constant $C$ in equation (15.13) changes abruptly. Fig. 15.16 shows the various absorption edges corresponding to ionisation of inner atomic shells. For a given element not all of these lie within the wavelength range covered in 'conventional' analysis (1–12 Å): for example, the K edges of heavy elements ($Z > 33$) occur below 1 Å, while the L and M edges of the lighter elements ($Z < 29$) lie beyond 12 Å.

Leroux (1961) gave empirical values of $C$ and $n$, enabling $\mu$ to be

Table 15.2. Mass attenuation coefficients for Kα lines (after Heinrich, 1987b).

| Absorber | | Emitter | | | | | | | | | | | | | | | | | | | | | | | | | |
|---|---|---|---|---|---|---|---|---|---|---|---|---|---|---|---|---|---|---|---|---|---|---|---|---|---|---|---|
| | Z | Na | Mg | Al | Si | P | S | Cl | Ar | K | Ca | Sc | Ti | V | Cr | Mn | Fe | Co | Ni | Cu | Zn |
| Li | 3 | 209 | 115 | 66 | 40 | 25 | 16 | 10 | 7 | 5 | 3 | 2 | 2 | 1 | 1 | 1 | 1 | 0 | 0 | 0 | 0 |
| Be | 4 | 570 | 322 | 189 | 115 | 72 | 46 | 31 | 21 | 14 | 10 | 7 | 5 | 4 | 3 | 2 | 2 | 1 | 1 | 1 | 1 |
| B | 5 | 1154 | 666 | 399 | 246 | 157 | 102 | 69 | 47 | 33 | 23 | 17 | 12 | 9 | 7 | 5 | 5 | 4 | 3 | 2 | 2 |
| C | 6 | 1929 | 1147 | 704 | 445 | 289 | 192 | 131 | 91 | 65 | 47 | 34 | 25 | 19 | 15 | 11 | 9 | 7 | 5 | 4 | 4 |
| N | 7 | 2917 | 1766 | 1101 | 706 | 463 | 311 | 213 | 149 | 106 | 76 | 56 | 42 | 31 | 24 | 19 | 14 | 11 | 9 | 7 | 6 |
| O | 8 | 4084 | 2512 | 1590 | 1032 | 685 | 464 | 320 | 225 | 161 | 117 | 86 | 64 | 48 | 37 | 29 | 22 | 18 | 14 | 11 | 9 |
| F | 9 | 5110 | 3189 | 2046 | 1344 | 901 | 616 | 429 | 303 | 218 | 159 | 117 | 88 | 66 | 51 | 39 | 31 | 24 | 19 | 16 | 13 |
| Ne | 10 | 6756 | 4268 | 2771 | 1841 | 1247 | 861 | 604 | 430 | 311 | 228 | 169 | 127 | 96 | 74 | 57 | 45 | 35 | 28 | 23 | 18 |
| Na | 11 | 574 | 5086 | 3337 | 2240 | 1532 | 1067 | 755 | 541 | 394 | 290 | 216 | 163 | 124 | 95 | 74 | 58 | 46 | 37 | 29 | 24 |
| Mg | 12 | 811 | 468 | 4168 | 2823 | 1948 | 1367 | 975 | 704 | 515 | 381 | 285 | 216 | 165 | 127 | 99 | 78 | 62 | 49 | 40 | 32 |
| Al | 13 | 1051 | 634 | 398 | 3282 | 2282 | 1614 | 1159 | 843 | 620 | 462 | 347 | 264 | 202 | 157 | 122 | 97 | 77 | 61 | 49 | 40 |
| Si | 14 | 1407 | 852 | 535 | 347 | 2774 | 1976 | 1429 | 1045 | 774 | 579 | 438 | 334 | 257 | 200 | 157 | 124 | 98 | 79 | 64 | 52 |
| P | 15 | 1728 | 1051 | 662 | 430 | 287 | 2230 | 1622 | 1193 | 889 | 668 | 507 | 389 | 301 | 235 | 184 | 146 | 117 | 94 | 76 | 62 |
| S | 16 | 2208 | 1348 | 852 | 555 | 371 | 255 | 1922 | 1422 | 1064 | 804 | 613 | 472 | 367 | 287 | 226 | 180 | 144 | 116 | 94 | 77 |
| Cl | 17 | 2584 | 1586 | 1005 | 656 | 439 | 302 | 212 | 1558 | 1172 | 890 | 682 | 527 | 411 | 322 | 255 | 203 | 163 | 132 | 107 | 87 |
| Ar | 18 | 2912 | 1796 | 1142 | 747 | 501 | 345 | 242 | 174 | 1246 | 950 | 731 | 567 | 444 | 349 | 277 | 222 | 178 | 144 | 117 | 96 |
| K | 19 | 3715 | 2302 | 1469 | 964 | 648 | 446 | 314 | 225 | 164 | 1152 | 890 | 693 | 544 | 430 | 342 | 274 | 221 | 179 | 146 | 120 |
| Ca | 20 | 4456 | 2774 | 1778 | 1170 | 788 | 543 | 383 | 274 | 200 | 149 | 1022 | 799 | 629 | 499 | 398 | 320 | 258 | 210 | 172 | 141 |
| Sc | 21 | 4817 | 3014 | 1940 | 1281 | 865 | 597 | 421 | 302 | 221 | 164 | 123 | 832 | 657 | 522 | 418 | 337 | 273 | 222 | 182 | 150 |
| Ti | 22 | 5415 | 3405 | 2200 | 1458 | 987 | 683 | 482 | 346 | 253 | 188 | 141 | 108 | 83 | 571 | 458 | 370 | 300 | 245 | 201 | 166 |
| V | 23 | 6029 | 3810 | 2472 | 1643 | 1116 | 773 | 547 | 393 | 288 | 213 | 161 | 123 | 95 | 74 | 498 | 403 | 328 | 269 | 221 | 183 |
| Cr | 24 | 6921 | 4395 | 2864 | 1910 | 1301 | 904 | 640 | 461 | 337 | 250 | 189 | 144 | 111 | 87 | 68 | 455 | 371 | 304 | 250 | 207 |
| Mn | 25 | 7606 | 4852 | 3175 | 2126 | 1452 | 1011 | 717 | 517 | 379 | 281 | 212 | 162 | 125 | 98 | 77 | 61 | 402 | 330 | 272 | 226 |
| Fe | 26 | 8612 | 5518 | 3626 | 2437 | 1669 | 1165 | 829 | 598 | 439 | 326 | 246 | 188 | 145 | 113 | 89 | 71 | 57 | 370 | 306 | 254 |
| Co | 27 | 9321 | 5999 | 3958 | 2669 | 1834 | 1284 | 915 | 661 | 486 | 362 | 273 | 208 | 161 | 126 | 99 | 79 | 64 | 51 | 329 | 274 |
| Ni | 28 | 10607 | 6856 | 4542 | 3074 | 2119 | 1488 | 1062 | 769 | 566 | 421 | 318 | 243 | 188 | 147 | 116 | 92 | 74 | 60 | 49 | 311 |
| Cu | 29 | 9469 | 7163 | 4764 | 3236 | 2238 | 1575 | 1127 | 818 | 602 | 449 | 339 | 260 | 201 | 157 | 124 | 99 | 79 | 64 | 52 | 43 |
| Zn | 30 | 8431 | 7826 | 5224 | 3561 | 2471 | 1744 | 1251 | 910 | 671 | 501 | 379 | 290 | 224 | 175 | 138 | 110 | 89 | 72 | 59 | 48 |

## 15.12 Mass attenuation coefficients

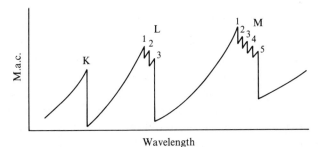

Fig. 15.16 Variation in m.a.c. with wavelength showing absorption edges (schematic).

calculated for any atomic number and wavelength. Heinrich (1966) fitted equation (15.13) to a more selectively chosen set of data, including some new measurements, and made $n$ somewhat dependent on $Z$. Having first derived $n$ from smoothed experimental data, values of $C$ were obtained graphically for each element from a smoothed plot of $C/Z^3$ versus $Z$. The comprehensive table of m.a.c. values given by Heinrich has been used extensively for calculating absorption corrections.

A revised tabulation produced by Heinrich (1987b) is based on the formula:

$$\mu = C(Z^4/A)\lambda^n\{1 - \exp[-(E+b)/a]\}, \qquad (15.14)$$

in which $C$ and $n$ are approximately constant but vary somewhat with $Z$. The main differences compared to the previous expression are the dependence on atomic weight $A$ as well as $Z$, as predicted theoretically (Guttmann and Wagenfeld, 1967) and the addition of the exponential term which takes account of the divergence of experimental data from simple dependence on $\lambda^n$ close to absorption edges. 'Best fit' values for the coefficients $C, n, a$ and $b$ in equation (15.14) were derived by Heinrich from a set of experimental and theoretical data. Selected m.a.c. values from Heinrich's revised tables are given in table 15.2. Accuracy is estimated to be $\pm 5\%$ except within $-5$ and $+20$ eV of absorption edges, between multiple L, M, etc. edges, and for energies below the $M_5$ edge.

In the low energy region (below 1 keV) the problems of measuring and interpolating m.a.c. values are especially severe. This question is discussed further in the context of light element analysis (chapter 18).

## 15.13 Layered samples

A useful extension of the application of the electron microprobe is to layered samples, as distinct from conventional samples, which are homogeneous on a scale of a few microns at least. The simplest form of layered sample consists of a thin surface layer on a solid substrate (fig. 15.17). Assuming that the film is composed of a pure element (differing from the substrate), its thickness can be derived from the observed X-ray intensity. This intensity is given by the integrated area of $\phi(\rho z)$ from the surface to a depth equal to the film thickness (fig. 15.18). The ratio of the intensity from the film to that from a solid sample of the same element (after correction for absorption) is equal to the ratio of the area just defined to that of the whole $\phi(\rho z)$ curve. This approach is applicable to films of thickness down to 1 nm or less.

Early procedures for surface film analysis (Sweeney, Seebold and Birks, 1960; Cockett and Davis, 1963; Hutchins, 1966) made use of empirical calibrations of intensity versus thickness. Subsequently methods based directly on $\phi(\rho z)$ functions were developed. For example, Philibert and Penot (1966) and Reuter (1972) used versions of the Philibert model (§15.5). Bishop and Poole (1973), on the other hand, devised a graphical method based on Monte-Carlo calculations. A problem which arises is that the usual $\phi(\rho z)$ models are not strictly applicable to a surface film on a substrate when these differ significantly in atomic number. The above authors employed forms of weighted averaging to overcome this difficulty.

For films containing more than one element, it is possible to determine both thickness and composition (Colby, 1968; Djurić and Cerović, 1969). Kyser and Murata (1974) applied Monte-Carlo calculations to binary alloy films on substrates, using a graphical iterative procedure to arrive at

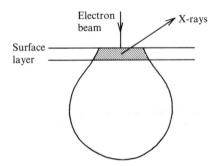

Fig. 15.17 Analysis of surface layer on solid substrate.

## 15.13 Layered samples

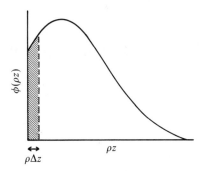

Fig. 15.18 X-ray intensity from surface layer of thickness $\Delta z$ given by area of shaded part of $\phi(\rho z)$ curve.

self-consistent results for the concentrations and thickness. Pouchou, Pichoir and Girard (1980) also used the Monte-Carlo method, whereas Pouchou and Pichoir (1985) applied the 'PAP' $\phi(\rho z)$ model (§15.8). For a general review of surface film analysis, see Scott (1985).

The ideas discussed above can be extended to include more complex samples with several layers. To obtain information on compositions and thicknesses of such layers, however, measurements must be made over a range of incident electron energies. The relative intensities of the relevant X-ray peaks vary as the $\rho z$ scale of $\phi(\rho z)$ changes with $E_0$, and it is possible in principle to deduce the thicknesses and compositions of the layers (Pouchou and Pichoir, 1984b, 1991). In practice, reliable results can be obtained only for a limited number of layers of contrasting compositions, and some degree of 'trial and error' is involved in finding a satisfactory solution.

The problem of the effect of differences in the electron scattering properties of layers differing in mean atomic number is even greater than for surface layers, and can be treated either by empirical modification of existing $\phi(\rho z)$ models or more fundamentally by Monte-Carlo calculations (Ammann and Karduck, 1990). An experimental study of the effect of heterogeneous layers using 'sandwich' samples has been reported by Brown and Chan (1990).

# 16

# Fluorescence corrections

## 16.1 Fluorescence excitation

Fluorescence occurs when the characteristic radiation of an element 'A' is excited by X-ray photons of higher energy than the critical excitation energy of A. Part of the continuous X-ray spectrum invariably satisfies this condition, hence 'continuum fluorescence' is always present. A correction for this effect, though usually small, and commonly ignored, strictly should be applied (§16.5). Fluorescence may also be excited by the characteristic radiation of other elements, subject to the above energy criterion. For $Z_A \leq 21$, fluorescent K-shell ionisation can be caused by the K radiation of any element of higher atomic number (fig. 16.1(a)), i.e. the condition for excitation is $Z_B - Z_A > 1$, where $Z_B$ is the atomic number of the exciting element. For $Z_A > 21$ the situation is somewhat different: when $Z_B - Z_A = 1$, only the K$\beta$ radiation of element B has sufficient energy to cause fluorescence, whereas when $Z_B - Z_A \geq 2$, both K$\alpha$ and K$\beta$ radiation satisfy the excitation condition (fig. 16.1(b)).

For fluorescence excitation by L lines the same principles apply. The L spectrum is considerably more complex than the K spectrum (fig. 1.2), hence there is a greater probability of excitation by some but not all of the L lines of a particular element, though for most purposes only the principal L$\alpha$ and L$\beta$ lines need be considered. It is also possible for L radiation to be excited by either the K or L lines of another element. Usually the L$\alpha$ line is used for analysis, in which case the appropriate excitation energy is that of the $L_3$ subshell. Fluorescence involving M radiation also occurs, but in practice can be neglected on the grounds that the effect is small (Büchner and Stienen, 1975).

## 16.2 Theory of characteristic fluorescence corrections

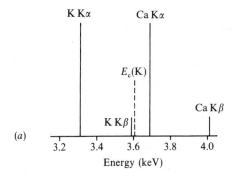

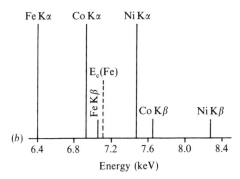

Fig. 16.1 Characteristic fluorescence: (a) excitation of K by Ca Kα and Kβ lines, (b) excitation of Fe by Co Kβ line (but not Kα) and by both Ni Kα and Kβ lines.

### 16.2 Theory of characteristic fluorescence corrections

In order to apply a correction for characteristic fluorescence it is necessary to calculate the intensity of the fluorescence radiation of element A ($I_f$) excited by the characteristic radiation of element B, and to obtain the ratio $I_f/I_A$, where $I_A$ is the intensity of 'A' radiation produced directly by electron bombardment. The fluorescence factor $F_f$ (§1.10) is given by: $F_f = 1/[1 + (I_f/I_A)]$.

The derivation of the equation for $I_f/I_A$ given below follows Castaing (1951) and refers to K–K fluorescence, though it can be adapted to fluorescence involving L lines (§16.4). In order to simplify the theory it is assumed initially that all primary (electron-excited) X-ray production takes place at the surface and half is absorbed in the sample. The fraction absorbed by A atoms is $C_A \mu_B^A / \mu_B$, where $\mu_B^A$ and $\mu_B$ are the m.a.c.s of pure A and the sample respectively for the radiation of the exciting element B, and

244                    16 Fluorescence corrections

$C_A$ is the mass concentration of element A. For the purpose of determining the m.a.c.s all the B radiation is assumed to be concentrated in the $K\alpha$ line.

A certain fraction of the radiation absorbed by A atoms gives rise to K-shell ionisation. This fraction may be deduced from consideration of the variation of the m.a.c. with energy in the vicinity of the K edge (fig. 16.2). Absorption on the low energy side of the edge arises from ionisation of shells other than K, and the additional absorption at the edge is associated purely with K-shell ionisation. The fraction of $\mu_K$ (the m.a.c. just on the high energy side of the edge) attributable to K-shell ionisation is thus $1 - 1/r$, or $(r-1)/r$, where $r$ is the 'absorption edge jump ratio' (the m.a.c. on the high energy side divided by that on the low energy side). Since the energy dependence of the contributions of all shells is nearly the same, this ratio applies anywhere on the high side of the edge. The fraction of all ionisations of A atoms that involve the K shell is thus $(r_A - 1)/r_A$. The probability of K-shell ionisation resulting in K(A) photon emission is equal to $\omega_K(A)$, the K-shell fluorescence yield (§1.5) of element A. Combining all the above factors we find that the rate of production of K(A) photons due to the fluorescence is:

$$I_f = 0.5\varepsilon I_B, \qquad (16.1)$$

where $\varepsilon = C_A(\mu_B^A/\mu_B)[(r_A - 1)/r_A]\omega_K(A) I_B$ is the intensity of K(B) radiation.

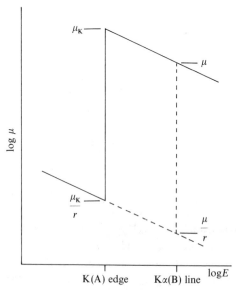

Fig. 16.2 Log–log plot of m.a.c. $\mu$ versus X-ray energy $E$ in vicinity of K absorption edge of element A.

## 16.2 Theory of characteristic fluorescence corrections

It remains to determine the ratio $I_A/I_B$ of the primary K(A) and K(B) intensities, in order to obtain $I_f/I_A$. Castaing (1951) used the following expression based on the Rosseland (1923) ionisation cross-section:

$$I_B/I_A = C_B \omega_K(B) A_A E_c(A) / C_A \omega_K(A) A_B E_c(B), \qquad (16.2)$$

where $A_A$ and $A_B$ are the atomic weights of elements A and B, and $E_c(A)$ and $E_c(B)$ are their critical excitation energies. This does not give the intensity ratio very accurately, especially when A and B differ considerably in atomic number, and Reed and Long (1963) proposed a modification based on the Green and Cosslett (1961) intensity formula (equation (13.8)), whereby $E_c(A)/E_c(B)$ in equation (16.2) is replaced by $[(U_B-1)/(U_A-1)]^{1.67}$. This reduces $I_B/I_A$ by an amount that increases with increasing difference in atomic number and is dependent on the incident electron energy. The expression $(U-1)^{1.67}$ is, however, an approximation for $U \ln U - U + 1$ (§13.3), which should be used in preference (Reed, 1990a). Combining equation (16.1) and the modified form of equation (16.2), we thus have:

$$I_f/I_A = 0.5 C_B(\mu_B^A/\mu_B)[(r_A-1)/r_A]\omega_K(B)(A_A/A_B) \\ [(U_B \ln U_B - U_B + 1)/(U_A \ln U_A - U_A + 1)]. \qquad (16.3)$$

### 16.2.1 The absorption term

The calculation of the absorption of the emerging fluorescence radiation is simplified by the initial assumption that primary X-ray production occurs entirely at the surface. The solid angle subtended by the annulus in fig. 16.3 is $2\pi \sin\phi \, d\phi$: therefore the intensity of the characteristic radiation emitted

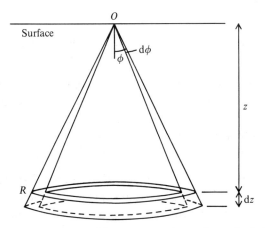

Fig. 16.3 Geometry of fluorescence absorption calculation.

246   16 Fluorescence corrections

by element B into this solid angle is $2\pi\sin\phi d\phi(I_B/4\pi)$, where $I_B$ is the total intensity of B radiation. In travelling to the annulus this radiation is attenuated by the factor $\exp(-\mu_B\rho z\sec\phi)$, where $\rho$ is the density and $\mu_B$ is the m.a.c. of the sample for B radiation. The fraction of the B radiation reaching the annulus that is absorbed in it is $\mu_B\rho\sec\phi dz$, and the rate of conversion of absorbed photons into K(A) photons may be written as $\varepsilon$ (see equation (16.1)). The attenuation of the A radiation in travelling to the surface is given by the factor $\exp(-\mu_A\rho\text{cosec}\psi z)$, where $\mu_A$ is the m.a.c. of the sample for this radiation and $\psi$ is the X-ray take off angle.

Combining these factors and integrating over $z$, we arrive at the intensity of the fluorescence radiation:

$$I_f = 0.5\varepsilon\mu_B\rho \int_0^\infty \int_0^{\pi/2} \exp[-(\mu_B\sec\phi + \mu_A\text{cosec}\psi)\rho z]\tan\phi d\phi dz I_B$$

$$= 0.5\varepsilon\mu_B \int_0^{\pi/2} [-\tan\phi/(\mu_B\sec\phi + \mu_A\text{cosec}\psi)]d\phi I_B$$

$$= 0.5\varepsilon \int_0^{\pi/2} \{-\sin\phi/[1 + (\mu_A/\mu_B)\text{cosec}\psi\cos\phi]\}d\phi I_B$$

$$= 0.5\varepsilon \int_1^0 \{1/[1 + (\mu_A/\mu_B)(\text{cosec}\psi)t]\}dt I_B$$

$$= 0.5\varepsilon[\ln(1+u)/u]I_B,$$

where $u = \mu_A\text{cosec}\psi/\mu_B$. The function $\ln(1+u)/u$ is plotted in fig. 16.4.

The finite depth of production of the primary radiation has two effects:

(1) the amount of fluorescence radiation generated is increased owing to the absorption of primary X-rays travelling upwards, and
(2) absorption of the fluorescence radiation is increased owing to the greater depth at which it is produced.

These act in opposite directions and tend to compensate each other.

Castaing assumed that the primary X-ray depth distribution follows Lenard's law (§13.8), thus: $I_B(z) = I_B(0)\exp(-\sigma\rho z)$, where $\sigma$ is the Lenard coefficient, which is related to $\sigma$ as used in the Philibert absorption

## 16.2 Theory of characteristic fluorescence corrections

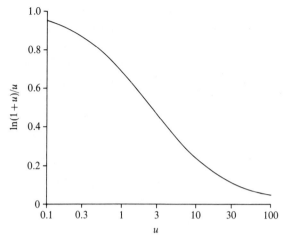

Fig. 16.4 Fluorescence absorption function, $\ln(1+u)/u$.

correction and can be obtained from equation (15.7). This leads to an additional term $\ln(1+v)/v$ (where $v = \sigma/\mu_B$). Combining the above expression with equations (16.1) and (16.2) we arrive at the final formula:

$$I_f/I_A = 0.5 C_B (\mu_B^A/\mu_B)[(r_A-1)/r_A]\omega_K(B)(A_A/A_B)$$
$$\times [(U_B \ln U_B - U_B + 1)/(U_A \ln U_A - U_A + 1)]$$
$$\times \{[\ln(1+u)]/u + [\ln(1+v)]/v\}. \qquad (16.4)$$

The exponential X-ray depth distribution is a rather crude approximation. Modified expressions derived from more realistic depth distribution functions have been proposed (Criss, 1968; Büchner and Stienen, 1975) but the effect on the results is small. Similarly the use of a depth distribution derived from Monte-Carlo calculations (Paduch and Barszcz, 1986) makes little difference to the size of the correction.

### 16.2.2 Fluorescence excited by Kβ radiation

In the derivation of the K–K fluorescence correction described above it is assumed that the exciting radiation is monochromatic. Ideally the calculation should be divided into two parts representing excitation by Kα and Kβ radiation, since the m.a.c.s required for substitution in equation (16.4) are different for the two lines. However, since less than 15% of the total K intensity is contained in the Kβ line, the monochromatic approximation does not cause significant errors in most cases.

As already noted in §16.1, for elements exceeding 21 in atomic number, only the $K\beta$ line of the element one higher in atomic number excites fluorescence. For calculating the correction in this case the m.a.c.s for B radiation in equation (16.4) should be those for the $K\beta$ line. Furthermore, $I_f/I_A$ must be multiplied by a factor $\delta$, equal to the $K\beta$ intensity of element B divided by the total K intensity (obtainable from Khan and Karimi (1980) for example).

### 16.2.3 Ternary and higher compounds

In ternary or higher compounds fluorescence may be excited by more than one element. For example, in a Cr–Fe–Ni alloy Cr is excited by both Fe and Ni. The value of $I_f/I_A$ required to take both contributions into account is obtained by adding the separate values calculated for each exciting element.

A further aspect of ternary systems is that the intensity of the radiation emitted by element B (Fe in the above example) is enhanced owing to fluorescence excited by the third element C (Ni). The usual correction formula does not allow for this. A rigorous calculation of this effect would be complicated, but an approximate result may be obtained by multiplying $I_f/I_A$ calculated for B–A fluorescence by $[1+(I_f/I_A)]$ calculated for C–B fluorescence. This effect is invariably small and can be neglected with little error.

### 16.3 K–K fluorescence corrections in practice

For calculating fluorescence corrections, data are required for m.a.c.s (§15.12) and other parameters. Springer (1967b) gave the following expression for $(r-1)/r$:

$$(r-1)/r = 0.924 - 0.00144Z. \tag{16.5}$$

Values for $\omega_K$ can be obtained from table A.6 (see appendix). An example of the calculation of a characteristic fluorescence correction is given in §17.2.2.

Fig. 16.5 shows the enhancement factor $1/F_f$ calculated for Fe in binary Ni/Fe alloys of varying Ni/Fe ratio. The non-linear variation in this factor with Fe concentration is attributable to the strong dependence of $\mu_B$ on $C_B$, the m.a.c. of Fe for Ni K radiation being much higher than that of Ni.

The size of the K–K fluorescence correction decreases as the difference between the atomic numbers of A and B increases (see fig. 16.6). Also, the correction increases with atomic number owing to the strong Z-dependence of the fluorescence yield (see fig. 1.4). For $Z_A < 20$ the correction is always small.

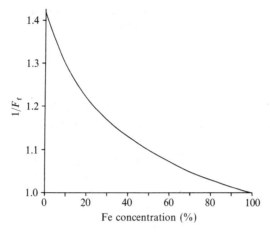

Fig. 16.5 Fluorescence enhancement factor, $1/F_f$, for Fe in Ni–Fe alloys ($E_0 = 25$ keV, $\psi = 40°$).

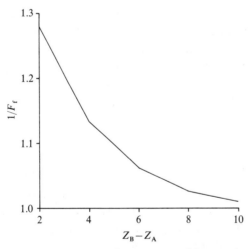

Fig. 16.6 Fluorescence enhancement factor versus difference in atomic number between exciting element ($Z_B$) and excited element ($Z_A$) for $Z_A = 22$ ($E_0 = 20$ keV).

## 16.4 Fluorescence involving L lines

The discussion so far has been confined to K lines. Fluorescence involving L lines also occurs and the following three forms are considered here: (1) L–L (the L lines of one element exciting the L shell of another), (2) K–L (the K lines of one element exciting the L shell of another) and (3) L–K (the

### 16.4.1 L–L fluorescence corrections

The derivation of the fluorescence correction given in §16.2 applies equally to all forms of characteristic fluorescence, with certain modifications. In fluorescence involving L lines the three separate L subshells are a complication, but as a first approximation the correction may be calculated as if there were a single L absorption edge and emission line (Reed, 1965). The excitation energy is taken to be that of the $L_3$ subshell and m.a.c.s for the $L\alpha$ line are used. The latter is a more serious approximation than the equivalent assumption in the case of the K shell, since the $L\alpha$ line contains only about 55% of the total L intensity. However, provided no large absorption edges are present between the main L lines of the exciting element B, the ratio $\mu_B^A/\mu_B$ is approximately the same for all lines and no serious error arises. If such edges are present, it is desirable to calculate the fluorescence generated by the lines above and below the edge separately. In L–L fluorescence the absorption edge jump ratio of the excited element A refers to the combined $L_1$, $L_2$ and $L_3$ edges. The total jump ratio $r_L$ is equal to the product of the jump ratios of each edge. The effective fluorescence yield is the weighted mean of the fluorescence yields of the individual subshells.

### 16.4.2 K–L and L–K fluorescence corrections

In calculating corrections for K–L and L–K fluorescence it is necessary to take into account the ratio of K and L intensities. The ionisation cross-sections of the L subshells exhibit the same energy dependence as the K shell (Burhop, 1940; Green and Cosslett, 1968). The Green–Cosslett intensity expression (equation (13.8)), or preferably the more rigorous form (equation (13.7)) can thus be used for both, the constant having different values $k_K$ and $k_L$ for K and L radiation. A value of 0.24 for $k_K/k_L$ has been obtained from experimental intensity measurements (Reed, 1965). For K–L fluorescence this factor is applied to equation (16.4), while for L–K fluorescence the reciprocal value (4.2) is substituted.

Büchner and Stienen (1975) described a somewhat different approach to the calculation of corrections for fluorescence involving L lines, in which a

## 16.5 Continuum fluorescence

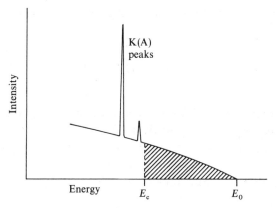

Fig. 16.7 Excitation of fluorescence by continuum radiation between $E_0$ and critical excitation energy $E_c$.

factor allowing for the effect of Coster–Kronig transitions (§A.7) on the relative L line intensities is included.

### 16.5 Continuum fluorescence

The energy criterion for fluorescence excitation is satisfied for that part of the continuum lying between $E_c$ and $E_0$ (fig. 16.7). Unlike characteristic fluorescence, this form of excitation is always present in both specimen and standard, and hence tends to cancel out. However, it is desirable that a correction should be applied, as described below.

#### 16.5.1 Continuum fluorescence in pure elements

The number of continuum photons, $dn$, between energies $E$ and $E+dE$ produced per incident electron is, according to Kramers (1923):

$$dn = aZ[(E_0 - E)/E]dE,$$

where $a$ is a constant, for which Dyson (1959) obtained the experimental value of $2.76 \times 10^{-9}$ photons s$^{-1}$ eV$^{-1}$.

Let us first consider a sample consisting of pure 'A'. Following Green and Cosslett (1961), the total number of photons capable of exciting fluorescence is:

$$n = aZ \int_{E_c}^{E_0} [(E_0 - E)/E]dE,$$

where $E_c$ is the critical excitation energy of A. Changing variables to $U$ ($= E/E_0$), we have:

$$n = aZE_c \int_1^{U_0} [(U_0 - U)/U] dU. \quad (16.6)$$

Assuming that primary X-ray production occurs entirely at the surface, half the total continuum intensity will be absorbed in the target. Of the absorbed photons, a fraction $(r-1)/r$ cause K-shell ionisation, where $r$ is the K absorption edge jump ratio of A. The generated intensity of continuum fluorescence is thus given by:

$$I_f = 0.5[(r_A - 1)/r_A]\omega_K(A)aZE_c \int_1^{U_0} [(U_0 - U)/U] dU, \quad (16.7)$$

$$= 0.5[(r_A - 1)/r_A]\omega_K(A)aZE_c(U_0 \ln U_0 - U_0 + 1). \quad (16.8)$$

The characteristic intensity produced by electron bombardment, $I_A$, as given by equation (13.7), shows the same dependence on $U_0$ as $I_f$. The ratio $I_f/I_A$ is therefore independent of the accelerating voltage and is given by:

$$I_f/I_A = 5.2 \times 10^{-12}(ac/R)[(r_A - 1)/r_A]AZE_c. \quad (16.9)$$

By making various approximations, equation (16.9) can be reduced to the simplified form:

$$I_f/I_A = 9.7 \times 10^{-8} Z^4. \quad (16.10)$$

For L radiation the constant in equation (16.10) takes the value $3.7 \times 10^{-9}$.

According to equation (16.10), the continuum fluorescence intensity as a fraction of the primary X-ray production (neglecting absorption) is about 2% for $Z = 20$, rising to 4% for $Z = 25$ and 8% for $Z = 30$.

### 16.5.2 Continuum fluorescence in compounds

For calculating continuum fluorescence corrections it is obviously necessary to be able to deal with compound samples. Absorption suffered by fluorescence radiation in emerging from the sample (neglected in the preceding section) must also be taken into account. Methods for calculating continuum fluorescence corrections have been derived by Hénoc (1962, 1968) and Springer (1967b, 1972). The treatment here follows the latter.

In the case of a pure element, the primary radiation is absorbed entirely

## 16.5 Continuum fluorescence

by the element itself, whereas in a compound the fraction absorbed by A atoms is $C_A \mu_c^A/\mu_c$, where $\mu_c^A$ and $\mu_c$ are the m.a.c.s of pure A and of the compound, for continuum radiation. In the absence of absorption edges, the ratio $\mu_c^A/\mu_c$ is effectively constant. If there are edges between $E_c$ and $E_0$, the calculation must be carried out in steps, as described later.

The absorption factor for fluorescence excited by monochromatic X-rays of energy $E$ is $\ln(1+u)/u$, where $u = (\mu_A/\mu_E)\text{cosec}\psi$, $\mu_A$ and $\mu_E$ being the m.a.c.s of the sample for A radiation and radiation of energy $E$ respectively (§16.2.1). In considering continuum fluorescence we are concerned with X-rays ranging in energy from $E_c$ to $E_0$, hence an integration is required. To a reasonably close approximation, $\mu_E$ is proportional to $E^{-3}$ hence $\mu_E = \mu_K E^{-3}$, where $\mu_K$ is the m.a.c. on the high energy side of the K absorption edge of A. Thus, equation (16.7) becomes:

$$I_f = HC_A(\mu_c^A/\mu_c)E_c I(w, U_0), \tag{16.11}$$

where $H = 0.5[(r_A - 1)/r_A]\omega_K(A)a\bar{Z}$, $w = (\mu_A/\mu_K)\text{cosec}\psi$, $\bar{Z}$ is the mean atomic number of the compound and:

$$I(w, U_0) = \int_1^{U_0} [(U_0 - U)/U]\{[\ln(1 + wU^3)]/wU^3\}dU.$$

For zero absorption ($w = 0$), $I(w, U_0) = (U_0 \ln U_0 - U_0 + 1)$, as in equation (16.8), hence the absorption factor, denoted by $h(w, U_0)$ is given by:

$$h(w, U_0) = I(w, U_0)/(U_0 \ln U_0 - U_0 + 1). \tag{16.12}$$

Springer (1967b) gave the following analytical solution for $I(w, U_0)$, which is equivalent to that derived in somewhat different form by Hénoc (1962):

$$I(w, U_0) = [w^{-1}U^{-3}\{(U/2) - (U_0/3)\}\ln(1 + wU^3) + U_0 \ln U \\ - (3/4)w^{-1/3}\ln(1 + w^{1/3}U) - \{(U_0/3) - (w^{-1/3}/4)\}\ln(1 + wU^3) \\ - (\sqrt{3}/2)w^{-1/3}\arctan\{(2/\sqrt{3})w^{1/3}U - 1/\sqrt{3}\}]_1^{U_0}.$$

Springer proposed the following alternative approximate expression, which is sufficiently accurate for practical purposes:

$$h(w, U_0) = \ln(1 + wU_0)/wU_0. \tag{16.13}$$

Combining equations (16.11) and (16.12), we have:

$$I_f = HC_A(\mu_c^A/\mu_c)E_c(U_0 \ln U_0 - U_0 + 1)h(w, U_0),$$

and, dividing by $I_A$ (obtained by multiplying equation (13.7) by $C_A$):

$$I_f/I_A = D_A E_c(\mu_c^A/\mu_c)h(w,U_0), \qquad (16.14)$$

in which $E_c$ is in keV and $D_A = 3.85 \times 10^{-6}[(r_A-1)/r_A]A\bar{Z}$. The value of the constant in this equation is derived from experimental measurements of continuum intensity (Reed, 1990a) and differs somewhat from that used by Springer ($4.34 \times 10^{-6}$).

The above expression enables continuum fluorescence corrections to be calculated for samples in which there are no absorption edges in the relevant energy range ($E_c$–$E_0$). The ratio $\mu_c^A/\mu_c$ is then effectively constant within this range and hence can be evaluated for any energy between $E_c$ and $E_0$, though it is convenient to substitute m.a.c. values for the high energy side of the absorption edge of A (i.e. $\mu_c^A$ and $\mu_c$).

We will now consider the more general case where absorption edges are present. Since these cause sudden changes in $\mu_c^A/\mu_c$ and $w$, the total continuum intensity must be obtained by summing separate integrals for each inter-edge region. The contribution $^n\triangle I_f$ for the region between the $n$th and $(n+1)$th edges can be obtained by integrating equation (16.11) from $E_n$ (the energy of the $n$th edge) to $E_0$ and from $E_{n+1}$ to $E_0$ and taking the difference (see fig. 16.8). Thus:

$$^n\triangle I_f = HC_A(\mu_n^A/\mu_n)[E_n I(w_n,U_n) - E_{n+1} I(w_{n+1},U_{n+1})],$$

where $U_n = E_0/E_n$, $U_{n+1} = E_0/E_{n+1}$, and $\mu_n^A$ is the m.a.c. of A on the high energy side of the $n$th edge, while $\mu_n$ is the m.a.c. of the sample for the same energy. In this expression, $w_n = \mu_A \text{cosec}\psi/\mu_n$ and $w_{n+1} = \mu_A \text{cosec}\psi/\mu'_{n+1}$, where $\mu'_{n+1}$ is the m.a.c. of the sample on the low energy side of the $(n+1)$th

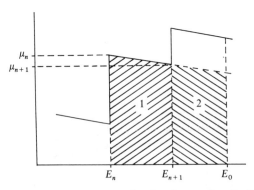

Fig. 16.8 Plot of m.a.c. versus energy, showing integration limits for continuum fluorescence calculation.

## 16.5 Continuum fluorescence

edge (see fig. 16.8). Now, substituting for $I(w,U)$ from equation (16.12):

$$^n\triangle I_f = HC_A(\mu_n^A/\mu_n)[E_n(U_n\ln U_n - U_n + 1)h(w_n,U_n) \\ - E_{n+1}(U_{n+1}\ln U_{n+1} - U_{n+1} + 1)h(w_{n+1},U_{n+1})].$$

The total continuum fluorescence intensity is obtained by summation over all the inter-edge energy band between $E_c$ and $E_0$. Dividing by $I_A$ as before, we thus obtain:

$$I_f/I_A = D_A(U_0\ln U_0 - U_0 + 1)^{-1}(\mu_n^A/\mu_n)[E_n(U_n\ln U_n - U_n + 1)h(w_n,U_n) \\ - E_{n+1}(U_{n+1}\ln U_{n+1} - U_{n+1} + 1)h(w_{n+1},U_{n+1})]. \quad (16.15)$$

For L radiation, $(r-1)/r$ becomes $(r_{L3} - 1)/r_{L3}r_{L2}r_{L1}$, and the energy used for calculating $U$ is the $L_3$ critical excitation energy. The m.a.c. on the high energy side of the $L_1$ edge is substituted for $\mu_c$.

The derivation above neglects the finite depth of production of continuum radiation. However, as shown by Green (1962, 1963b), this has an approximately equal effect on both $I_f$ and $I_A$ and can therefore be neglected to a reasonable approximation.

Heinrich (1987a) proposed an alternative approach whereby the relevant energy range is divided into fixed steps, for each of which the fluorescence contribution is calculated as for characteristic fluorescence (the exciting radiation being assumed to be monochromatic). The shape of the continuum is taken into account by means of weighting factors which may be derived from the Kramers equation or any other preferred expression. It is not necessary to assume the cube law dependence of $\mu$ on $E$, and absorption edges are automatically taken into account. The use of finite integration steps entails a degree of approximation: however, with ten or more steps the error is negligible.

### 16.5.3 Continuum fluorescence corrections in practice

Continuum fluorescence corrections may be calculated from equation (16.15), which simplifies to equation (16.14) when there are no absorption edges between $E_c$ and $E_0$. The absorption factor $h(w,U)$) may be calculated from the simplified expression given above (equation (16.13)). Values of the m.a.c. on the high energy side of the K absorption edge are given in table 16.1.

The correction may be applied by means of a factor $F_f$ analogous to that used for characteristic fluorescence, given by: $F_f = 1/(1 + I_f/I_A)$. When both kinds of fluorescence occur, the two values of $I_f/I_A$ are added. As noted by Springer (1967b), the size of the correction is strongly influenced by

Table 16.1. M.a.c.s for high energy side of K absorption edge (derived from data given by Heinrich, 1987b).

| Absorber | | Emitter | | | | | | | | | | | | | | | | | | | | | | |
|---|---|---|---|---|---|---|---|---|---|---|---|---|---|---|---|---|---|---|---|---|---|---|---|---|
| | | Na | Mg | Al | Si | P | S | Cl | K | Ca | Sc | Ti | V | Cr | Mn | Fe | Co | Ni | Cu | Zn |
| 6 | C | 1773 | 1025 | 611 | 377 | 239 | 156 | 105 | 49 | 35 | 25 | 18 | 14 | 10 | 8 | 6 | 4 | 3 | 3 | 2 |
| 7 | N | 2689 | 1584 | 960 | 600 | 385 | 253 | 171 | 81 | 58 | 42 | 31 | 23 | 17 | 13 | 10 | 8 | 6 | 5 | 4 |
| 8 | O | 3776 | 2267 | 1393 | 881 | 572 | 380 | 258 | 124 | 89 | 64 | 47 | 35 | 27 | 20 | 16 | 12 | 10 | 8 | 6 |
| 9 | F | 4735 | 2880 | 1798 | 1152 | 756 | 507 | 347 | 169 | 121 | 88 | 65 | 49 | 37 | 28 | 22 | 17 | 13 | 11 | 8 |
| 11 | Na | 7409 | 4618 | 2953 | 1936 | 1297 | 886 | 617 | 309 | 223 | 164 | 122 | 92 | 70 | 54 | 42 | 33 | 26 | 21 | 17 |
| 12 | Mg | 746 | 5726 | 3698 | 2448 | 1655 | 1141 | 800 | 406 | 295 | 218 | 163 | 123 | 94 | 73 | 57 | 45 | 35 | 28 | 23 |
| 13 | Al | 968 | 569 | 4276 | 2855 | 1947 | 1352 | 956 | 491 | 359 | 267 | 200 | 152 | 117 | 90 | 71 | 56 | 44 | 35 | 28 |
| 14 | Si | 1296 | 765 | 468 | 3455 | 2374 | 1661 | 1182 | 615 | 453 | 338 | 255 | 194 | 149 | 116 | 91 | 72 | 57 | 46 | 37 |
| 15 | P | 1594 | 944 | 579 | 368 | 2671 | 1881 | 1347 | 709 | 525 | 393 | 298 | 228 | 176 | 137 | 108 | 85 | 68 | 55 | 44 |
| 16 | S | 2038 | 1213 | 746 | 474 | 311 | 2223 | 1602 | 853 | 634 | 478 | 363 | 279 | 216 | 169 | 133 | 106 | 84 | 68 | 55 |
| 17 | Cl | 2388 | 1428 | 881 | 562 | 369 | 249 | 1752 | 943 | 704 | 533 | 407 | 314 | 244 | 191 | 151 | 120 | 96 | 78 | 63 |
| 19 | K | 3438 | 2076 | 1290 | 827 | 545 | 369 | 256 | 1219 | 919 | 701 | 540 | 419 | 328 | 258 | 205 | 164 | 132 | 107 | 87 |
| 20 | Ca | 4128 | 2505 | 1563 | 1005 | 664 | 450 | 312 | 158 | 1055 | 808 | 624 | 486 | 381 | 302 | 240 | 193 | 155 | 126 | 103 |
| 21 | Sc | 4466 | 2724 | 1707 | 1101 | 729 | 495 | 344 | 174 | 127 | 841 | 652 | 510 | 401 | 318 | 254 | 204 | 165 | 134 | 109 |
| 22 | Ti | 5024 | 3080 | 1939 | 1254 | 833 | 566 | 394 | 199 | 146 | 109 | 711 | 557 | 440 | 350 | 280 | 225 | 183 | 149 | 122 |
| 23 | V | 5599 | 3450 | 2181 | 1416 | 942 | 642 | 447 | 227 | 166 | 124 | 93 | 605 | 479 | 382 | 306 | 247 | 201 | 164 | 134 |
| 24 | Cr | 6433 | 3984 | 2529 | 1648 | 1100 | 751 | 524 | 266 | 195 | 145 | 110 | 84 | 539 | 430 | 346 | 280 | 228 | 186 | 153 |
| 25 | Mn | 7074 | 4402 | 2807 | 1836 | 1229 | 841 | 588 | 299 | 219 | 163 | 124 | 95 | 73 | 466 | 375 | 304 | 248 | 203 | 167 |
| 26 | Fe | 8016 | 5012 | 3210 | 2107 | 1415 | 970 | 680 | 347 | 254 | 190 | 143 | 110 | 85 | 67 | 421 | 342 | 279 | 229 | 188 |
| 27 | Co | 8682 | 5454 | 3508 | 2311 | 1557 | 1071 | 751 | 384 | 282 | 210 | 159 | 122 | 95 | 74 | 59 | 367 | 300 | 247 | 204 |
| 28 | Ni | 9887 | 6239 | 4030 | 2665 | 1801 | 1242 | 873 | 448 | 329 | 246 | 186 | 143 | 111 | 87 | 68 | 55 | 341 | 281 | 232 |
| 29 | Cu | 8832 | 6525 | 4232 | 2809 | 1904 | 1317 | 928 | 478 | 351 | 262 | 199 | 152 | 118 | 93 | 73 | 58 | 47 | 292 | 242 |
| 30 | Zn | 9616 | 7134 | 4645 | 3095 | 2105 | 1460 | 1031 | 532 | 392 | 293 | 222 | 170 | 132 | 104 | 82 | 65 | 53 | 43 | 264 |

## 16.6 Fluorescence near phase boundaries

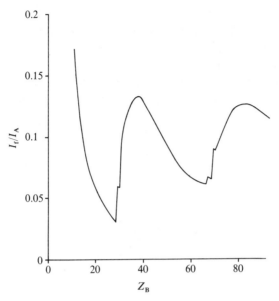

Fig. 16.9 Relative continuum fluorescence intensity ($I_f/I_A$) calculated for trace of Zn ($Z=30$) in element B ($E_0 = 30$ keV, $\psi = 75°$) (after Springer (1972)).

absorption effects. Fig. 16.9 shows the relative size of the continuum fluorescence correction contribution for a trace element (A) as a function of the atomic number $Z_B$ of the matrix. For $Z_B < Z_A$ the continuum fluorescence intensity increases rapidly with decreasing $Z_B$, because the ratio $\mu_c^A/\mu_c^B$ increases more rapidly (owing to the strong dependence of $\mu_c^B$ on $Z_B$) than the continuum intensity (which is proportional to $Z_B$) decreases. Thus, the effect of absorption predominates.

Although the continuum fluorescence effect can evidently be quite significant, the correction required is always smaller because the effect exists in both specimen and standard. Calculations for a wide range of compositions typically show corrections of up to about 5% (relative), with about 20% requiring corrections exceeding 1% (Springer and Rosner, 1969). An example of a continuum fluorescence correction calculation is given in §17.2.3.

## 16.6 Fluorescence near phase boundaries

The fluorescence correction in its usual form is calculated on the assumption that the volume within which fluorescence is excited is homogeneous and of the same composition as the analysed point. Errors may be significant when there are inhomogeneities on a scale comparable with the distance penetrated by the exciting radiation (e.g. of the order of 10 μm). These are largest when primary (electron-excited) X-ray production takes place entirely within a small inclusion of phase 1, with nearly all of the fluorescence excitation occurring in the surrounding phase (2), especially when the concentration of element A is low in phase 1 and high in phase 2 (fig. 16.10).

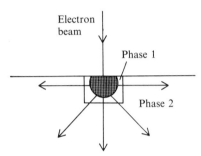

Fig. 16.10 Analysis of small inclusion of phase 1, with fluorescence excited in surrounding phase (2).

The characteristic fluorescence correction as derived in §16.2 may readily be modified for the limiting case where all primary X-ray production occurs in phase 1 and all fluorescence in phase 2. In equation (16.1), $C_A$ is replaced by $C_{A2}$ (the concentration of A in phase 2) and $I_B$ now represents the K(B) intensity excited by electrons in phase 1. In equation (16.2) the intensity ratio $I_B/I_A$ refers to a phase containing concentrations $C_A$ and $C_B$ of A and B. It is more useful to calculate the ratio of the fluorescence intensity to $I_{A0}$, the intensity obtained from pure A, hence we put $C_A = 1$ and $C_B$ becomes $C_{B1}$. Equation (16.4) may thus be used to calculate $I_f/I_{A0}$, with the following substitutions: $C_B \to C_{B1} C_{A2}$, $\mu_B \to \mu_{B2}$ (the m.a.c. of phase 2 for B radiation). The appropriate values of $u$ and $v$ are those for phase 2.

The above approach can be used to estimate the maximum boundary fluorescence effect in particular cases. The largest effects occur when phases 1 and 2 are both pure elements and are close in atomic number. For example, if A is Fe and B is Ni, the calculated value of $I_f/I_{A0}$ is 0.13, i.e. there appears to be 13% Fe in a small inclusion of pure Ni embedded in pure Fe. This indicates the possibility of quite large errors due to uncorrected boundary fluorescence.

The comparable expression for continuum fluorescence can be obtained

## 16.6 Fluorescence near phase boundaries

by suitable adaptation of the equations derived in §16.5.2, leading to the expression:

$$I_f/I_{A0} = D_A E_c C_{A2}(\mu_c^A/\mu_{c2})\ln(1+w_2 U_0)/w_2,$$

where $w_2 = (\mu_{A2}/\mu_{c2})\mathrm{cosec}\psi$ and $D_A$ is as defined in §16.5.2, with $\bar{Z} = \bar{Z}_1$. For example, if phase 1 is pure Fe and phase 2 pure Ni (the inverse of the case considered above), then when the beam is on the Fe inclusion, continuum fluorescence excitation of the adjacent Ni takes place and the calculated value of $I_f/I_{A0}$ is 0.033. The effect is thus considerably smaller than for characteristic fluorescence of Fe by Ni, but is not negligible.

Another geometrical configuration that is amenable to calculation is a semi-infinite vertical interface between two phases (fig. 16.11). Neglecting the finite size of the primary X-ray source, the methods of calculating characteristic and continuum fluorescence for small inclusions described above can be adapted simply by applying a factor of 0.5 (assuming the probe to be located in phase 1 adjacent to the interface), which represents the reduction in the solid angle subtended by the excited phase.

The rigorous calculation of the dependence of the fluorescence intensity on the distance $x$ of the probe from the interface involves numerical integrations (Maurice, Seguin and Hénoc, 1966; Hénoc, Heinrich and Zemskoff, 1969; Bastin, van Loo, Vosters and Vrolijk, 1983). However, a reasonably close approximation is given by a simple exponential (Reed, 1964), thus: $I_f(x) = I_f(0)\exp(-\tau x)$, where $I_f(0)$ is the fluorescence intensity for $x=0$, which may be calculated as above. For characteristic excitation, $\tau = 2.4\mu_{B1}\rho_1$. For example, when phase 1 is pure Ni, $\tau = 1250\,\mathrm{cm}^{-1}$, hence $I_f(x)$ falls to 10% of $I_f(0)$ at $x = 18\,\mu\mathrm{m}$. The corresponding expression for continuum excitation is $\tau = 2.4\mu_{c1}\rho_1/U_0$, which gives a value of $5910\,\mathrm{cm}^{-1}$ for pure Fe, and for continuum excitation of Ni by Fe, the fluorescence

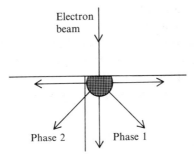

Fig. 16.11 Analysis of phase 1 close to vertical interface with phase 2.

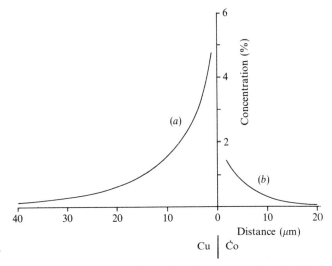

Fig. 16.12 Experimental plots of boundary fluorescence for Cu–Co couple (after Bastin et al., (1983)): (a) apparent Co concentration with beam on Cu (caused by combined characteristic and continuum fluorescence); (b) apparent Cu concentration with beam on Co (caused by continuum fluorescence only).

intensity falls to 10% in 4 μm. The distance over which continuum fluorescence is significant is thus considerably smaller than in the case of characteristic fluorescence.

Measurements of fluorescence effects in the vicinity of artificial interfaces (Reed and Long, 1963; Maurice et al., 1966; Hénoc et al., 1969; Bastin et al., 1983) agree quite well with the calculations (see fig. 16.12).

Fluorescence effects for other geometries have been considered by various authors. Cox, Love and Scott (1979) derived an analytical expression applicable to the case of a surface layer differing in composition from the underlying substrate, while Bastin et al. (1983) and Armstrong and Buseck (1985) demonstrated the application of numerical integration to such samples. Bastin et al. also treated the case of a hemispherical inclusion of one phase embedded in another, while Armstrong and Buseck considered isolated thin films and spherical particles. Nockolds, Nasir, Cliff and Lorimer (1980) derived an analytical expression for films sufficiently thin for the assumption of uniform primary X-ray production as a function of depth to be made.

### *16.6.1 Fluorescence uncertainty*

In the circumstances described above, the operator would probably be aware of the presence of phase 2, but in opaque specimens it is possible to have an interface or particle below the surface giving rise to analytical errors, of which he is unaware. The term 'fluorescence uncertainty' describes this potential source of error.

Fluorescence uncertainty and errors due to inhomogeneity, and inadequacies in the correction even for homogeneous samples, decrease with decreasing X-ray take-off angle $\psi$ owing to the absorption of the emerging fluorescence radiation. This may be used as an argument in favour of a low take-off angle, but if fluorescence is to be reduced significantly, $\psi$ must be less than 20°, which results in absorption corrections becoming unduly large. Hence, on balance, a high take-off angle is preferable.

# 17
# Matrix corrections in practice

## 17.1 Introduction

The uncorrected concentration ($C'$) of each element is given by:

$$C' = C_0(I/I_0),$$

where $I$ and $I_0$ are the characteristic X-ray intensities emitted by specimen and standard respectively and $C_0$ is the concentration of the element concerned in the standard. The intensities are assumed to have been corrected for dead time, background, etc. Matrix corrections may be represented by the factors $F$ for specimen and $F_0$ for standard, thus:

$$C = C'(F/F_0),$$

where $C$ is the true concentration. These factors are each the product of separate factors – $F_a$ (absorption), $F_f$ (fluorescence), $F_b$ (backscattering) and $F_s$ (stopping power), which may be calculated by methods described in chapters 15, 16, 14 and 13 respectively. An example of such a calculation is given in §17.2. The above factors are dependent on the composition, which is not known until the true values of the correction factors have been determined. This difficulty is overcome by iteration (§17.3).

Matrix corrections are, of course, calculated by computer in practice and the user is often not fully aware of the processes involved in producing the final result. In the following sections a complete correction calculation is followed step by step, for illustrative purposes. Computer programs for matrix corrections are discussed in §17.5.

Instead of applying the rather complicated correction procedures described in chapters 13–16, it is possible to use a much simpler approximate approach whereby the corrections are combined in a single parameter – the 'alpha coefficient' – representing the effect which each element has on the X-ray emission of every other element (§17.4). This is

## 17.2 Example of correction calculation

In order to illustrate the calculation of corrections, we now consider a simple example, namely an alloy containing 75% Fe and 25% Ni, analysed for Fe and Ni at an accelerating voltage of 25 kV, with an X-ray take-off angle ($\psi$) of 40°, using pure elements as standards. It is assumed for the purpose of this exercise that the true composition is known at the start, thereby avoiding the need for iteration.

### 17.2.1 Absorption

Mass attenuation coefficients for pure Fe and Ni can be obtained from table 15.2 and values for the alloy are calculated using the mass concentrations:

| Absorber | Emitter | |
|---|---|---|
| | Fe | Ni |
| Fe | 71 | 370 |
| Ni | 92 | 60 |
| Alloy | 76 | 292 |

From these m.a.c. values, $\chi$ is obtained by multiplying by cosec $\psi$ (1.556). The parameter $h$ in the Philibert absorption correction (§15.5) is equal to $1.2A/Z^2$, and the mean value for the alloy is calculated using mass concentrations. For substitution in equation (15.5), $h/(1+h)$ is required. The relevant values are as follows:

| Sample | $h$ | $h/(1+h)$ |
|---|---|---|
| Fe | 0.099 | 0.090 |
| Ni | 0.090 | 0.083 |
| Alloy | 0.097 | 0.088 |

The parameter $\sigma$ is a function of $E_0$ and $E_c$, and can be calculated from equation (15.7):

| Emitter | $E_c$ (keV) | $\sigma$ |
|---|---|---|
| Fe | 7.11 | 2540 |
| Ni | 8.33 | 2653 |

264     17 Matrix corrections in practice

Now equation (15.5) may be used to obtain $F_a$ values:

| Element | Sample | $F_a$ |
|---|---|---|
| Fe | Fe | $[1+(110/2540)][1+0.090(110/2540)] = 1.047$ |
|  | Alloy | $[1+(118/2540)][1+0.088(118/2540)] = 1.051$ |
| Ni | Ni | $[1+(93/2653)][1+0.083(93/2653)] = 1.038$ |
|  | Alloy | $[1+(454/2653)][1+0.088(454/2653)] = 1.189$ |

The correction factors for Fe and Ni are thus 1.004 (1.051/1.047) and 1.145 (1.189/1.038) respectively. The absorption of Fe K$\alpha$ radiation is slight and nearly equal in standard and alloy – hence the correction is small. In the case of Ni, absorption is significant in the alloy (owing to the Fe K absorption edge) but not in the standard, and the correction is therefore relatively large.

### 17.2.2 Characteristic fluorescence

From consideration of the relevant energies, it is apparent that Fe K$\alpha$ radiation is excited by Ni K radiation, but not vice-versa: hence a characteristic fluorescence correction is required only for Fe in the alloy. The calculation follows the procedure given in §16.2, where A is Fe and B is Ni. The data required are as follows:

$(r_A - 1)/r_A = 0.887$ (from equation (16.5)),
$\omega_K(B) = 0.406$ (from table A.6),
$\mu_B^A = 370$, $\mu_A = 76$, $\mu_B = 292$ (see previous section),
$U_A \ln U_A - U_A + 1 = 1.904$, $U_B \ln U_B - U_B + 1 = 1.297$,
$\sigma = 2540$ (see previous section).

For the absorption term (§16.2.1), we require the parameters $u$ and $v$:

$u = \mu_A \mathrm{cosec}\psi / \mu_B = 76 \times 1.556/292 = 0.405$
$v = \sigma/\mu_B = 2540/292 = 8.70$.

Hence:

$$[\ln(1+u)]/u = 0.840, \quad [\ln(1+v)]/v = 0.261.$$

From equation (16.4):

$$I_f/I_A = 0.5 \times 0.25 \times (370/292) \times 0.887 \times 0.406 \times (55.9/58.7)$$
$$\times (1.297/1.904) \times (0.840 + 0.261) = 0.041.$$

### 17.2.3 Continuum fluorescence

The calculation follows the procedure given in §16.5.2. Only a single-stage calculation is necessary for the pure elements and for Ni in the alloy, since there are no absorption edges between $E_c$ and $E_0$. In the case of Fe in the alloy, however, the calculation must be carried out in two stages, covering the energy ranges: (1) from 7.11 keV ($E_c$ for Fe) to 8.33 keV ($E_c$ for Ni), and (2) from 8.33 keV to 25 keV. For calculating $D_A$ in equations (16.14) and (16.15) the following data are required:

|    | Z  | A    | $(r-1)/r$ |
|----|----|------|-----------|
| Fe | 26 | 55.9 | 0.887     |
| Ni | 28 | 58.7 | 0.884     |

The following values of $D_A$ are then obtained:

| Sample | Line  | $D_A$ |
|--------|-------|-------|
| Fe     | Fe Kα | $3.85 \times 10^{-6} \times 0.887 \times 55.9 \times 26 = 4.96 \times 10^{-3}$ |
| Ni     | Ni Kα | $3.85 \times 10^{-6} \times 0.884 \times 58.7 \times 28 = 5.59 \times 10^{-3}$ |
| Alloy  | Fe Kα | $3.85 \times 10^{-6} \times 0.887 \times 55.9 \times 26.5 = 5.06 \times 10^{-3}$ |
| Alloy  | Ni Kα | $3.85 \times 10^{-6} \times 0.884 \times 58.7 \times 26.5 = 5.29 \times 10^{-3}$ |

The m.a.c.s required for calculating the absorption factors are shown in fig. 17.1. The absorption factor $h(w, U_0)$ is now calculated for each case, using equation (16.13). Three separate values are required for Fe in the alloy, for different energy ranges (see below).

| Sample | Line  | $w$ | | $U_0$ | $h(w,U_0)$ |
|--------|-------|-----|---|-------|------------|
| Fe     | Fe Kα | $71 \times 1.556/421 = 0.262$ | | 3.52 | 0.708 |
| Ni     | Ni Kα | $60 \times 1.556/341 = 0.274$ | | 3.00 | 0.730 |
| Alloy  | Fe Kα | (1) $76 \times 1.556/334 = 0.354$ | | 3.52 | 0.650 |
|        |       | (2) $76 \times 1.556/222 = 0.533$ | | 3.00 | 0.597 |
|        |       | (3) $76 \times 1.556/294 = 0.402$ | | 3.00 | 0.656 |
| Alloy  | Ni Kα | $292 \times 1.556/294 = 1.545$ | | 3.00 | 0.373 |

Equation (16.14) can be used to obtain $I_f/I_A$ for the cases where only a single-stage calculation is required:

| Sample | Line  | $I_f/I_A$ |
|--------|-------|-----------|
| Fe     | Fe Kα | $4.96 \times 10^{-3} \times 7.11 \times (421/421) \times 0.708 = 0.025$ |
| Ni     | Ni Kα | $5.59 \times 10^{-3} \times 8.33 \times (341/341) \times 0.730 = 0.034$ |
| Alloy  | Ni Kα | $5.29 \times 10^{-3} \times 8.33 \times (341/294) \times 0.373 = 0.019$ |

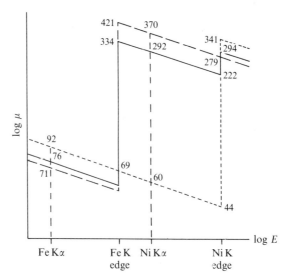

Fig. 17.1 Log–log plot of m.a.c. ($\mu$) for pure Fe (long dashes), pure Ni (short dashes), and Fe–Ni alloy (solid line), as function of X-ray energy, with values used in continuum fluorescence calculation.

For Fe K$\alpha$ in the alloy, we require values of $U_0 \ln U_0 - U_0 + 1$, which are 1.904 and 1.297 for Fe and Ni respectively. Equation (16.15) is used to calculate the contribution of the continuum between 7.11 and 8.33 keV:

$$(I_f/I_A)_1 = 5.06 \times 10^{-3} \times (1/1.904) \times (421/334)$$
$$\times [(7.11 \times 1.904 \times 0.650) - (8.33 \times 1.297 \times 0.597)] = 0.008$$

The second stage covers energies from 8.33 to 25 keV and in this case the second term in the bracket is zero:

$$(I_f/I_A)_2 = 5.06 \times 10^{-3} \times (1/1.904) \times (279/294)$$
$$\times 8.33 \times 1.297 \times 0.656 = 0.018.$$

Adding both contributions we have: $I_f/I_A = 0.026$.

In this case only, characteristic fluorescence is also excited, for which $I_f/I_A = 0.041$ (see previous section). Adding this to the continuum contribution gives $I_f/I_A = 0.067$.

The fluorescence factors, given by $F_f = 1/[1 + (I_f/I_A)]$, are as follows:

| Element | Sample | $F_f$ |
|---|---|---|
| Fe | Fe | 0.976 |
|    | Alloy | 0.937 |
| Ni | Ni | 0.968 |
|    | Alloy | 0.981 |

## 17.2 Example of correction calculation

The continuum fluorescence effect is very similar in the alloy and the pure element standards, hence the only important correction is that for characteristic fluorescence for Fe in the alloy.

### 17.2.4 Backscattering

Values of $R$, the factor by which the X-ray intensity is reduced due to electron backscattering, are obtained from the polynomial in §14.9.2. For the alloy, mass concentration averaging is used. The factor $F_b$ is equal to $1/R$. The values obtained are as follows:

| Element | Sample | $R$ | $F_b$ |
|---------|--------|-------|-------|
| Fe      | Fe     | 0.866 | 1.155 |
|         | Ni     | 0.854 |       |
|         | Alloy  | 0.863 | 1.159 |
| Ni      | Fe     | 0.874 |       |
|         | Ni     | 0.864 | 1.157 |
|         | Alloy  | 0.871 | 1.148 |

The corrections are small (less than 1%) because of the small difference in atomic number between Fe and Ni.

### 17.2.5 Stopping power

The stopping power correction is calculated from the Bethe expression (equation (13.16)), with $J$ obtained from equation (13.17) and $\bar{E}$ from equation (13.15). The data required are as follows:

| Line  | Element | $Z/A$ | $J$ | $\bar{E}$ | $S$   |
|-------|---------|-------|-----|-----------|-------|
| Fe Kα | Fe      | 0.466 | 332 | 16055     | 1.879 |
|       | Ni      | 0.477 | 362 | 16055     | 1.882 |
| Ni Kα | Fe      | 0.466 | 332 | 16665     | 1.896 |
|       | Ni      | 0.477 | 362 | 16665     | 1.900 |

The stopping power factor $F_s$ is equal to $S$ and for the alloy is calculated from the pure element values, using mass concentration averaging:

| Element | Sample | $F_s$ |
|---------|--------|-------|
| Fe      | Fe     | 1.879 |
|         | Alloy  | 1.880 |
| Ni      | Ni     | 1.900 |
|         | Alloy  | 1.897 |

The stopping power corrections are very small because of the small difference in atomic number between Fe and Ni.

### 17.2.6 Total correction factors

The combined factor $F$ is the product of the individual factors derived above, and the total correction factor (that by which the uncorrected concentration $C'$ must be multiplied to give the corrected concentration $C$) is equal to $F/F_0$, where $F$ refers to the alloy and $F_0$ to the standard. The values obtained are as follows:

| Element | Sample | $F$ | $F/F_0$ |
|---|---|---|---|
| Fe | Fe | 2.218 | |
| | Alloy | 2.146 | 0.968 |
| Ni | Ni | 2.209 | |
| | Alloy | 2.540 | 1.150 |

For Fe the main correction is fluorescence, which makes the uncorrected concentration too high. In the case of Ni, absorption is predominant and its effect is opposite, the uncorrected concentration being too low.

## 17.3 Iteration

As already observed, matrix correction factors are dependent on the composition of the specimen, which is not known initially. The usual approach to this problem is to apply an iterative procedure, whereby estimated concentrations are used in the correction factor calculations and, having applied the corrections thus obtained, the calculations are repeated. As many iterations are executed as necessary to achieve convergence, when the concentrations do not change significantly between successive calculations. Various forms of iteration can be used, as described below. The main motive for using the more sophisticated procedures is to ensure satisfactory convergence even in extreme cases.

### 17.3.1 Simple iteration

In 'simple iteration' the normalised corrected concentrations obtained in the first cycle of correction calculations are used for recalculating the correction factors in the second cycle, and so on. In the first calculation, 'raw' uncorrected concentrations are used. Usually convergence is quite rapid, but if the corrections are large, the corrected concentrations obtained in successive cycles may oscillate and fail to converge, or only do so in an excessively large number of cycles. The only correction that is ever large enough to cause such difficulties is that for absorption. The worst case is when a strongly absorbing element is calculated by difference. Thus, the uncorrected concentration of an element with an emission line strongly absorbed by an element estimated by difference (e.g. Mg K$\alpha$ absorbed by O) is much lower than the true concentration, and in the first cycle the correction is overestimated due to the high assumed concentration of the absorbing element. This, in turn, leads to underestimation of the correction in the next cycle, and so on. Large and possibly non-convergent oscillations then occur. In such circumstances other iteration procedures are preferable (see below). Non-convergence is less likely for complete analyses or those in which residual elements are estimated stoichiometrically, since normalisation then tends to assist convergence.

### 17.3.2 Hyperbolic iteration

Criss and Birks (1966) proposed an iterative procedure based on the 'alpha coefficient' approximation (§17.4), the calibration curve of $C'$ versus $C$ for a binary compound being represented by a hyperbolic expression (equation (17.4)). This formula can be used simply as an aid to convergence, with no constraint placed on the method used to derive the relationship between X-ray intensity and composition.

For each element the value of $C'$ that would be obtained for a given assumed composition is calculated from first principles and $\alpha$ is deduced from equation (17.5), which is then used to calculate $C$ from the observed value of $C'$. This procedure is repeated using a new assumed composition derived from the latest $C$ values, and in this way progressively improved $\alpha$ values are obtained.

For binary compounds convergence is very rapid, since $\alpha$, as determined from the usual correction procedures, is almost independent of the composition. Ternary and high-order systems are treated as pseudo-binary, i.e. all elements other than the one under consideration are lumped together. However, the relative proportions of these elements may change from one cycle to the next, causing $\alpha$ to change, possibly substantially. Therefore this form of iteration will not necessarily perform well for multi-element analyses with large corrections. Performance is improved if normalisation is carried out before each iteration (Heinrich, 1972; Springer, 1976). This also helps to avoid difficulties which otherwise arise when absorption is less in the specimen than in the standard (Pouchou and Pichoir, 1987).

### 17.3.3 Wegstein iteration

Of various general methods for accelerating convergence, one which has proved successful in the present context is that of Wegstein, in which the following expression is used to derive the concentration $\bar{C}_{n+1}$ for substitution in the correction formula in the $(n+1)$th iteration:

$$\bar{C}_{n+1} = C_n - [(C_n - C_{n-1})(C_n - \bar{C}_n)]/(C_n - C_{n-1} - \bar{C}_n + \bar{C}_{n-1}), \quad (17.1)$$

where $C_{n-1}$ and $C_n$ are the corrected concentrations obtained in the $(n-1)$th and $n$th iterations respectively, and $\bar{C}_{n-1}$ and $\bar{C}_n$ are the concentrations substituted in the correction formula in the same iterations. Two cycles of simple iteration are completed before using equation (17.1).

This method is significantly better than simple iteration for analyses with large corrections (Reed and Mason, 1967; Beaman and Isasi, 1970; Springer, 1976). Non-convergence is practically eliminated and the number of iterations is usually reduced, though this advantage is offset by the greater complexity of the calculations.

## 17.4 Alpha coefficients

The approximation:

$$C'_A = \alpha_A C_A/(\alpha_A C_A + \alpha_B C_B) \quad (17.2)$$

was proposed by Castaing (1951) for atomic number corrections in binary compounds and was further developed by Poole and Thomas (1963, 1966). Originally, $C'_A$ in equation (17.2) represented the apparent concentration as derived from the ratio of intensities *generated in the sample* (absorption being treated separately).

For a binary system, $C_B = 1 - C_A$, hence equation (17.2) may be written:

$$C_A/C'_A = \alpha_{AB} + (1 - \alpha_{AB})C_A, \quad (17.3)$$

where $\alpha_{AB} = \alpha_B/\alpha_A$, or alternatively:

$$(1 - C'_A)/C'_A = \alpha_{AB}(1 - C_A)/C_A. \quad (17.4)$$

From inspection of equation (17.3), it is apparent that $\alpha_{AB}$ represents the correction factor for a vanishingly small concentration of A in B. For $\alpha_{AB} < 1$: $C_A < C'_A$, whereas if $\alpha_{AB} > 1$: $C_A > C'_A$ (see fig. 17.2). The latter is typical of absorption corrections, where absorption is usually greater in the

## 17.4 Alpha coefficients

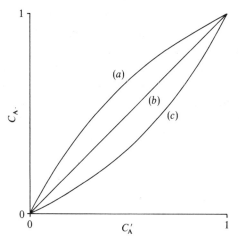

Fig. 17.2 Apparent concentration of element A ($C'_A$) versus true concentration ($C_A$) in binary A–B alloy: (a) $\alpha > 1$, (b) $\alpha = 1$, (c) $\alpha < 1$.

sample than in the standard, and the uncorrected concentration is lower than the true concentration.

Ziebold and Ogilvie (1964) noted that experimental data for binary alloys give linear plots of $C_A/C'_A$ versus $C_A$, thereby justifying the use of alpha coefficients to represent the complete correction, and further showed this to be consistent with the properties of the Philibert absorption correction (Ziebold and Ogilvie, 1966). The characteristic fluorescence correction behaves non-linearly (fig. 16.5), but the absolute errors involved in using alpha coefficients to include fluorescence effects are usually not large.

For a ternary compound containing elements A, B and C, it is reasonable to assume that the effective alpha coefficient can be obtained by linear combination of binary coefficients, thus:

$$\alpha_{ABC} = (C_B \alpha_{AB} + C_C \alpha_{AC})/(C_B + C_C). \tag{17.5}$$

Experimental results for Cu–Ag–Au alloys were adduced by Ziebold and Ogilvie (1964) in support of this assumption. Generalising equation (17.5) for any number of elements, we have:

$$\alpha_A = \sum C_i \alpha_{Ai}/\sum C_i. \tag{17.6}$$

This approach has the advantage that corrections for multi-component compounds can be derived from empirical data for binary standards, which are easier to obtain.

Bence and Albee (1968) obtained alpha coefficients for geological materials consisting of mixed oxides (e.g. silicates) from measurements on samples of known composition. Gaps in the experimental data were filled by using calculated factors. In this procedure, the concentrations are expressed as the percentage by weight of the oxide. These alpha coefficients are valid only for an X-ray take-off angle of 52.5° and an accelerating voltage of 15 kV. Albee and Ray (1970) used ZAF procedures to extend the range of oxides covered. Laguitton *et al.* (1975) calculated alpha coefficients for various ternary alloy systems. Armstrong (1988) calculated alpha coefficients for $\psi = 40°$, using various methods and fitting the results to second-order polynomials, thereby allowing for variation in the coefficients with composition. This considerably improves the agreement between calculated corrections and experimental data.

The enormous increase in available computing power has reduced the usefulness of the alpha coefficient method, given that corrections can now be calculated by the more rigorous methods described in previous chapters quickly enough for most purposes, except possibly for mapping, where the counting time per point is small.

## 17.5 Correction programs

Desirable features in a computer program for calculating corrections include flexibility and convenience for the user, though to some extent these are incompatible, since the availability of a wide range of options as regard operating conditions, type of analysis, etc., requires the user to make a lot of choices. Practical correction programs represent a compromise between flexibility and convenience, and in some cases are tailored to specific requirements, in order to limit the number of options.

A program of general applicability must allow for all elements, with a choice of K, L or M lines, and have the capacity to handle a large number of elements per analysis. Allowance must be made for standards being either pure elements or compounds. It should be possible to calculate corrections for any accelerating voltage (within a reasonable range, e.g. 5–50 kV) and X-ray take-off angle. Since the latter is usually fixed for a given instrument, its value can be entered initially and thereafter remain constant. Where a scanning electron microscope is used for quantitative analysis, however, it is desirable to be able to allow for varying specimen tilt.

In calculating the correction factors, it should be possible to include one or more elements in the assumed composition which are not amongst those actually determined (e.g. oxygen in oxides and silicates). The concentration

## 17.5 Correction programs

of such residual elements can be estimated either on the basis of stoichiometry or by difference, i.e. by subtracting the sum of the concentrations of the other elements present from 100%.

One of the iteration methods described in §17.3 can be used, with a convergence test, so that iteration is stopped when the difference between successive results is less than a certain amount. For most purposes it is better not to normalise the results to give a sum of exactly 100%, since the value of the total is a useful (though not infallible) test of the quality of the analysis, and reveals errors such as the omission of a significant element.

It is desirable to give the error for each concentration, derived from counting statistics. Concentrations below the limit of detection (defined, for example, as 3 s.d. of the background count) can be eliminated from the final analysis. It is advantageous to have the correction factors listed, enabling the user to note cases where the corrections are severe and likely to affect the accuracy of the results. It is sometimes useful to express the analysis in forms other than elemental weight per cent. For silicates etc. it is often desirable to express the results as the weight per cent of the oxides. In some cases atomic concentrations are useful, since they facilitate the assessment of stoichiometry.

Examples of published correction programs are 'COR 2' (Hénoc, Heinrich and Myklebust, 1973), which is a rigorous program originally intended for off-line use, and 'FRAME' (Yakowitz, Myklebust and Heinrich, 1973), which uses simplified methods of calculation and is intended for on-line application. Proprietary software as provided with current commercial systems is not always fully accessible to the user but generally incorporates some combination of correction methods described in the preceding chapters.

# 18
# Light element analysis

## 18.1 Introduction

The wavelengths and energies of the K$\alpha$ lines of the 'light' elements ($Z < 10$) are given in table 18.1. Most of these lines lie beyond the wavelength range of the crystals ordinarily used in w.d. spectrometers, but synthetic multilayers of large interplanar spacing (§6.3) enable the coverage to be extended to Be ($Z = 4$). For detecting long wavelength X-rays, proportional counters can be fitted with ultra-thin entrance windows (§7.2). The same considerations apply to the L and M lines of heavier elements lying in the long wavelength region, though these are less commonly used for analysis.

Standard Si(Li) detectors used for e.d. analysis have beryllium windows which severely absorb X-rays of energy below about 1 keV and hence are unsuitable for light element analysis. However, reasonable sensitivity is achieved by fitting an ultra-thin window or having none at all (§9.4). Spectral resolution is poor compared with w.d. spectrometers, but is sufficient to resolve the K$\alpha$ lines of different light elements. Spectral interferences are more troublesome, however, because there are often L or M lines of heavier elements in the same region. In this connection the superior resolution of the w.d. spectrometer is advantageous.

It might be expected that the intensities of the K$\alpha$ lines of light elements would be low because of their low fluorescence yields. However, the overvoltage ratio $U_0$ ($= E_0/E_c$) is usually considerably higher, and, since X-ray intensity increases quite rapidly with $U_0$ (§13.3), the effect of low fluorescence yield is counteracted to a large extent. However, the detection efficiency is less than for 'normal' elements and count-rates are generally lower. Peak to background ratios are also lower, partly because of the rapid rise in the continuum intensity with decreasing energy (§1.6), resulting in higher detection limits.

## 18.1 Introduction

Table 18.1. *Data for Kα lines of light elements (after White and Johnson, (1970)).*

| Element | Z | Wavelength (Å) | Energy (eV) | Crit. exc. energy (eV) |
|---|---|---|---|---|
| Be | 4 | 114.0 | 109 | 112 |
| B  | 5 | 67.6  | 183 | 192 |
| C  | 6 | 44.7  | 277 | 284 |
| N  | 7 | 31.6  | 392 | 400 |
| O  | 8 | 23.6  | 525 | 532 |
| F  | 9 | 18.3  | 677 | 687 |

Light element analysis is subject to various difficulties which either do not exist or are less acute in 'conventional' analysis. This applies both to the experimental determination of peak intensities and to the calculation of matrix corrections. Hence, special procedures are required if quantitative light element analysis is to achieve an accuracy approaching that obtainable for heavier elements.

Light element K spectra are subject to significant chemical effects, including wavelength shifts and changes in shape (fig. 18.1). These are much

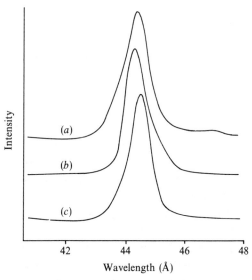

Fig. 18.1 Profiles of C Kα peak for different materials, showing chemical bonding effect: (*a*) $Fe_3C$, (*b*) SiC, (*c*) $Mo_2C$ (Bastin and Heijligers, 1986a).

more marked than for heavier elements because K$\alpha$ lines are produced by K–L transitions and for $Z<10$ the L shell contains electrons that are involved in bonding. Also the effect of a given energy shift is more apparent when the line concerned has a low energy. In order to avoid errors due to chemical effects, specimen and standard should preferably be closely similar chemically. If not, then a correction must be applied, as described in §18.4.

After peak intensity measurements from specimen and standard have been carried out, it remains to calculate the matrix corrections. The severity of the absorption of long wavelength X-rays in emerging from the sample requires that special attention be paid to the model upon which the absorption correction is based. Further, errors in m.a.c.s have a greater effect than for 'normal' wavelengths, yet are known less accurately. Empirically determined values are therefore often used in preference to data from tables.

## 18.2 Long wavelength w.d. spectrometry

Of the crystals commonly used in w.d. spectrometers, only the acid phthalates have sufficiently large spacing to be relevant in the present context. Thus, TAP ($2d=25.9$ Å) can be used for the K$\alpha$ lines of F and O, though for the latter the Bragg angle is beyond the upper limit of some spectrometers. Furthermore, compared to evaporated multilayers, the intensity obtained is typically lower by about a factor of about 10.

Multilayers, of both the soap film (§6.3.1) and evaporated (§6.3.2) types, are available with $2d$ values up to at least 150 Å. Considerably higher intensities are obtainable with the latter type, at the expense of somewhat inferior resolution. Fig. 18.2 shows the Bragg angle for the K$\alpha$ lines of light elements as a function of $2d$. Lead stearate ($2d=100$ Å) can be used for B, C, N and O, though the Bragg angle for O is below the minimum available for some spectrometers. The Be K$\alpha$ line requires a larger $2d$ value, which can be obtained either with an evaporated multilayer or a soap film with a greater chain length (see table 6.2). For O, an evaporated multilayer with $2d=60$ Å is commonly used; this also covers the F K$\alpha$ line.

Count-rates for light elements (except Be) obtainable with evaporated multilayers are typically in the region of $10^6$ counts s$^{-1}$ per $\mu$A probe current (normalised to an elemental concentration of 100% and with an accelerating voltage of 10 kV), and thus are not much lower than for 'ordinary' elements using true crystals (see table 6.3). Maximum efficiency is obtained over a limited wavelength range: in fact it is desirable to use a

18.2 Long wavelength w.d. spectrometry

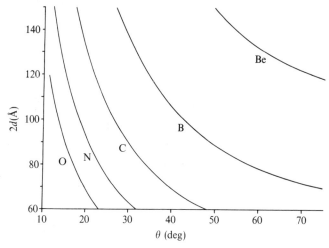

Fig. 18.2 Bragg angle ($\theta$) versus $2d$ for light element K$\alpha$ Lines.

different multilayer for each element. (The above remarks about intensity assume that the optimum choice is made.) Soap film multilayers such as lead stearate give intensities typically 5–10 times lower, but varying less with wavelength.

Usually multilayer diffracting elements are the same size as, and thus can be directly substituted for, conventional crystals. Higher intensities can be obtained by increasing the size, where the spectrometer design allows (Kawabe, Takagi, Saito and Tagata, 1988).

Compared to crystals used in the 'conventional' wavelength range, the resolution of synthetic multilayers is considerably worse (see fig. 18.3), though the K$\alpha$ lines of light elements are resolved easily. Inadequate resolution of interfering lines from heavier elements, however, sometimes requires the application of deconvolution techniques in order to obtain the true peak intensity.

Such interferences are sometimes caused by high order reflections, in which case it is desirable to use pulse height analysis (§11.4) to minimise their effect (fig. 18.4). (A useful aspect of evaporated multilayers is that they give negligible reflections of order higher than 2.) For pulse height analysis, the width of the pulse height distribution is important. The counter voltage should be no higher than necessary to obtain sufficient pulse height, since the pulse height distribution tends to broaden with increasing voltage. It is advantageous to use a low density counter gas, such as argon with up to 75% methane (instead of the usual 10%), which gives a narrower

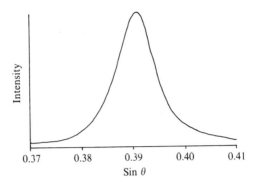

Fig. 18.3 Peak profile of O Kα recorded with evaporated multilayer ($2d = 60$ Å).

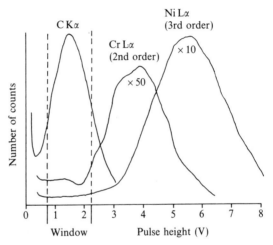

Fig. 18.4 Use of pulse height analysis to discriminate against high order reflections (after Ruste (1979)).

distribution for long wavelength X-rays (§7.3). Alternatively, the density of the gas may be reduced by decreasing its pressure, which has the additional benefit that absorption of high energy X-rays is reduced, thereby discriminating against high order reflections (Rehbach, Karduck and Burchard, 1985).

## 18.3 Low energy e.d. spectrometry

The Kα lines of light elements are completely absorbed by the beryllium window fitted to standard Si(Li) detectors, but can be detected if a thin window is substituted, or the window is removed completely (§9.4). The detection efficiency is still somewhat less than for higher energies, owing to

## 18.3 Low energy e.d. spectrometry

absorption in the layers of gold and silicon in front of the active region of the detector, but reasonable count-rates are obtainable.

The energy resolution at low energies is governed mainly by the electronic noise originating in the detector and preamplifier, rather than statistical fluctuations in the ionisation processes, which are important at higher energies (§9.5). Since the former could be reduced in principle by technical improvements, there is a prospect of enhancing the energy resolution at low energies. Such improvements have indeed occurred in recent years and enable light element K$\alpha$ peaks to be separated better from each other and from interfering peaks of heavier elements. A further benefit is that the tail of the gaussian noise distribution (which is centred on the zero of the energy scale) has less effect on B K$\alpha$ (183 eV), which can now be well resolved (fig. 18.5). The Be K$\alpha$ peak can also be resolved if the noise peak is sufficiently narrow (Lowe, 1989).

Spectrum processing techniques developed for e.d. analysis in the 'normal' energy range, described in chapter 12, can be applied to light elements. Chemical effects, which are important in w.d. analysis, are much less significant in e.d. analysis because of the greater width of the peaks. The peak to background ratios obtained for light elements are low, partly because of the inherently broad peaks in e.d. spectra, but also because of the additional contribution to the background intensity of higher-energy X-ray pulses displaced by incomplete charge collection (§9.6). Detection limits are thus considerably higher than in the case of w.d. analysis.

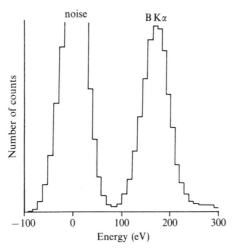

Fig. 18.5 B K$\alpha$ peak recorded with windowless Si(Li) detector (after Statham and Nashishibi (1988)).

## 18.4 Peak intensity measurements

Assuming that the peak of interest is adequately resolved from other peaks in the spectrum, the peak intensity can be measured with a w.d. spectrometer in the usual fashion, that is by setting the spectrometer to the maximum intensity and counting for a suitable time. However, the position of the peak is liable to change as a function of chemical bonding: for example, Bastin and Heijligers (1986a) observed the position of the C K$\alpha$ peak to vary from 44.34 to 44.53 Å in different carbides. In general the size of the peak shift decreases with increasing atomic number and is thus less important for O K$\alpha$. If the same wavelength setting is used for a specimen as for a standard which is chemically different, serious errors may ensue. These can be avoided by using a 'peak seek' routine to determine the correct setting for each, though this is impracticable where the peak is small.

For quantitative analysis, it is strictly the *area* of the peak that is required, since this represents the total emission. In conventional analysis, peak height is proportional to area, but not in the case of light elements, owing to variations in peak width and shape. Peak areas can be measured by stepping the spectrometer across the whole of the peak profile and summing the counts. However, this is inconvenient for routine analysis and a more practical approach is to measure the intensity at the maximum point and apply a 'profile factor' (Weisweiler, 1975b), or 'area/peak factor' (Bastin and Heijligers, 1986a) to obtain the area. This factor has to be determined experimentally in the first instance by integration, as described above, and can be assumed thereafter to remain constant (for the same compound). It should be noted that determinations of integrated peak areas are particularly susceptible to the effects of carbon contamination, owing to the long measuring time required, and anti-contamination measures should be applied (§18.7).

For metal carbides the C K$\alpha$ profile factor varies considerably: values ranging from 0.715 to 1.048 (relative to $Fe_3C$) have been reported (Brown, Schwaab and Von Rosenstiel, 1984; Bastin and Heijligers, 1986a). The values may depend somewhat on the spectrometer resolution (the above data were obtained with a lead stearate multilayer). The factor is found to be correlated with the position of the metal in the periodic table. The spread of values for O in oxides is smaller: Bastin and Heijligers (1989) found a range of 0.974–1.015 relative to $Fe_2O_3$ (using an evaporated multilayer with $2d = 60$ Å). For B in borides, a range similar to that for carbides was found by Bastin and Heijligers (1986c).

In measuring the B K$\alpha$ peak the last-named authors observed variations in peak position and shape with orientation of the sample relative to the spectrometer, caused by the polarisation of the X-rays emitted from crystalline samples, the recorded intensity being dependent on the orientation of the plane of polarisation relative to the diffracting planes. This is especially noticeable in the case of $TiB_2$, for example. The effects on peak position and profile factor are found to be correlated, hence peak intensities can be corrected by applying a factor derived from the measured peak position using a previously determined calibration plot.

## 18.5 Background corrections

In light element analysis, peak to background ratios are typically considerably lower than for 'normal' elements. Further, measuring background by offsetting the spectrometer on each side of the peak is often unreliable, owing to factors such as large steps caused by absorption edges, the greater width of the peaks, and the presence of interfering peaks of heavier elements. In some cases the background is enhanced by 'total X-ray reflection', which occurs when long wavelength X-rays are incident on the diffracting element of the spectrometer at a low angle (Rehbach and Karduck, 1990).

Such difficulties can be avoided by measuring the intensity at the wavelength of the peak concerned on pure samples of the other elements present and calculating the average for the analysed material (Ruste, 1979). An alternative approach is to measure the intensity at the peak position on a sample containing none of the light element concerned, but otherwise closely similar in composition. An example would be to use a carbon-free steel to determine the background for C K$\alpha$ in other steels.

## 18.6 Conducting coating

The conducting coating necessary for electrically non-conducting samples has effects that are more acute for light element analysis than for 'normal' elements. One factor is absorption suffered by the emerging X-rays as they pass through the coating. For example, absorption by carbon is especially severe in the case of O K$\alpha$, which lies above the C K absorption edge: a 20 nm carbon layer reduces the O K$\alpha$ intensity by about 8% (assuming a take-off angle of 40°). The coating also reduces the generated intensity owing to its effect on the incident electrons, which increases as the accelerating voltage decreases (§11.10), becoming more serious at the low accelerating voltages commonly used for light element analysis.

These effects can be minimised by coating specimens and standards simultaneously in order to obtain a uniform thickness. Armstrong (1988) adopted the practice of coating a sample of $Al_2O_3$ with each batch of specimens and measuring the O K$\alpha$ intensity, from which the thickness of carbon is deduced and used to estimate the correction for the specimen intensities. This obviates the need for repeated recoating of the standards.

It is advantageous to use a coating material other than carbon in some cases. For example, Weisweiler (1974) used copper on the grounds that this absorbs O K$\alpha$ less than carbon, and measured the Cu K$\alpha$ intensity to obtain an estimate of the thickness in order to apply a correction for absorption in the coating. Similarly Bizouard and Charpentier (1979) used nickel, whereas Willich and Obertop (1990) used a gold layer a few nanometres thick and measured the Au M$\alpha$ intensity in order to monitor the thickness.

## 18.7 Contamination

Contamination (consisting mainly of carbon) deposited at the point of impact of the electron beam (§2.6) is only a minor problem in conventional analysis, but is much more significant for light elements, since it causes enhancement of the C K$\alpha$ intensity and reduction in the intensity of other lines due to absorption. Various measures can be applied in order to minimise the contamination rate, including the substitution of a turbo pump for the usual oil diffusion pump, using a cold trap to condense hydrocarbons etc. present in the specimen chamber, and directing a gas jet at the point of impact of the beam. Cleanliness of the specimen is also important. The effect of contamination can be minimised by continuously moving the sample during analysis, though this obviously entails a sacrifice of effective spatial resolution.

This problem has been investigated by Rehbach *et al.* (1985), who found the gas jet to be most effective (fig. 18.6). Air is a suitable gas for this purpose. It should be directed at the specimen via a fine nozzle, the flow rate being controlled with a valve so as to limit the pressure rise in the specimen chamber to an acceptable level. The air jet not only inhibits carbon deposition, but also removes carbon already present. The C K$\alpha$ intensity thus tends to decrease with time after placing a new spot under the beam, and may take a considerable time to reach a plateau. For determining low concentrations of carbon it is therefore desirable to include a suitable delay in the measuring routine (Duerr and Ogilvie, 1972). Even with the precautions just described, the spurious C K$\alpha$ signal cannot be eliminated completely and it is difficult to obtain meaningful results for carbon concentrations below about 0.1%.

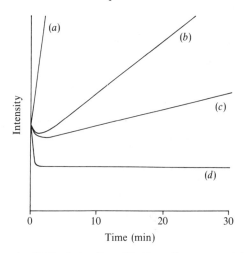

Fig. 18.6 Change in C Kα intensity with time (brass sample), using different anti-contamination measures: (a) none, (b) cold finger, (c) air jet, (d) cold finger and air jet (after Rehbach et al. (1985)).

## 18.8 Absorption corrections

Qualitatively the depth distribution of X-ray production, $\phi(\rho z)$, has essentially the same form for light elements as for heavier elements (§15.1). However, owing to the fact that the overvoltage ratio $U_0$ ($=E_0/E_c$) is generally much higher for light elements (e.g. for C, $U_0=35$ when $E_0=10$ keV), the variation in the ionisation cross-section $Q$ with electron energy ($E$) between $E_0$ and $E_c$ is much more significant for light elements (see fig. 18.7). Consequently, X-ray production is concentrated more towards the ends of the electron trajectories, modifying the shape of $\phi(\rho z)$. Also, the value $\phi_0$ of $\phi(\rho z)$ at the surface is higher (for a given matrix), because of the greater contribution by electrons scattered from within the sample, which have lost significant energy. (This trend can be seen in fig. 15.12.)

When absorption is severe the only X-rays that escape absorption are those generated near the surface (fig. 18.8). The shape of $\phi(\rho z)$ in the near-surface region is therefore critical for light elements, whereas for shorter wavelengths this part of the $\phi(\rho z)$ curve is relatively unimportant. It follows that the simplified Philibert formula (§15.5), which is derived from a $\phi(\rho z)$ function that falls to zero at the surface, is inappropriate for light element analysis.

Duncumb and Melford (1966) and Love, Cox and Scott (1974) found that even the 'full' Philibert model in which $\phi(\rho z)$ is more accurately represented in the near-surface region did not agree well with experimental

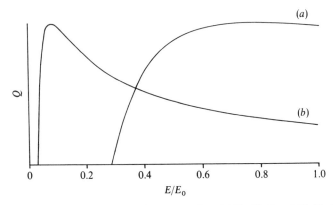

Fig. 18.7 Ionisation cross-section ($Q$) versus $E/E_0$: (a) Fe K, $E_0 = 25$ keV, (b) C K, $E_0 = 10$ keV.

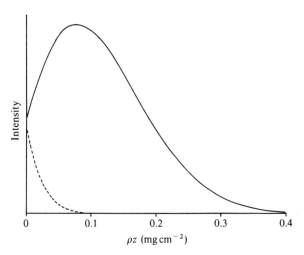

Fig. 18.8 Depth distribution of emergent (dashed line) and generated (solid line) B K$\alpha$ X-rays, AlB$_2$ sample (after Bastin and Heijligers (1986c)).

data for carbon compounds, owing to the effect of high $U$ (see above). Various modified versions of the Philibert model have been proposed and give improved results for light elements (e.g. Reuter, 1972; Ruste, 1979). However, correction methods developed more recently, based on other types of $\phi(\rho z)$ model (§§15.6–15.8), perform better for light element analysis.

An important source of information about $\phi(\rho z)$ is experimental data obtained by the 'tracer method' (§15.2). For light elements such measurements are particularly difficult, because of the heavy absorption of the

## 18.8 Absorption corrections

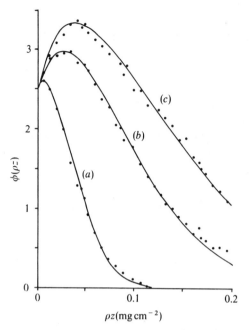

Fig. 18.9 Experimental $\phi(\rho z)$ curves for C K$\alpha$ in Ag obtained by tracer method (dots), compared to curves calculated from a gaussian model, for different accelerating voltages: (a) 5 kV, (b) 10 kV, (c) 15 kV (after Karduck and Rehbach (1985)).

emitted X-rays and the shallow depth of production for low incident electron energies. Karduck and Rehbach (1985) used very thin tracer layers to obtain more reliable data for light elements than previously. Some typical results are shown in fig. 18.9.

### 18.8.1 'Thin film' model

Duncumb and Melford (1966) proposed an extreme simplification of the absorption correction based on the assumption that only X-rays generated in the immediate surface layer contribute to the measured intensity when absorption is severe. If this layer is sufficiently thin, $\phi(\rho z)$ can be assumed to be constant, with a value $\phi_0$. The effective thickness of the layer is proportional to the 'escape depth' of the X-rays, given by $1/\chi$ (where $\chi = \mu \operatorname{cosec} \psi$). Hence:

$$C'/C = \phi_0(\mathrm{sp})\chi(\mathrm{st})/\phi_0(\mathrm{st})\chi(\mathrm{sp}), \qquad (18.1)$$

where $C'$ and $C$ are the uncorrected and 'true' concentrations (§1.10) and 'sp' and 'st' refer to specimen and standard. Methods of obtaining $\phi_0$ are described in §15.9.

The thin film model is valid only for an incident electron energy ($E_0$) sufficiently high that the depth of electron penetration is much greater than $1/\chi$. If a series of measurements of $C'$ with increasing $E_0$ is made, then the corrected concentration, $C$, as derived from equation (18.1), should tend towards a constant value, which may be assumed to be the 'true' concentration.

### *18.8.2 Monte-Carlo models*

The depth distribution function $\phi(\rho z)$ which is required for absorption corrections can be generated by simulating electron trajectories using the Monte-Carlo method (§14.6). This approach is particularly useful for light element analysis, in view of the difficulty in obtaining experimental $\phi(\rho z)$ data. Duncumb and Melford (1966) showed that much better agreement with experimental measurements of the C K$\alpha$ intensity from diamond was obtained by using absorption factors derived from Monte-Carlo calculations than with the standard Philibert correction, reflecting the greater realism of the $\phi(\rho z)$ function in the former case (fig. 18.10).

Commonly used Monte-Carlo models suffer from certain shortcomings when applied to light elements, related primarily to the low excitation energies of these elements. Karduck and Rehbach (1988) developed a model specifically for light elements, incorporating various features

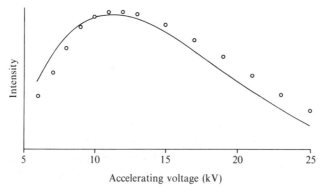

Fig. 18.10 C K$\alpha$ intensity from diamond as function of accelerating voltage: circles – experimental points, line – curve calculated from Monte-Carlo model (after Duncumb and Melford (1966)).

including modification of the stopping power at low energies (§13.4.2) and allowance for the contribution of fast secondary electrons (Reimer and Krefting, 1976), which do not normally contribute significantly to X-ray production. Though not commonly used directly for correction calculations, the results of Monte-Carlo calculations provide a useful input for adjustment of the parameters in the 'phi-rho-z' models which are used for this purpose.

## 18.9 Specimen tilt

Errors due to the surface of the specimen not being at the 'correct' angle are more serious for light elements than for 'normal' elements because absorption is more severe. It is therefore important to pay attention to specimen mounting arrangements and if necessary apply a correction for tilt as described below.

When the surface is flat (though possibly tilted) the optical microscope can be used to determine the tilt angle by moving the specimen a known distance and noting the change in the $z$-axis setting required to stay in sharp focus. A correction to $\psi$ can then be applied. (The shape of $\phi(\rho z)$ is also affected by specimen tilt, but this and the effect on the backscatter factor can be neglected for small angles.)

An approach which is applicable to cases where there may be local variations in tilt has been described by Mackenzie (1990). This entails determining the difference in the intensity obtained when the specimen is rotated through 180° by reversing the (rectangular) holder in its carrier. The tilt angle can be determined from such experimental data to better than $\pm 0.2°$. The 'correct' intensity may be obtained by averaging the values measured in each position (for small tilt angles).

## 18.10 Mass attenuation coefficients

For calculating absorption corrections, knowledge of the m.a.c.s of the elements present for the relevant wavelengths is required. For severe absorption the relative error in the absorption factor is equal to about twice that in $\chi$. Light element analysis is thus particularly sensitive to errors in m.a.c.s.

### 18.10.1 Experimental m.a.c. data

The usual way of determining m.a.c.s is to place an absorbing film of known thickness in a beam of X-rays of known wavelength and measure the attenuation factor. For long wavelength X-rays the absorber must be very thin and it is difficult to measure its thickness accurately. Some such measurements have been made, however. For example, Weisweiler (1969) used optical transmission photometry and interferometry, as well as weighing, to measure the thickness of carbon films for m.a.c. measurements.

Lurio and Reuter (1975) developed a method of m.a.c. measurement in which an absorbing film is deposited on a solid substrate in which X-rays are generated by proton bombardment. Absorption in the film is determined by comparing the intensities measured with and without the film, which has a negligible effect on the protons. The film thickness is measured by Rutherford backscattering. This method has been applied to the determination of m.a.c.s for B and C K$\alpha$ radiation (Lurio and Reuter, 1977; Lurio, Reuter and Keller, 1977). A similar method using electron excitation has been employed by Küchler (1990).

It is much easier to determine the mass thickness of an absorber in gaseous form. Henke and Elgin (1970), for example, obtained m.a.c. values for 11 elements for wavelengths between 8 and 114 Å, using a cell containing the elements (or compounds of them) in the form of gases placed in the X-ray path of a fluorescence spectrometer.

### 18.10.2 M.a.c. tables

Even for 'normal' wavelengths, compiling m.a.c. tables presents difficulties, owing to the limited accuracy and incomplete coverage of experimental data (§15.12). For long wavelengths the experimental data are even more inadequate. Also, the interpolation formulae normally used are not necessarily valid. Various tabulations have nevertheless been compiled for light element analysis, of which the most commonly used are those of Henke and Elgin (1970) and Henke *et al.* (1982), from which table 18.2 is derived.

### 18.10.3 Determining m.a.c.s from microprobe measurements

In view of the difficulties mentioned above, it is appropriate to consider other ways of determining m.a.c.s, including obtaining them from electron microprobe measurements. Thus, given a valid method for calculating the absorption correction, the m.a.c. for a sample of known composition may be estimated by finding the value that gives the 'right' result. The accuracy of the m.a.c. values obtained in this way is, of course, dependent on the validity of the absorption correction model used. Thus, for example,

## 18.10 Mass attenuation coefficients

Table 18.2. *M.a.c.s for Kα lines of light elements, after Henke et al. (1982).*

| Absorber | Emitter | | | | | |
|---|---|---|---|---|---|---|
| | Be | B | C | N | O | F |
| 1 H | 8950 | 1730 | 462 | 149 | 573 | 245 |
| 2 He | 42200 | 10700 | 3350 | 1140 | 455 | 201 |
| 3 Li | 112000 | 31600 | 10300 | 3810 | 1600 | 735 |
| 4 Be | 6500 | 60600 | 22000 | 8880 | 3920 | 1880 |
| 5 B | 11200 | 3350 | 37000 | 15800 | 7420 | 3680 |
| 6 C | 19200 | 6350 | 2350 | 25500 | 12400 | 6370 |
| 7 N | 36400 | 11200 | 4220 | 1810 | 17300 | 9160 |
| 8 O | 54000 | 16500 | 6040 | 2530 | 1200 | 12400 |
| 9 F | 73700 | 23300 | 8750 | 3700 | 1770 | 922 |
| 10 Ne | 100000 | 35400 | 13600 | 5620 | 2580 | 1300 |
| 11 Na | 122000 | 43900 | 17000 | 7330 | 3520 | 1830 |
| 12 Mg | 127000 | 59500 | 23900 | 11000 | 5170 | 2620 |
| 13 Al | 107000 | 64000 | 31000 | 13800 | 6720 | 3410 |
| 14 Si | 115000 | 84000 | 36800 | 16500 | 8790 | 4540 |
| 15 P | 9050 | 66200 | 41300 | 20600 | 10500 | 5530 |
| 16 S | 12800 | 74200 | 47900 | 24900 | 13000 | 7010 |
| 17 Cl | 16100 | 7550 | 50800 | 27500 | 14100 | 7590 |
| 18 Ar | 18000 | 9370 | 45600 | 29500 | 16200 | 8920 |
| 19 K | 21000 | 11300 | 5680 | 35900 | 19400 | 10600 |
| 20 Ca | 22200 | 13000 | 6840 | 35600 | 22000 | 12400 |
| 21 Sc | 25000 | 14200 | 7460 | 3990 | 23600 | 13300 |
| 22 Ti | 27500 | 15300 | 8090 | 4360 | 22100 | 14500 |
| 23 V | 30700 | 16700 | 8840 | 4790 | 24300 | 15800 |
| 24 Cr | 40200 | 20700 | 10600 | 5630 | 3140 | 15900 |
| 25 Mn | 41800 | 22200 | 11500 | 6160 | 3470 | 16100 |
| 26 Fe | 54300 | 27600 | 13900 | 7190 | 4000 | 2330 |
| 27 Co | 57100 | 30900 | 15600 | 8020 | 4410 | 2570 |
| 28 Ni | 64400 | 35700 | 18200 | 9340 | 5120 | 2940 |
| 29 Cu | 74400 | 40400 | 18700 | 9980 | 5920 | 3190 |
| 30 Zn | 73200 | 44200 | 23500 | 12000 | 6550 | 3720 |

Kohlhaas and Scheiding (1970) and Weisweiler (1975c) used the standard Philibert model, and the m.a.c. values obtained cannot be regarded as having absolute validity, though they may give correct results in the conditions under which they were determined. On the other hand, if a more realistic $\phi(\rho z)$ model is used (e.g. Bastin and Heijligers, 1986b), the m.a.c. values obtained can be assumed to be more accurate.

There are significant advantages in deriving m.a.c. values from measurements made at different accelerating voltages. Fig. 18.10 shows how intensity typically varies with $E_0$: the rise in the generated intensity with

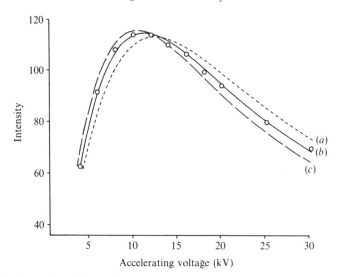

Fig. 18.11 Intensity of O Kα from YBa$_2$Cu$_3$O$_7$ versus accelerating voltage: circles – experimental points, lines – curves calculated from 'PAP' model, using m.a.c. values of (a) 5150, (b) 6150, (c) 7150 (after Mackenzie (1991)).

increasing $E_0$ is counteracted by the rapid increase in absorption, causing the observed intensity to pass through a maximum. Kyser (1972) derived the following relationship between the incident electron energy, $E_p$ (in eV), corresponding to the maximum intensity and $\chi$, based on a gaussian $\phi(\rho z)$ model: $\chi = (2660/E_p)^{1.68}$

Given an accurate expression for the dependence of the generated intensity on $E_0$ (§13.3) and a valid absorption correction model, it is possible to deduce an m.a.c. value by finding the best fit to the plot of intensity versus $E_0$ (see fig. 18.11). A program developed for this purpose, based on the 'PAP' absorption model (§15.8) has been described by Pouchou and Pichoir (1988). An accuracy of $\pm 2\%$ is obtainable in cases where there is a large amount of absorption (Mackenzie, 1991). The relative error is greater for small m.a.c. values, but in such cases the correction is also smaller. Fig. 18.12 shows some m.a.c. values for pure elements determined by this method, which differ significantly from tabulated values, probably because of chemical bonding effects.

In the conventional absorption correction procedure, the m.a.c. of the specimen for a given line is calculated from tabulated values for pure elements, requiring knowledge of the elemental concentrations (iteration being used to overcome the difficulty that these are not known initially). Applying the method described above to the determination of the m.a.c. of

## 18.10 Mass attenuation coefficients

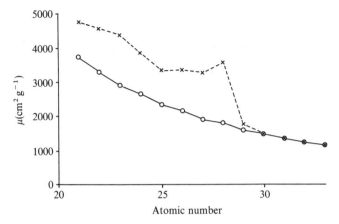

Fig. 18.12 Empirical m.a.c. values for L$\alpha$ lines determined by variable accelerating voltage measurements (Pouchou and Pichoir, 1988): crosses – experimental points, circles – values given by Heinrich (1987b).

the specimen avoids the need for calculation and eliminates the dependence of the accuracy of the result for one element on the accuracy with which the concentrations of other elements are known. This is especially advantageous for light element analysis, where the absorption correction is often very sensitive to the concentrations of other elements which absorb strongly. The same approach can also be applied to the intensity *ratio* between specimen and standard. In this case the m.a.c. and concentration can both be obtained simultaneously (Mackenzie, 1991). The above discussion neglects the dependence of the shape of $\phi(\rho z)$ on composition, but this has a relatively weak effect and can be taken into account easily.

# Appendix: Origin of characteristic X-rays

## A.1 Atomic structure

In the Rutherford–Bohr model of the atom, electrons orbit around the nucleus. In the later theory of wave mechanics, orbits are described by *wave functions* representing the probability of finding an electron at a given location. However, the concept of an 'orbital' is still useful and is easier to visualise than an abstract mathematical function. An orbital which represents the spatial probability distribution is defined by quantum numbers as described below. According to Pauli's exclusion principle, only one electron can have a given set of quantum numbers.

The *principal* quantum number is the main determinant of the binding energy and distance from the nucleus of an orbital electron. The inner shells, designated K, L, M, etc., correspond to $n=1,2,3,\ldots$ respectively, the K shell having the greatest binding energy and being closest to the nucleus. The other quantum numbers have a relatively small effect on energy and cause the shells (other than the K shell) to be split into subshells.

The *azimuthal* quantum number $l$ takes values $0,1,2,\ldots,(n-1)$, for a given value of $n$ and represents the angular momentum of the orbital electron. The letters s, p, d, f, etc., used in optical spectroscopy correspond to $l=0,1,2,3$, etc. Thus, for the 3d orbital $n=3$ and $l=2$. A further quantum number $s$ represents the *spin* of the electron and takes values $+1/2$ and $-1/2$. The *total angular momentum* is described by another quantum number $j$, equal to $l \pm s$ (only positive values allowed). In the case of the L shell ($n=2$), three different combinations of quantum number are possible, giving three subshells:

| Subshell | $l$ | $j$ |
|---|---|---|
| $L_1$ | 0 | 1/2 |
| $L_2$ | 1 | 1/2 |
| $L_3$ | 1 | 3/2 |

The number of subshells is equal to $2n-1$: for example the M shell ($n=3$) has five subshells. A complete list of inner shells and their quantum configurations is given in table A.1.

In order to determine the number of electrons occupying each subshell it is necessary to consider yet another quantum number, $m_j$, which is related to the

Table A.1. *Inner electron shells.*

| Shell | n | l | spectroscopic designation | j | Population | Subshell number |
|---|---|---|---|---|---|---|
| K | 1 | 0 | 1s | 1/2 | 2 | — |
| L | 2 | 0 | 2s | 1/2 | 2 | 1 |
|   | 2 | 1 | 2p | 1/2 | 2 | 2 |
|   | 2 | 1 | 2p | 3/2 | 4 | 3 |
| M | 3 | 0 | 3s | 1/2 | 2 | 1 |
|   | 3 | 1 | 3p | 1/2 | 2 | 2 |
|   | 3 | 1 | 3p | 3/2 | 4 | 3 |
|   | 3 | 2 | 3d | 3/2 | 4 | 4 |
|   | 3 | 2 | 3d | 5/2 | 6 | 5 |
| N | 4 | 0 | 4s | 1/2 | 2 | 1 |
|   | 4 | 1 | 4p | 1/2 | 2 | 2 |
|   | 4 | 1 | 4p | 3/2 | 4 | 3 |
|   | 4 | 2 | 4d | 3/2 | 4 | 4 |
|   | 4 | 2 | 4d | 5/2 | 6 | 5 |
|   | 4 | 3 | 4f | 5/2 | 6 | 6 |
|   | 4 | 3 | 4f | 7/2 | 8 | 7 |
| O | 5 | 0 | 5s | 1/2 | 2 | 1 |
|   | 5 | 1 | 5p | 1/2 | 2 | 2 |
|   | 5 | 1 | 5p | 3/2 | 4 | 3 |
|   | 5 | 2 | 5d | 3/2 | 4 | 4 |
|   | 5 | 2 | 5d | 5/2 | 6 | 5 |

spatial quantisation of the angular momentum. The different orientations taken up by the angular momentum vector in the presence of an applied magnetic field are represented by the values $j, j-1, \ldots 0, \ldots -j$, of $m_j$. Normally no such field is present and states with different $m_j$ but the same $n$, $l$ and $j$ are 'degenerate' – that is they have the same energy. Hence, from Pauli's exclusion principle it follows that the number of electrons occupying a given energy level is $2j+1$: thus the $L_1$, $L_2$ and $L_3$ subshells have populations of 2, 2 and 4 respectively. The populations of other shells are listed in table A.1.

The shell structure is built up by electrons occupying vacant orbitals in order of energy. The K shell, being the most tightly bound, is filled first, and contains two electrons. Next the L shell starts to fill and is complete at $Z=10$. The M shell starts to fill next, but after $Z=18$ the 4s orbital of the N shell has a lower energy than the 3d M-shell orbital and both M and N shells are partially occupied up to $Z=30$, when the M shell is complete. Similar behaviour occurs with other shells at higher atomic numbers, as shown in fig. A.1.

### A.2 Characteristic X-ray emission

Electromagnetic radiation is emitted when an orbital electron undergoes a transition between one state (as defined by a set of quantum numbers) and

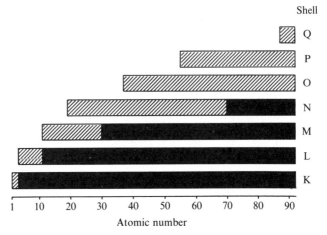

Fig. A.1 Occupancy of atomic shells as function of atomic number: partially filled – shaded, completely filled – black.

another, these states having different energies. The photon energy of the emitted radiation is equal to the difference between these energies and is related to frequency ($v$) thus: $E = hv$, where $h$ is Planck's constant. The wavelength ($\lambda$) is given by: $\lambda = c/v$, where $c$ is the velocity of light, hence $E = hc/\lambda$. The usual unit of energy is the electron volt (eV), defined as the energy acquired by an electron when its potential is increased by 1 V, and the practical unit of wavelength is the Ångstrom unit (1 Å $= 10^{-10}$ m). Using these units, the relationship between energy and wavelength is: $E\lambda = 12396$.[1]

X-ray line spectra are produced by electron transitions between inner shells, which are normally fully occupied. A necessary prerequisite for characteristic X-ray emission is the removal of an inner-shell electron, which provides a vacancy or 'hole' into which an electron from a shell further from the nucleus can 'fall'. The vacancy may be produced in various ways, including bombardment with electrons or other charged particles such as protons, or absorption of an X-ray photon. In any case the energy of the incident particle or photon must be greater than the binding energy of the electron concerned. The minimum energy required to produce the vacancy is the 'critical excitation energy' ($E_c$).

The energy of a characteristic X-ray line is equal to the difference between the energy of the atom in its initial and final states. In the case of the $K\alpha_1$ line, for example, the inner shell of the atom in its initial state is complete except for a vacancy in the K shell. In the final state the vacancy is in the $L_3$ shell. The energy of the emitted photons is equal to the difference in energy between two levels.

The discussion in §A.1 refers to single electron orbitals, whereas X-ray line emission involves vacancies in otherwise full inner shells. However, Pauli's 'vacancy principle' states that such a vacancy is equivalent to a single electron

---

[1] This is the value generally used; the updated value of 12398.5 recommended by Jenkins, Manne, Robin and Senemand (1991) is not significantly different from a practical viewpoint.

## A.2 Characteristic X-ray emission

orbiting an atomic core comprised of the nucleus and complete inner electron shells, enabling the transitions which give rise to X-rays to be treated on the same basis.

Not all transitions are allowed by the rules of quantum theory, according to which the following conditions must be satisfied: $\triangle n \neq 0$, $\triangle l = \pm 1$, $\triangle j = \pm 1$ or 0 (but if $j = 0$ initially, then $\triangle j = 0$ is forbidden). These are the conditions for 'electric dipole' transitions, which give rise to the principal emission lines.
Other types of transition can occur, but with much lower probability. Examples of 'forbidden' lines produced by such transitions are: $K\beta_4$ (K–N$_4$) and $L\beta_9$ (L$_1$–M$_5$).

The atomic states which are relevant to characteristic X-ray production can be represented as horizontal lines on an energy level diagram, in which the zero of the energy scale corresponds to an atom in its rest state with no electrons missing. The energies of the levels associated with the absence of one electron from a shell are positive. The highest level is that of the K shell, since a K electron requires the greatest amount of energy to remove it. Transitions of 'holes' from higher to lower levels giving rise to X-ray emission are described by vertical lines in the energy level diagram. Characteristic X-ray energies may be deduced from the difference in energy between such levels (fig. A.2).

In the usual (Siegbahn) system of line nomenclature, the Greek letters $\alpha$, $\beta$ and $\gamma$ refer to groups of lines of similar wavelength, in order of decreasing

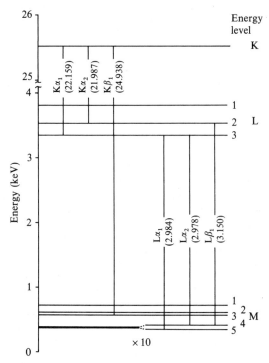

Fig. A.2 Energy level diagram showing K, L and M lines of Ag, and transitions for principal characteristic line (with energies given in keV).

Table A.2. *Initial and final levels for characteristic lines.*

| Line | Initial | Final | Line | Initial | Final |
|---|---|---|---|---|---|
| $K\alpha_1$ | K | $L_3$ | $L\gamma_1$ | $L_2$ | $N_4$ |
| $K\alpha_2$ | K | $L_2$ | $L\gamma_2$ | $L_1$ | $N_{2,3}$ |
| $K\beta_1$ | K | $M_3$ | $L\gamma_3$ | $L_1$ | $N_3$ |
| $K\beta_2$ | K | $N_{2,3}$ | $L\eta$ | $L_2$ | $M_1$ |
| $K\beta_3$ | K | $M_2$ | $Ll$ | $L_3$ | $M_1$ |
| $L\alpha_1$ | $L_3$ | $M_5$ | $M\alpha_1$ | $M_5$ | $N_7$ |
| $L\alpha_2$ | $L_3$ | $M_4$ | $M\alpha_2$ | $M_5$ | $N_6$ |
| $L\beta_1$ | $L_2$ | $M_4$ | $M\beta_1$ | $M_4$ | $N_6$ |
| $L\beta_2$ | $L_3$ | $N_5$ | $M\gamma_1$ | $M_3$ | $N_5$ |
| $L\beta_3$ | $L_1$ | $M_3$ | $M\gamma_2$ | $M_3$ | $N_4$ |
| $L\beta_4$ | $L_1$ | $M_2$ | $M\delta$ | $M_2$ | $N_4$ |
| $L\beta_5$ | $L_3$ | $O_{4,5}$ | $M\varepsilon$ | $M_3$ | $O_5$ |
| $L\beta_{15}$ | $L_3$ | $N_4$ | $M\zeta_1$ | $M_5$ | $N_3$ |

intensity, while numerical subscripts distinguish the lines within each group, also in decreasing intensity order. There are also some 'odd' lines which fall outside this scheme. Since Greek letters are unsuitable for computers, it is common practice to substitute the equivalent Roman letter (a for $\alpha$, b for $\beta$, etc.). The principal characteristic lines and the corresponding initial and final energy levels are listed in table A.2.

The Siegbahn system was devised before the origin of characteristic X-rays was understood: a more logical approach is to identify the line by the transition which produces it, as recommended by the International Union of Pure and Applied Chemistry (Jenkins et al., 1991). Thus, for example, $K\alpha_1$ becomes $K-L_3$. However, the older and more familiar system is used here.

### A.3 Wavelengths and energies of X-ray lines

Moseley (1913, 1914) noted that the square roots of the frequencies of the $K\alpha$ lines of different elements lie on a straight line when plotted against atomic number ('Moseley's law'). Using energy in place of frequency, this relationship can be expressed by the equation: $E = a(Z-b)^2$, where $a$ and $b$ are constants. This is a consequence of the dependence of the electron binding energy on the nuclear charge. The constant $b$ is approximately 1 for the $K\alpha$ lines and represents the reduction in the effective nuclear charge caused by the single remaining K electron. For $L\alpha$ lines, $b$ has a value of approximately 7.

Moseley's law is not sufficiently accurate to give line energies with high precision: for this purpose it is necessary to use a polynomial expression (e.g. Springer and Nolan, 1976). However, for calculating line energies as required for matrix corrections it is preferable to use tabulated data, since a fitted polynomial may occasionally cause a line to be placed on the wrong side of an absorption edge. Energies and wavelengths can be obtained from tables such as those of White and Johnson (1970). Values for the principal K, L and M lines are listed in tables A.3, A.4 and A.5 respectively.

Table A.3. *Wavelengths (in Å), energies and critical excitation energies (in keV) of K lines (after White and Johnson (1970)).*

| Element | Z | $K\alpha_1$ $\lambda$ | $K\alpha_1$ $E$ | $K\beta_1$ $\lambda$ | $K\beta_1$ $E$ | $E_c$ |
|---|---|---|---|---|---|---|
| Be | 4  | 114.0 | 0.109 |       |       | 0.112 |
| B  | 5  | 67.60 | 0.183 |       |       | 0.192 |
| C  | 6  | 44.70 | 0.277 |       |       | 0.284 |
| N  | 7  | 31.60 | 0.392 |       |       | 0.400 |
| O  | 8  | 23.62 | 0.525 |       |       | 0.532 |
| F  | 9  | 18.32 | 0.677 |       |       | 0.687 |
| Ne | 10 | 14.61 | 0.848 |       |       | 0.867 |
| Na | 11 | 11.91 | 1.041 | 11.62 | 1.067 | 1.071 |
| Mg | 12 | 9.890 | 1.253 | 9.570 | 1.295 | 1.303 |
| Al | 13 | 8.340 | 1.486 | 7.982 | 1.553 | 1.560 |
| Si | 14 | 7.126 | 1.739 | 6.778 | 1.829 | 1.840 |
| P  | 15 | 6.158 | 2.013 | 5.804 | 2.136 | 2.143 |
| S  | 16 | 5.372 | 2.307 | 5.032 | 2.464 | 2.470 |
| Cl | 17 | 4.729 | 2.621 | 4.403 | 2.815 | 2.819 |
| Ar | 18 | 4.193 | 2.957 | 3.886 | 3.190 | 3.202 |
| K  | 19 | 3.742 | 3.312 | 3.454 | 3.589 | 3.607 |
| Ca | 20 | 3.359 | 3.690 | 3.090 | 4.012 | 4.037 |
| Sc | 21 | 3.032 | 4.088 | 2.780 | 4.460 | 4.488 |
| Ti | 22 | 2.750 | 4.508 | 2.514 | 4.931 | 4.964 |
| V  | 23 | 2.505 | 4.949 | 2.284 | 5.426 | 5.463 |
| Cr | 24 | 2.291 | 5.411 | 2.085 | 5.946 | 5.988 |
| Mn | 25 | 2.103 | 5.894 | 1.910 | 6.489 | 6.536 |
| Fe | 26 | 1.937 | 6.398 | 1.757 | 7.057 | 7.110 |
| Co | 27 | 1.790 | 6.924 | 1.621 | 7.648 | 7.708 |
| Ni | 28 | 1.659 | 7.471 | 1.500 | 8.263 | 8.330 |
| Cu | 29 | 1.542 | 8.040 | 1.392 | 8.904 | 8.979 |
| Zn | 30 | 1.436 | 8.630 | 1.295 | 9.570 | 9.659 |
| Ga | 31 | 1.341 | 9.241 | 1.208 | 10.26 | 10.37 |
| Ge | 32 | 1.255 | 9.874 | 1.129 | 10.98 | 11.10 |
| As | 33 | 1.177 | 10.53 | 1.057 | 11.72 | 11.86 |
| Se | 34 | 1.106 | 11.21 | 0.992 | 12.49 | 12.56 |
| Br | 35 | 1.041 | 11.91 | 0.933 | 13.29 | 13.47 |

Table A.4. *Wavelengths (in Å), energies and critical excitation energies$^a$ (in keV) of L lines (after White and Johnson (1970)).*

| Element | Z | L$\alpha_1$ $\lambda$ | L$\alpha_1$ E | L$\beta_1$ $\lambda$ | L$\beta_1$ E | $E_c$ |
|---|---|---|---|---|---|---|
| Ga | 31 | 11.29 | 1.098 | 11.02 | 1.125 | 1.117 |
| Ge | 32 | 10.44 | 1.188 | 10.18 | 1.218 | 1.217 |
| As | 33 | 9.671 | 1.282 | 9.414 | 1.317 | 1.323 |
| Se | 34 | 8.990 | 1.379 | 8.736 | 1.419 | 1.434 |
| Br | 35 | 8.375 | 1.480 | 8.125 | 1.526 | 1.553 |
| Kr | 36 | 7.817 | 1.586 | 7.576 | 1.636 | 1.677 |
| Rb | 37 | 7.318 | 1.694 | 7.076 | 1.752 | 1.806 |
| Sr | 38 | 6.863 | 1.806 | 6.624 | 1.871 | 1.941 |
| Y | 39 | 6.449 | 1.922 | 6.212 | 1.995 | 2.079 |
| Zr | 40 | 6.071 | 2.042 | 5.836 | 2.124 | 2.222 |
| Nb | 41 | 5.724 | 2.166 | 5.492 | 2.257 | 2.370 |
| Mo | 42 | 5.407 | 2.293 | 5.177 | 2.394 | 2.523 |
| Ru | 44 | 4.846 | 2.558 | 4.621 | 2.683 | 2.837 |
| Rh | 45 | 4.597 | 2.696 | 4.374 | 2.834 | 3.002 |
| Pd | 46 | 4.368 | 2.838 | 4.146 | 2.990 | 3.172 |
| Ag | 47 | 4.154 | 2.984 | 3.935 | 3.150 | 3.350 |
| Cd | 48 | 3.956 | 3.133 | 3.738 | 3.316 | 3.537 |
| In | 49 | 3.772 | 3.286 | 3.555 | 3.487 | 3.730 |
| Sn | 50 | 3.600 | 3.443 | 3.385 | 3.662 | 3.928 |
| Sb | 51 | 3.439 | 3.604 | 3.226 | 3.843 | 4.132 |
| Te | 52 | 3.289 | 3.769 | 3.077 | 4.029 | 4.341 |
| I | 53 | 3.149 | 3.937 | 2.937 | 4.220 | 4.558 |
| Xe | 54 | 3.017 | 4.109 | 2.806 | 4.417 | 4.781 |
| Cs | 55 | 2.892 | 4.286 | 2.684 | 4.619 | 5.011 |
| Ba | 56 | 2.776 | 4.465 | 2.568 | 4.827 | 5.246 |
| La | 57 | 2.666 | 4.650 | 2.459 | 5.041 | 5.483 |
| Ce | 58 | 2.562 | 4.839 | 2.356 | 5.261 | 5.723 |
| Pr | 59 | 2.463 | 5.033 | 2.259 | 5.488 | 5.962 |
| Nd | 60 | 2.370 | 5.229 | 2.167 | 5.721 | 6.208 |
| Sm | 62 | 2.200 | 5.635 | 1.998 | 6.204 | 6.716 |
| Eu | 63 | 2.121 | 5.845 | 1.920 | 6.455 | 6.979 |
| Gd | 64 | 2.047 | 6.056 | 1.847 | 6.712 | 7.242 |
| Tb | 65 | 1.977 | 6.272 | 1.777 | 6.977 | 7.514 |
| Dy | 66 | 1.909 | 6.494 | 1.711 | 7.246 | 7.788 |
| Ho | 67 | 1.845 | 6.719 | 1.648 | 7.524 | 8.066 |
| Er | 68 | 1.784 | 6.947 | 1.587 | 7.809 | 8.356 |
| Tm | 69 | 1.727 | 7.179 | 1.530 | 8.100 | 8.648 |
| Yb | 70 | 1.672 | 7.414 | 1.476 | 8.400 | 8.942 |
| Lu | 71 | 1.620 | 7.654 | 1.424 | 8.708 | 9.247 |
| Hf | 72 | 1.570 | 7.898 | 1.374 | 9.021 | 9.556 |
| Ta | 73 | 1.522 | 8.145 | 1.327 | 9.342 | 9.875 |
| W | 74 | 1.476 | 8.396 | 1.282 | 9.671 | 10.20 |
| Re | 75 | 1.433 | 8.651 | 1.239 | 10.01 | 10.53 |
| Os | 76 | 1.391 | 8.910 | 1.197 | 10.35 | 10.87 |

Table A.4. (cont).

| Element | Z | $L\alpha_1$ | | $L\beta_1$ | | |
|---|---|---|---|---|---|---|
| | | $\lambda$ | $E$ | $\lambda$ | $E$ | $E_c$ |
| Ir | 77 | 1.351 | 9.174 | 1.158 | 10.71 | 11.21 |
| Pt | 78 | 1.313 | 9.441 | 1.120 | 11.07 | 11.56 |
| Au | 79 | 1.276 | 9.712 | 1.084 | 11.44 | 11.92 |
| Hg | 80 | 1.241 | 9.987 | 1.049 | 11.82 | 12.28 |
| Tl | 81 | 1.207 | 10.27 | 1.015 | 12.21 | 12.66 |
| Pb | 82 | 1.175 | 10.55 | 0.983 | 12.61 | 13.04 |
| Bi | 83 | 1.144 | 10.84 | 0.952 | 13.02 | 13.42 |

[a] Values given are for the $L_3$ subshell and apply to the $L\alpha_1$ line only.

Table A.5. Wavelengths (in Å), energies and critical excitation energies[a] (in keV) of $M\alpha_1$ lines (after White and Johnson (1970)).

| Element | Z | $\lambda$ | $E$ | $E_c$ |
|---|---|---|---|---|
| Sm | 62 | 11.47 | 1.081 | |
| Eu | 63 | 10.96 | 1.131 | |
| Gd | 64 | 10.46 | 1.185 | |
| Tb | 65 | 10.00 | 1.240 | |
| Dy | 66 | 9.590 | 1.293 | |
| Ho | 67 | 9.200 | 1.347 | |
| Er | 68 | 8.820 | 1.405 | |
| Tm | 69 | 8.480 | 1.462 | |
| Yb | 70 | 8.149 | 1.521 | |
| Lu | 71 | 7.840 | 1.581 | |
| Hf | 72 | 7.539 | 1.644 | |
| Ta | 73 | 7.252 | 1.709 | 1.743 |
| W | 74 | 6.983 | 1.775 | 1.815 |
| Re | 75 | 6.729 | 1.842 | 1.890 |
| Os | 76 | 6.478 | 1.914 | 1.968 |
| Ir | 77 | 6.262 | 1.980 | 2.049 |
| Pt | 78 | 6.047 | 2.050 | 2.134 |
| Au | 79 | 5.840 | 2.123 | 2.220 |
| Hg | 80 | 5.648 | 2.195 | 2.313 |
| Tl | 81 | 5.460 | 2.270 | 2.406 |
| Pb | 82 | 5.286 | 2.345 | 2.502 |
| Bi | 83 | 5.118 | 2.422 | 2.602 |
| Th | 90 | 4.138 | 2.996 | 3.324 |
| U | 92 | 3.910 | 3.170 | 3.545 |

[a] Values not given are very close to the line energy and for practical purposes can be assumed to be the same.

### A.4 Relative intensities

Certain predictions about the relative intensities of characteristic X-ray lines can be made. Thus, in the case of the $K\alpha_1$ and $K\alpha_2$ lines the populations of the relevant final energy levels ($L_3$ and $L_2$) are in the ratio 2:1, and the observed intensity ratio is in good agreement. For other groups of lines a simplified treatment can be applied, even though different subshells of the same shell are involved, by using the 'Burger–Dorgelo' rule (Compton and Allison, 1935). In the case of the $L\alpha_1$, $L\alpha_2$ and $L\beta_1$ lines, all of which derive from L–M transitions, the $L_2$ and $L_3$ levels are first assumed to be merged into a single level, as shown in fig. A.3(a). From the populations of the $M_4$ and $M_5$ levels it may be deduced that the ratio $L\alpha_1/(L\alpha_2+L\beta_1)$ is equal to 6/4. It is then assumed that the $L_2$ and $L_3$ levels are separated, while the $M_4$ and $M_5$ levels are merged, as in fig. A.3(b), from which it is inferred that $(L\alpha_1+L\alpha_2)/L\beta_1=4/2$. Combining these results, it follows that the intensities of the $L\alpha_1$, $L\alpha_2$ and $L\beta_1$ lines are in the ratio 9:1:5, which is in reasonable agreement with experiment.

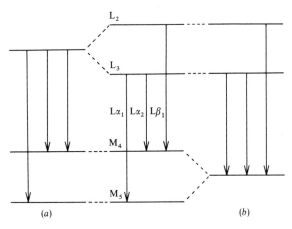

Fig. A.3 Simplified energy level diagrams used in Burger–Dorgelo calculation of relative line intensities (see text).

It should be noted that observed intensities are influenced by incomplete occupancy of the relevant levels. For example, the $K\beta_1/K\alpha_1$ ratio is fairly constant for atomic numbers above about 20, but decreases quite rapidly for lower Z values owing to incomplete filling of the $M_3$ subshell (the $K\beta_1$ line being produced by the K–$M_3$ transition). Similar behaviour occurs in the case of L and M lines. Relative line intensities are given by White and Johnson (1970), but in some cases these are only estimated or interpolated values and therefore are not necessarily accurate.

### A.5 Satellite lines

Satellite lines do not correspond to any transition identifiable in the energy level diagram and are therefore known as 'non-diagram' lines. The $K\alpha$ line has several satellites on the high energy side, designated $K\alpha'$, $K\alpha_3$, $K\alpha_4$, etc., with intensities of up to 5% of the main line in the case of Al, for example. These lines are only visible when a w.d. spectrometer is used and even then are usually not fully resolved, appearing as a 'tail' on the short wavelength side of peaks. The $K\beta_1$ line and the principal L and M lines also have satellites.

Satellites are produced by transitions occurring in doubly ionised atoms: for example, a transition from a KL state (with one K and one L electron missing) to an LL state (two L electrons missing). The additional vacancy reduces the screening of the charge on the nucleus, thereby increasing the binding energies of the electrons and shifting the line to a slightly higher energy. Double ionisation may be produced directly by electron bombardment if the incident electrons have sufficient energy, or as a result of the Auger effect (see §A.6).

## A.6 Auger effect and fluorescence yield

Inner-shell ionisation does not inevitably lead to characteristic X-ray emission. An alternative possibility is a 'radiationless' transition, whereby the energy released as the electron 'falls' into the initial vacancy is used to eject another electron from the atom. This is known as the 'Auger effect'. Such a transition can be described in terms of the shells containing the initial and final vacancies, e.g. $K-L_2L_3$. When this occurs, the atom is left in a doubly ionised state. Satellite lines with energies slightly different from the principal lines in the spectrum result from radiative transitions occurring in atoms which are already doubly ionised (§A.5).

The 'fluorescence yield' ($\omega$) is defined as the probability of ionisation of a given shell being followed by characteristic X-ray emission, as opposed to a radiationless Auger transition, and is independent of the means by which the initial vacancy was produced. It can be shown theoretically that the probability of an Auger transition is approximately independent of $Z$ (Wentzel, 1927), whereas the probability of a radiative transition varies as $Z^4$. The fluorescence yield is thus given approximately by: $\omega = Z^4/(a + Z^4)$. The constant $a$ has a value of about $10^6$ for the K shell. The same expression is applicable to other shells, with different values of $a$. Better accuracy can be obtained with more complicated semi-empirical expressions, such as that proposed by Burhop (1955): $[\omega/(1-\omega)]^{-\frac{1}{4}} = -A + BZ + CZ^3$. For practical purposes, however, it is probably best to use tabulated values based on experimental data, such as those given in tables A.6 (K shell) and A.7 (L shell).

## A.7 Coster–Kronig transitions

In multiple shells a special kind of radiationless transition is possible, involving the transfer of a vacancy from one shell to another. This is known as a 'Coster–Kronig' (C–K) transition. For example, an initial vacancy in the $L_1$ subshell may move to the $L_3$ subshell and the energy released be used to eject an $M_4$ electron. This is described as an $L_1-L_3M_4$ transition. The C–K process is a special case of the Auger effect and applies to the L and M shells but not the K shell.

For a C–K transition to occur, the ejected outer electron must have a binding energy less than the energy difference between the subshells between which the initial vacancy moves. Thus, in the above case the condition is: $[E(L_1) - E(L_3)] > E(M_4)$, where $E$ represents binding energy. This condition is satisfied for $50 > Z > 73$. A list of the atomic number ranges for which different L-shell C–K transitions occur is given in table A.8.

Table A.6. *Fluorescence yields for the K shell (after Krause (1979))*.

| Element | Z | $\omega_K$ | Element | Z | $\omega_K$ |
|---|---|---|---|---|---|
| B  | 5  | 0.0017 | Cr | 24 | 0.275 |
| C  | 6  | 0.0028 | Mn | 25 | 0.308 |
| N  | 7  | 0.0052 | Fe | 26 | 0.340 |
| O  | 8  | 0.0083 | Co | 27 | 0.373 |
| F  | 9  | 0.013  | Ni | 28 | 0.406 |
| Ne | 10 | 0.018  | Cu | 29 | 0.440 |
| Na | 11 | 0.023  | Zn | 30 | 0.474 |
| Mg | 12 | 0.030  | Ga | 31 | 0.507 |
| Al | 13 | 0.039  | Ge | 32 | 0.535 |
| Si | 14 | 0.050  | As | 33 | 0.562 |
| P  | 15 | 0.063  | Se | 34 | 0.589 |
| S  | 16 | 0.078  | Br | 35 | 0.618 |
| Cl | 17 | 0.097  | Kr | 36 | 0.643 |
| Ar | 18 | 0.118  | Rb | 37 | 0.667 |
| K  | 19 | 0.140  | Sr | 38 | 0.690 |
| Ca | 20 | 0.163  | Y  | 39 | 0.710 |
| Sc | 21 | 0.188  | Zr | 40 | 0.730 |
| Ti | 22 | 0.214  | Nb | 41 | 0.747 |
| V  | 23 | 0.243  | Mo | 42 | 0.765 |

Table A.7. *Fluorescence yields for the L subshells (after Krause (1979))*.

| Element | Z | $\omega_{L1}$ | $\omega_{L2}$ | $\omega_{L3}$ |
|---|---|---|---|---|
| Zn | 30 | 0.002 | 0.011 | 0.012 |
| Ga | 31 | 0.002 | 0.012 | 0.013 |
| Ge | 32 | 0.002 | 0.013 | 0.015 |
| As | 33 | 0.003 | 0.014 | 0.016 |
| Se | 34 | 0.003 | 0.016 | 0.018 |
| Br | 35 | 0.004 | 0.018 | 0.020 |
| Kr | 36 | 0.004 | 0.020 | 0.022 |
| Rb | 37 | 0.005 | 0.022 | 0.024 |
| Sr | 38 | 0.005 | 0.024 | 0.026 |
| Y  | 39 | 0.006 | 0.026 | 0.028 |
| Zr | 40 | 0.007 | 0.028 | 0.031 |
| Nb | 41 | 0.009 | 0.031 | 0.034 |
| Mo | 42 | 0.010 | 0.034 | 0.037 |
| Ru | 44 | 0.012 | 0.040 | 0.043 |
| Rh | 45 | 0.013 | 0.043 | 0.046 |
| Pd | 46 | 0.014 | 0.047 | 0.049 |
| Ag | 47 | 0.016 | 0.051 | 0.052 |
| Cd | 48 | 0.018 | 0.056 | 0.056 |
| In | 49 | 0.020 | 0.061 | 0.060 |
| Sn | 50 | 0.037 | 0.065 | 0.064 |
| Sb | 51 | 0.039 | 0.069 | 0.069 |
| Te | 52 | 0.041 | 0.074 | 0.074 |

Table A.7. (cont.)

| Element | Z | $\omega_{L1}$ | $\omega_{L2}$ | $\omega_{L3}$ |
|---------|----|-------|-------|-------|
| I  | 53 | 0.044 | 0.079 | 0.079 |
| Xe | 54 | 0.046 | 0.083 | 0.085 |
| Cs | 55 | 0.049 | 0.090 | 0.091 |
| Ba | 56 | 0.052 | 0.096 | 0.097 |
| La | 57 | 0.055 | 0.103 | 0.104 |
| Ce | 58 | 0.058 | 0.110 | 0.111 |
| Pr | 59 | 0.061 | 0.117 | 0.118 |
| Nd | 60 | 0.064 | 0.124 | 0.125 |
| Sm | 62 | 0.071 | 0.140 | 0.139 |
| Eu | 63 | 0.075 | 0.149 | 0.147 |
| Gd | 64 | 0.079 | 0.158 | 0.155 |
| Tb | 65 | 0.083 | 0.167 | 0.164 |
| Dy | 66 | 0.089 | 0.178 | 0.174 |
| Ho | 67 | 0.094 | 0.189 | 0.182 |
| Er | 68 | 0.100 | 0.200 | 0.192 |
| Tm | 69 | 0.106 | 0.211 | 0.201 |
| Yb | 70 | 0.112 | 0.222 | 0.210 |
| Lu | 71 | 0.120 | 0.234 | 0.220 |
| Hf | 72 | 0.128 | 0.246 | 0.231 |
| Ta | 73 | 0.137 | 0.258 | 0.243 |
| W  | 74 | 0.147 | 0.270 | 0.255 |
| Re | 75 | 0.144 | 0.283 | 0.268 |
| Os | 76 | 0.130 | 0.295 | 0.281 |
| Ir | 77 | 0.120 | 0.308 | 0.294 |
| Pt | 78 | 0.114 | 0.321 | 0.306 |
| Au | 79 | 0.107 | 0.334 | 0.320 |
| Hg | 80 | 0.107 | 0.347 | 0.333 |
| Tl | 81 | 0.107 | 0.360 | 0.347 |
| Pb | 82 | 0.112 | 0.373 | 0.360 |
| Bi | 83 | 0.117 | 0.387 | 0.373 |
| Th | 90 | 0.161 | 0.479 | 0.463 |
| U  | 92 | 0.176 | 0.467 | 0.489 |

Table A.8. *Limiting atomic numbers for Coster–Kronig transitions of the type* $L_x L_y - M_z$.

| x | y | z=1 | 2 | 3 | 4 | 5 |
|---|---|-----|-----|-----|-----|-----|
| 1 | 2 | <31 | <36 | <37 | <40 | <40 |
| 1 | 3 | <31 | <36 | <37 | <50 | <50 |
|   |   |     |     |     | >73 | >77 |
| 2 | 3 |     |     |     | <30 | <30 |
|   |   |     |     |     |     | >90 |

Table A.9. *Coster–Kronig yields for the L shell (after Krause (1979))*.

| Element | Z | $f_{1,2}$ | $f_{1,3}$ | $f_{2,3}$ |
|---|---|---|---|---|
| Zn | 30 | 0.29 | 0.54 | 0.03 |
| Ga | 31 | 0.29 | 0.53 | 0.03 |
| Ge | 32 | 0.28 | 0.53 | 0.05 |
| As | 33 | 0.28 | 0.53 | 0.06 |
| Se | 34 | 0.28 | 0.52 | 0.08 |
| Br | 35 | 0.28 | 0.52 | 0.09 |
| Kr | 36 | 0.27 | 0.52 | 0.10 |
| Rb | 37 | 0.27 | 0.52 | 0.11 |
| Sr | 38 | 0.26 | 0.52 | 0.12 |
| Y | 39 | 0.26 | 0.52 | 0.13 |
| Zr | 40 | 0.26 | 0.52 | 0.13 |
| Nb | 41 | 0.10 | 0.61 | 0.14 |
| Mo | 42 | 0.10 | 0.61 | 0.14 |
| Ru | 44 | 0.10 | 0.61 | 0.15 |
| Rh | 45 | 0.10 | 0.60 | 0.15 |
| Pd | 46 | 0.10 | 0.60 | 0.15 |
| Ag | 47 | 0.10 | 0.59 | 0.15 |
| Cd | 48 | 0.10 | 0.59 | 0.16 |
| In | 49 | 0.10 | 0.59 | 0.16 |
| Sn | 50 | 0.17 | 0.27 | 0.16 |
| Sb | 51 | 0.17 | 0.28 | 0.16 |
| Te | 52 | 0.18 | 0.28 | 0.16 |
| I | 53 | 0.18 | 0.28 | 0.15 |
| Xe | 54 | 0.19 | 0.28 | 0.15 |
| Cs | 55 | 0.19 | 0.28 | 0.15 |
| Ba | 56 | 0.19 | 0.28 | 0.15 |
| La | 57 | 0.19 | 0.29 | 0.15 |
| Ce | 58 | 0.19 | 0.29 | 0.15 |
| Pr | 59 | 0.19 | 0.29 | 0.15 |
| Nd | 60 | 0.19 | 0.30 | 0.15 |
| Sm | 62 | 0.19 | 0.30 | 0.15 |
| Eu | 63 | 0.19 | 0.30 | 0.15 |
| Gd | 64 | 0.19 | 0.30 | 0.15 |
| Tb | 65 | 0.19 | 0.30 | 0.15 |
| Dy | 66 | 0.19 | 0.30 | 0.14 |
| Ho | 67 | 0.19 | 0.30 | 0.14 |
| Er | 68 | 0.19 | 0.30 | 0.14 |
| Tm | 69 | 0.19 | 0.29 | 0.14 |
| Yb | 70 | 0.19 | 0.29 | 0.14 |
| Lu | 71 | 0.19 | 0.28 | 0.14 |
| Hf | 72 | 0.18 | 0.28 | 0.14 |
| Ta | 73 | 0.18 | 0.28 | 0.13 |
| W | 74 | 0.17 | 0.28 | 0.13 |
| Re | 75 | 0.16 | 0.33 | 0.13 |
| Os | 76 | 0.16 | 0.39 | 0.13 |
| Ir | 77 | 0.15 | 0.45 | 0.13 |

## A.7 Coster–Kronig transitions

Table A.9. (cont.)

| Element | Z | $f_{1,2}$ | $f_{1,2}$ | $f_{2,3}$ |
|---|---|---|---|---|
| Pt | 78 | 0.14 | 0.50 | 0.12 |
| Au | 79 | 0.14 | 0.53 | 0.12 |
| Hg | 80 | 0.13 | 0.56 | 0.12 |
| Tl | 81 | 0.13 | 0.57 | 0.12 |
| Pb | 82 | 0.12 | 0.58 | 0.12 |
| Bi | 83 | 0.11 | 0.58 | 0.11 |
| Th | 90 | 0.09 | 0.57 | 0.11 |
| U | 92 | 0.08 | 0.57 | 0.17 |

The probability of a vacancy moving from one subshell to another is given by the C–K yield. In the case of the L shell there are three such yields – $f_{1,2}$, $f_{1,3}$ and $f_{2,3}$ – in which the subscripts refer to the initial and final levels. Values for these yields are given in table A.9.

The intensities of L lines involving the $L_2$ and $L_3$ levels are enhanced by the transfer of vacancies from $L_1$ to $L_2$ and from both $L_1$ and $L_2$ to $L_3$. The enhancement factors are not constant, however, because they depend on the relative numbers of initial vacancies in the subshells, which vary with the excitation conditions. In the case of electron excitation, the ionisation cross-sections depend on the ratio $E_0/E_c$, and their relative size is a function of $E_0$. Further, the relative excitation ratios are significantly different for X-ray fluorescence.

The enhancement factors are given by the expressions:

$$1 + f_{1,2} n_1/n_2, \text{ for } L_2,$$

and

$$1 + (f_{1,3} + f_{1,2} f_{2,3}) n_1/n_3 + f_{2,3} n_2/n_3, \text{ for } L_3,$$

where $n_1$, $n_2$ and $n_3$ are the production rates of vacancies in the $L_1$, $L_2$ and $L_3$ subshells. For electron excitation these are approximately proportional to the electron populations of these levels, which are 2, 2 and 4 respectively. Enhancement factors can thus be estimated using C–K yields from table A.9: in the case of Ta ($Z = 73$), for example, the calculated values are 1.18 for $L_2$ and 1.22 for $L_3$. More rigorous estimation requires knowledge of the ionisation cross-sections for the electron energy concerned.

# References[1]

Agarwal B. K. (1991) *X-ray Spectroscopy*, 2nd edition (Berlin: Springer).
Alberti G., Clerici R. & Zambra A. (1979) *Nucl. Instr. Meth.* **158**, 425.
Albee A. L. & Ray L. (1970) *Anal. Chem.* **42**, 1408
Almasi G. S., Blair J., Ogilvie R. E. & Schwarz R. J. (1965) *J. Appl. Phys.* **36**, 1848.
Alves M. A. F., Policarpo A. J. P. L. & Santos M. C. M. (1973) *Nucl. Instr. Meth.* **111**, 413.
Ammann N. & Karduck P. (1990) *Microbeam Analysis* – 1990, p. 150.
Ancey M., Bastenaire F. & Tixier R. (1977) *J. Phys. D* **11**, 817.
Archard G. D. (1953) *J. Sci. Instr.* **30**, 352.
Archard G. D. (1961) *J. Appl. Phys.* **32**, 1505.
Archard G. D. & Mulvey T. (1963) *ICXOM* 3, p. 393.
Armigliato A., Bentini G. G. & Ruffini G. (1976) *J. Microsc.* **108**, 31.
Armstrong J. T. (1988) *Microbeam Analysis* – 1988, p. 469.
Armstrong J. T. & Buseck P. R. (1985) *X-ray Spectrom.* **14**, 172.
August H.-J. & Wernisch J. (1990) *Scanning* **12**, 14.
August H.-J. & Wernisch J. (1991a) *Scanning* **13**, 207.
August H.-J. & Wernisch J. (1991b) *X-ray Spectrom.* **20**, 131.
Axel P. (1954) *Rev. Sci. Instr.* **25**, 391.
Baines M., Dean E. M. & Wilson J. M. (1975) *J. Phys. E* **8**, 305.
Bakker C. J. & Segrè E. (1951) *Phys. Rev.* **81**, 489.
Ball M. D. & McCartney D. G. (1981) *J. Microsc.* **124**, 57.
Barbi N. C., Skinner D. P. & Blinder S. (1976) *MAS-11*, paper no. 8.
Barkla C. G. & Sadler C. A. (1909) *Phil. Mag.* **17**, 739.
Bartosek J., Masek J., Adams F. & Hoste J. (1972) *Nucl. Instr. Meth.* **104**, 221.
Bassett R. & Mulvey T. (1969) *ICXOM* 5, p. 225.
Bastin G. F. & Heijligers H. J. M. (1986a) *X-ray Spectrom.* **15**, 135.
Bastin G. F. & Heijligers H. J. M. (1986b) *X-ray Spectrom.* **15**, 143.
Bastin G. F. & Heijligers H. J. M. (1986c) *Microbeam Analysis* – 1986, p. 285.
Bastin G. F. & Heijligers H. J. M. (1989) *Microbeam Analysis* – 1989, p. 207.
Bastin G. F. & Heijligers H. J. M. (1991) In *Electron Probe Quantitation*, ed. K. F. J. Heinrich & D. E. Newbury (New York: Plenum), p. 145.
Bastin G. F., Heijligers H. J. M. & van Loo F. J. J. (1986) *Scanning* **8**, 45.

---

[1] For details of conference proceedings (indicated here by title or initials), see following sections.

Bastin G. F., van Loo F. J. J., Vosters P. J. C. & Vrolijk J. G. A. (1983) *Scanning* **5**, 172.
Bastin G. F., van Loo, F. J. J. & Heijligers H. J. M. (1984) *X-ray Spectrom.* **13**, 91.
Baun W. L. (1969) *Rev. Sci. Instr.* **40**, 1101.
Bawdekar V. S. (1975) *IEEE Trans. Nucl. Sci.* **22**, 282.
Beaman D. R. & Isasi J. A. (1970) *Anal. Chem.* **42**, 1540.
Beaman D. R., Isasi J. A., Birnbaum H. K. & Lewis R. (1972) *J. Phys. E* **5**, 767.
Belk J. A. (1966) *ICXOM* 4, p. 120.
Bence A. E. & Albee A. L. (1968) *J. Geol.* **76**, 382.
Berger S. L. & Seltzer S. M. (1964) In *Studies of Penetration of Charged Particles in Matter*, Nat. Res. Council publ. 1133 (Washington: Nat. Acad. Sci.), p. 205.
Bethe H. A. (1930) *Ann. Phys. Leipz.* **5**, 325.
Bethe H. A. & Ashkin J. (1953) *Experimental Nuclear Physics* (New York: John Wiley).
Bethe H. A., Rose M. E. & Smith L. P. (1938) *Proc. Amer. Phil. Soc.* **78**, 573.
Birks L. S. & Seal R. T. (1957) *J. Appl. Phys.* **28**, 541.
Bishop H. E. (1965) *Proc. Phys. Soc.* **85**, 855.
Bishop H. E. (1966a) *ICXOM* 4, p. 153.
Bishop H. E. (1966b) *ICXOM* 4, p. 112.
Bishop H. E. (1966c) Ph.D. thesis, Univ. Cambridge.
Bishop H. E. (1967) *Brit. J. Appl. Phys.* **18**, 703.
Bishop H. E. (1968) *J. Phys. D* **1**, 673.
Bishop H. E. (1974) *J. Phys. D.* **7**, 2009.
Bishop H. E. & Poole D. M. (1973) *J. Phys. D.* **6**, 1142.
Bizouard H. & Charpentier F. (1979) *J. Microsc. Spectrosc. Electr.* **4**, 149.
Bloch, F. (1933) *Zeit. Phys.* **81**, 363.
Blodgett K. B. (1935) *J. Amer. Chem. Soc.* **57**, 1007.
Blodgett K. B. & Langmuir I. (1937) *Phys. Rev.* **51**, 964.
Bloomer R. N. (1957) *Brit. J. Appl. Phys.* **8**, 83.
Blum F. & Brandt M. P. (1973) *X-ray Spectrom.* **2**, 121.
Böcker J. & Hehenkamp T. (1977) *Mikrochim. Acta Suppl.* **7**, 209.
Bombelka E. & Richter F.-W. (1988) *X-ray Spectrom.* **17**, 23.
Borovskii I. B. (1953) *Coll. Probl. Metall.*, 135.
Borovskii I. B. & Rydnik V. I. (1968) In *Quantitative Electron Probe Microanalysis*, ed. K. F. J. Heinrich, N.B.S. spec. publ. 298 (Washington: U.S. Dept. Comm.), p. 35.
Bostrom T. E. & Nockolds C. E. (1989) *Microbeam Analysis* – 1989, p. 233.
Bothe W. (1927) *Handb. Phys.* (Berlin: Springer) **24**, 18.
Bowman M. J. & Hardie D. I. (1972) *J. Phys. E* **5**, 9.
Bragg W. H. & Bragg W. L. (1913) *Proc. Roy. Soc. A* **88**, 428.
Bragg W. L., James R. W. & Bosanquet C. H. (1921) *Phil. Mag.* **41**, 309.
Broers A. N. (1967) *J. Appl. Phys.* **38**, 1991.
Brombach J. D. (1978) *X-ray Spectrom.* **7**, 81.
Brown D. B. & Ogilvie R. E. (1966) *J. Appl. Phys.* **37**, 4429.
Brown D. B., Wittry D. B. & Kyser D. F. (1969) *J. Appl. Phys* **40**, 1627.
Brown J. D. (1969) *Adv. Electronics & Electron Phys. Suppl.* **6**, 45.
Brown J. D. & Chan A. (1990) *ICXOM* 12, p. 60.
Brown J. D. & Packwood R. H. (1982) *X-ray Spectrom.* **11**, 187.
Brown J. D. & Parobek L. (1972) *ICXOM* 6, p. 163.
Brown J. D. & Parobek L. (1973) *Adv. X-ray Anal.* **16**, 198.

Brown J. D. & Parobek L. (1974) *Adv. X-ray Anal.* **17**, 479.
Brown J. D. & Parobek L. (1976) *X-ray Spectrom.* **5**, 36.
Brown J. D., Schwaab P. & Von Rosenstiel A. P. (1984) *J. de Phys. Coll.* **C2**, 609.
Büchner A. R. & Stienen J. P. M. (1975) *Mikrochim. Acta Suppl.* 6, 227.
Burek A. J. & Blake R. L. (1973) *Adv. X-ray Anal.* **16**, 37.
Burger P., Lampert M. O., Henck R. & Kemmer J. (1984) *IEEE Trans. Nucl. Sci.* **31**, 344.
Burhop E. H. S. (1940) *Proc. Camb. Phil. Soc.* **36**, 43.
Burhop E. H. S. (1955) *J. Phys. Rad.* **16**, 625.
Burkhalter P. G., Brown J. D. & Myklebust R. L. (1966) *Rev. Sci. Instr.* **37**, 1267.
Bussolati C., Manfredi P. F., Marioli D. & Krasowski R. (1978) *Nucl. Instr. Meth.* **156**, 553.
Cairns J. A., Desborough C. L. & Holloway D. F. (1970) *Nucl. Instr. Meth.* **88**, 239.
Caldwell D. O. (1955) *Phys. Rev.* **100**, 291.
Campbell A. J. & Gibbons R. (1966) In *The Electron Microprobe*, ed. T. D. McKinley, K. F. J. Heinrich & D. B. Wittry (New York: Wiley), p. 75.
Caruso A. J. & Kim H. H. (1968) *Rev. Sci. Instr.* **39**, 1059.
Castaing R. (1951) Ph.D. thesis, Univ. Paris.
Castaing R. (1960) *Adv. Electronics & Electron Phys.* **13**, 317.
Castaing R. (1967) *Phys. Bull.* **18**, 93.
Castaing R. & Descamps J. (1955) *J. Phys. Rad.* **16**, 304.
Castaing R. & Hénoc J. (1966) *ICXOM* 4, p. 120.
Cazaux J. (1986) *J. Appl. Phys.* **59**, 1418.
Cazaux J. & Le Gressus C. (1991) *Scanning Microsc.* **5**, 17.
Charles M. W. (1971) *J. Appl. Phys.* **42**, 3329.
Charles M. W. (1972) *J. Phys. E* **5**, 95.
Charles M. W. & Cooke B. A. (1968) *Nucl. Instr. Meth.* **61**, 31.
Cobet U. & Traub F. (1971) *Exp. Tech. Phys.* **19**, 479.
Cockett G. H. & Davis C. D. (1963) *Brit. J. Appl. Phys.* **14**, 813.
Cohen D. D. (1987) *X-ray Spectrom.* **16**, 237.
Colby J. W. (1968) *Adv. X-ray Anal.* **11**, 287.
Compton A. H. & Allison S. K. (1935) *X-rays in Theory and Experiment* (New York: Van Nostrand).
Conru H. W. & Laberge P. C. (1975) *J. Phys. E* **8**, 136.
Cosslett V. E. (1964) *Brit. J. Appl. Phys.* **15**, 107.
Cosslett V. E. & Duncumb P. (1956) *Nature* **177**, 1172.
Cosslett V. E. & Thomas R. N. (1964a) *Brit. J. Appl. Phys.* **15**, 1283.
Cosslett V. E. & Thomas R. N. (1964b) *Brit. J. Appl. Phys.* **15**, 883.
Cosslett V. E. & Thomas R. N. (1965) *Brit. J. Appl. Phys.* **16**, 779.
Coulon J. & Zeller C. (1973) *Compt. Rend. Acad. Sci. Paris B* **276**, 215.
Covell D. F., Sandomire M. M. & Eichen M. S. (1960) *Anal. Chem.* **32**, 1086.
Cox M. G. C., Love G. & Scott V. D. (1979) *J. Phys. D* **12**, 1441.
Cox C. E., Lowe B. G. & Sareen R. A. (1988) *IEEE Trans. Nucl. Sci.* **35**, 28.
Crawford C. K. (1979) *Scanning Electron Microsc./1979/1*, 19.
Crewe A. V., Eggenberger D. N., Wall J. & Welter L. M. (1968) *Rev. Sci. Instr.* **39**, 576.
Criss J. (1968) In *Quantitative Electron Probe Microanalysis*, ed. K. F. J. Heinrich, N.B.S. spec. publ. 298 (Washington: U.S. Dept. Comm.), p. 53.

Criss J. & Birks L. S. (1966) In *The Electron Microprobe*, ed. T. D. McKinley K. F. J. Heinrich & D. B. Wittry (New York: Wiley), p. 217.
Curgenven L. & Duncumb P. (1971) *T.I. Res. Lab.* rep. no. 303.
Danguy L. & Quivy R. (1956) *J. Phys. Rad.* **16**, 320.
Darlington E. F. H. St. G. (1975) *J. Phys. D* **8**, 85.
Darwin C. G. (1914) *Phil. Mag.* **27**, 675.
den Boggende A. J. F., Brinkman A. C. & de Graaff W. (1969) *J. Phys. E* **2**, 701.
Derian J. C. & Castaing R. (1966) *ICXOM* 4, p. 193.
Deslattes R. D., Simson B. G. & La Villa R. E. (1966) *Rev. Sci. Instr.* **37**, 596.
Ditsman S. A. (1960) *Izv. Akad. Nauk SSSR Ser, Fiz.* **24**, 367.
Djurić & Cerović (1969) *ICXOM* 5, p. 99.
Dolby R. M. & Cosslett V. E. (1960) *ICXOM* 2, p. 351.
Drescher H., Reimer L. & Seidel H. (1970) *Zeit. Angew. Phys.* **29**, 331.
DuMond J. & Youtz J. P. (1935) *Phys. Rev.* **48**, 703.
Duerr J. S. & Ogilvie R. E. (1972) *Anal. Chem.* **44**, 2361.
Duncumb P. (1960) *ICXOM* 2, p. 365.
Duncumb P. (1963) *ICXOM* 3, p. 431.
Duncumb P. (1991) pers. comm.
Duncumb P. & Cosslett V. E. (1957) *ICXOM* 1, p. 374.
Duncumb P. & Da Casa C. (1967) Paper presented at Conf. on Electron Probe Microanalysis, London.
Duncumb P. & Melford D. A. (1960) *ICXOM* 2, p. 358.
Duncumb P. & Melford D. A. (1966) *ICXOM* 4, p. 240.
Duncumb P. & Reed S. J. B. (1968) In *Quantitative Electron Probe Microanalysis*, ed. K. F. J. Heinrich, N.B.S. spec. publ. 298 (Washington: U.S. Dept. Comm.) p. 133.
Duncumb P. & Shields P. K. (1966) In *The Electron Microprobe*, ed. T. D. McKinley, K. F. J. Heinrich & D. B. Wittry (New York: Wiley), p. 284.
Duncumb P., Shields-Mason P. K. & Da Casa C. (1969) *ICXOM* 5, p. 146.
Dürr G., Hofer W. O., Schulz F. & Wittmaack K. (1971) *Zeit. Phys.* **246**, 312.
Dyson N. A. (1959) *Proc. Phys. Soc.* **73**, 924.
Dyson N. A. (1974) *Nucl. Instr. Meth.* **114**, 131.
Elad E. (1972) *Trans. IEEE Nucl. Sci.* **19**, 403.
Elion H. A. & Ogilvie R. E. (1962) *Rev. Sci. Instr.* **33**, 753.
Everhart T. E. (1960) *J. Appl. Phys.* **31**, 1483.
Everhart T. E. & Thornley R. F. M. (1960) *J. Sci. Instr.* **37**, 246.
Fairstein E. & Hahn J. (1965) *Nucleonics* **23**, 50.
Fano U. (1947) *Phys. Rev.* **72**, 26.
Fioratti M. P. & Piermattei S. R. (1971) *Nucl. Instr. Meth.* **96**, 605.
Fiori C. E., Myklebust R. L., Heinrich K. F. J. & Yakowitz H. (1976) *Anal. Chem.* **48**, 233.
Fiori C. E., Myklebust R. L. & Gorlen K. (1981) In *Energy Dispersive X-ray Spectrometry*, ed. K. F. J. Heinrich, D. E. Newbury, R. L. Myklebust & C. E. Fiori, N.B.S. spec. publ. 604 (Washington: U.S. Dept. Comm.), p. 233.
Fitzgerald R., Keil K. & Heinrich K. F. J. (1968) *Science* **159**, 528.
Fontijn L. A., Bok A. B. & Kornet J. G. (1969) *ICXOM* 5, p. 261.
Fredriksson K. (1966) *ICXOM* 4, p. 305.
Friskney C. A. & Haworth C. W. (1967) *J. Appl. Phys.* **38**, 3796.
Frost M. T., Harrowfield I. R. & Zuiderwyk M. (1981) *J. Phys. E* **14**, 597.

Futamato M., Nakazawa M., Usami K., Hosoki S. & Kawabe U. (1980) *J. Appl. Phys.* **51**, 3869.
Gallagher W.J. & Cipolla S.J. (1974) *Nucl. Instr. Meth.* **125**, 405.
Gaber M. (1987) *X-ray Spectrom.* **16**, 17.
Gedcke D.A., Ayers J.B. & DeNee P.B. (1978) *Scanning Electron Micr./1978/2*, 581.
Gehrke R.J. & Davies R.C. (1975) *Anal. Chem.* **47**, 1537.
Gennai N., Murata K. & Shimizu R. (1971) *Jap. J. Appl. Phys.* **10**, 491.
Golstein J.I., Newbury D.E., Echlin P., Joy D.C., Fiori C.E. & Lifshin E. (1981) *Scanning Electron Microscopy and X-ray Microanalysis* (New York: Plenum).
Goldstein J.I., Majeske F.J. & Yakowitz H. (1967) *Adv. X-ray Anal.* **10**, 431.
Golijanin D.M., Wittry D.B. & Sun S. (1989) *Microbeam Analysis – 1989*, p. 186.
Goudsmit S. & Saunderson J.L. (1940) *Phys. Rev.* **57**, 24.
Goulding F.S. (1966) *Nucl. Instr. Meth.* **43**, 1.
Goulding F.S. (1977) *Nucl. Instr. Meth.* **142**, 213.
Goulding F.S., Walton J.T. & Malone D.F. (1969) *Nucl. Instr. Meth.* **71**, 273.
Greaves C. (1970) *Vacuum* **20**, 332.
Green M. (1962) Ph.D. thesis, Univ. Cambridge.
Green M. (1963a) *Proc. Phys. Soc.* **82**, 204.
Green M. (1963b) *ICXOM* 3, p. 361.
Green M. (1964) *Proc. Phys. Soc.* **83**, 435.
Green M. & Cosslett V.E. (1961) *Proc. Phys. Soc.* **78**, 1206.
Green M. & Cosslett V.E. (1968) *J. Phys. D* **1**, 425.
Gui-Nian D. & Turner K.E. (1989) *X-ray Spectrom.* **18**, 57.
Guttmann A.J. & Wagenfeld H. (1967) *Acta Cryst.* **22**, 334.
Haine M.E. & Einstein P.A. (1952) *Brit. J. Appl. Phys.* **3**, 40.
Haine M.E., Einstein P.A. & Borcherds P.H. (1958) *Brit. J. Appl. Phys.* **9**, 482.
Hall T.A. (1968) In *Quantitative Electron Probe Microanalysis*, ed. K.F.J. Heinrich, N.B.S. spec. publ. 298 (Washington: U.S. Dept. Comm.), p. 269.
Hall T.A. (1977) *J. Microsc.* **110**, 103.
Ham W.R. (1910) *Phys. Rev.* **30**, 1.
Hanna G.C., Kirkwood D.H.W. & Pontecorvo B. (1949) *Phys. Rev.* **75**, 985.
Hay H.J. (1985) *Nucl. Instr. Meth.* **B10/11**, 624.
Heckel J. & Jugelt P. (1984) *X-ray Spectrom.* **13**, 159.
Hehenkamp T. & Böcker J. (1974) *Mikrochim. Acta suppl.* **5**, 29.
Heinrich K.F.J. (1962) *Rev. Sci. Instr.* **33**, 884.
Heinrich K.F.J. (1963) *Adv. X-ray Anal.* **6**, 291.
Heinrich K.F.J. (1964) *Adv. X-ray Anal.* **7**, 325.
Heinrich K.F.J. (1966) In *The Electron Microprobe*, ed. T.D. McKinley, K.F.J. Heinrich & D.B. Wittry (New York: Wiley), p. 296.
Heinrich K.F.J. (1967) *EPASA* 2, paper no. 7.
Heinrich K.F.J. (1968) In *Quantitative Electron Probe Microanalysis* ed. K.F.J. Heinrich, NBS spec. publ. 298 (Washington: US Dept Comm.) p. 5.
Heinrich K.F.J. (1972) *Anal. Chem.* **44**, 350.
Heinrich K.F.J. (1987a) *Microbeam Analysis – 1987*, p. 24.
Heinrich K.F.J. (1987b) *ICXOM* 11, p. 67.
Heinrich K.F.J., Vieth D. & Yakowitz H. (1966) *Adv. X-ray Anal.* **9**, 208.
Hendricks R.W. (1969) *Rev. Sci. Instr.* **40**, 1216.
Hendricks R.W. (1972) *Nucl. Instr. Meth.* **102**, 309.

Henke B. L. (1965) *Adv. X-ray Anal.* **8**, 269.
Henke B. L. & Elgin R. L. (1970) *Adv. X-ray Anal.* **13**, 639.
Henke B. L., Lee P., Tanaka T. J., Shimabukuro R. L. & Fujikawa B. K. (1982) *Atom. Data Nucl. Data Tables* **27**, 1.
Henke B. L., Uejio B. L., Yamada H. T. & Tackaberry R. E. (1986) *Opt. Eng.* **25**, 937.
Hénoc J. (1962) Ph.D. thesis, Univ. Paris.
Hénoc J. (1968) In *Quantitative Electron Probe Microanalysis*, ed. K. F. J. Heinrich, N.B.S. spec. publ. 298 (Washington: U.S. Dept. Comm.), p. 197.
Hénoc J., Heinrich K. F. J. & Myklebust R. L. (1973) *A Rigorous Correction Procedure for Quantitative Electron Probe Microanalysis (COR)*, NBS tech. note 769 (Washington: U.S. Dept. Comm.).
Hénoc J., Heinrich K. F. J. & Zemskoff A. (1969) *ICXOM* 5, p. 187.
Herrmann R. & Reimer L. (1984) *Scanning* **6**, 20.
Hillier J. (1947) U.S. patent no. 2419029.
Hohn F. J. (1985) *Scanning Electron Microsc./1985/4*, 1327.
Holland L., Laurenson L., Baker P. N. & Davis M. J. (1972) *Nature* **238**, 36.
Howes J. H. & Allsworth F. L. (1986) *IEEE Trans. Nucl. Sci.* **33**, 283.
Huang L. Y. & Lei H. N. (1981) *Scanning Electron Microsc./1981/1*, 21.
Hunger H.-J. & Küchler L. (1979) *Phys. Stat. Sol. A* **59**, 35.
Hutchins G. A. (1966) In *The Electron Microprobe*, ed. T. D. McKinley, K. F. J. Heinrich & D. B. Wittry (New York: Wiley), p. 390.
Ingram P. & Shelburne J. D. (1980) *Scanning Electron Microsc./1980/2*, 285.
Iwanczyk J. S., Dabrowski A. J., Huth G. C., Bradley J. G., Conley J. M. & Albee A. L. (1986) *IEEE Trans. Nucl. Sci.* **33**, 355.
Iwanczyk J. S., Dabrowski A. J., Huth G. C. & Economou T. E. (1985) *Appl. Phys. Lett.* **46**, 606.
Jaklevic J. M. & Goulding F. S. (1971) *IEEE Trans. Nucl. Sci.* **18**, 187.
Jaklevic J. M. & Goulding F. S. (1972) *IEEE Trans. Nucl. Sci.* **19**, 384.
Jaklevic J. M., Goulding F. S. & Landis D. A. (1972) *IEEE Trans. Nucl. Sci.* **19**, 392.
Jenkins R., Croke J. F. & Wastberg R. G. (1972) *X-ray Spectrom.* **1**, 59.
Jenkins R., Manne R., Robin R. & Senemand C. (1991) *X-ray Spectrom.* **20**, 149.
Johann H. H. (1931) *Zeit. Phys.* **69**, 185.
Johansson T. (1932) *Naturwiss.* **20**, 758.
Johansson T. (1933) *Zeit. Phys.* **82**, 507.
Johansson S. A. E. & Campbell J. L. (1988) *Proton-induced X-ray Emission* (New York: Wiley).
Jones M. P., Gavrilovic J. & Beaven C. H. J. (1966) *Trans. Inst. Mining Metall.* **75**, 273.
Joy D. C. & Luo S. (1989) *Scanning* **11**, 176.
Joy D. C., Romig A. D. & Goldstein J. I., eds. (1986) *Principles of Analytical Electron Microscopy*, (New York: Plenum).
Kanaya K. & Ono S. (1978) *J. Phys. D* **11**, 1495.
Kandiah K. (1966) In *Radiation Measurements in Nuclear Power* (London: Inst. Phys.), p. 420.
Kandiah K. (1971) *Nucl. Instr. Meth.* **95**, 289.
Kandiah K., Smith A. J. & White G. (1975) *IEEE Trans. Nucl. Sci.* **22**, 2058.
Kandiah K., Stirling A., Trotman D. L. & White G. (1968) *Int. Symp. Nucl. Electron.*, Versailles, AERE, Harwell, rept. no. R5852.
Kanter H. (1957) *Ann. Phys. Lpz.* **20**, 144.

Karduck P. & Rehbach W. (1985) *Mikrochim. Acta Suppl.* **11**, 289.
Karduck P. & Rehbach W. (1988) *Microbeam Analysis* – 1988, p. 277.
Karlovac N. & Gedcke D. A. (1973) *EPASA* 8, paper no. 17.
Kawabe K., Takagi S., Saito M. & Tagata S. (1988) *Microbeam Analysis* – 1988, p. 341.
Kaye G. W. C. (1909) *Phil. Trans. Roy. Soc. A* **209**, 123.
Keith H. D. & Loomis T. C. (1976) *X-ray Spectrom.* **5**, 93.
Kemmer J. (1980) *Nucl. Instr. Meth.* **169**, 499.
Kerrick D. M., Eminhizer L. B. & Villaume J. F. (1973) *Amer. Mineral.* **58**, 920.
Khan M. R. & Karimi M. (1980) *X-ray Spectrom.* **9**, 32.
Kimoto S. & Hashimoto H. (1966) In *The Electron Microprobe*, ed. T. D. McKinley, K. F. J. Heinrich & D. B. Wittry (New York: Wiley), p. 480.
Kirianenko A., Maurice F., Calais D. & Adda Y. (1963) *ICXOM* 3, p. 559.
Kirkpatrick P. & Hare D. G. (1934) *Phys. Rev.* **46**, 831.
Kirkpatrick P. & Wiedmann L. (1945) *Phys. Rev.* **67**, 321.
Kitazawa T., Shuman H. & Somlyo A. P. (1983) *Ultramicrosc.* **11**, 251.
Knoll M. (1935) *Zeit. Tech. Phys.* **16**, 467.
Kohlhaas E. & Scheiding F. (1970) *Mikrochim. Acta Suppl.* **4**, 131.
Kramers H. A. (1923) *Phil. Mag.* **46**, 836.
Krause M. O. (1979) *J. Phys. Chem. Ref. Data* **8**, 307.
Küchler L. (1990) *ICXOM* 12, 147.
Kulenkampff H. & Spyra W. (1954) *Zeit. Phys.* **137**, 416.
Kyser D. F. (1972) *ICXOM* 6, p. 147.
Kyser D. F. & Murata K. (1974) *IBM J. Res. Dev.* **18**, 352.
Lábár J. (1987) *X-ray Spectrom.* **16**, 33.
Lafferty J. N. (1951) *J. Appl. Phys.* **22**, 229.
Laguitton D., Rouseau R. & Claisse F. (1975) *Anal. Chem.* **47**, 2174.
Landis D. A., Goulding F. S. & Jaklevic J. M. (1970) *Nucl. Instr. Meth.* **87**, 211.
Landis D. A., Goulding F. S. & Jarrett B. V. (1972) *Nucl. Instr. Meth.* **101**, 127.
Langmuir D. B. (1937) *Proc. Inst. Radio Eng.* **25**, 977.
Lantto V. (1979) *J. Phys. D* **12**, 1181.
Lawson W. H. (1967) *J. Sci. Instr.* **44**, 917.
Le Poole J. B. (1964) In *Proc. 3rd Eur. Conf. on Electron Microsc., Prague* (Prague: Czech. Acad. Sci) p. 439.
Lenard P. (1895) *Ann. Phys. Lpz.* **56**, 255.
Leroux J. (1961) *Adv. X-ray Anal.* **5**, 153.
Lifshin E., Ciccarelli M. F. & Bolon R. B. (1975) In *Practical Scanning Electron Microscopy, Electron and Ion Microprobe Analysis*, ed. J. I. Goldstein & H. Yakowitz (New York: Plenum), p. 263.
Lineweaver J. L. (1963) *J. Appl. Phys.* **34**, 1786.
Lins S. J. & Kukuk H. S. (1960) In *Trans. 7th Natl. Symp. on Vac. Technol.* (New York: Pergamon) p. 333.
Livingston H. S. & Bethe H. A. (1937) *Rev. Mod. Phys.* **9**, 245.
Llacer J., Haller E. E. & Cordi R. C. (1977) *IEEE Trans. Nucl. Sci.* **24**, 53.
Long J. V. P. & Cosslett V. E. (1957) *ICXOM* 1, p. 435.
Love G. & Scott V. D. (1978a) *J. Phys. D* **11**, 7.
Love G. & Scott V. D. (1978b) *J. Phys. D* **11**, 1369.
Love G. & Scott V. D. (1980) *ICXOM* 9, 28.
Love G., Cox M. G. C. & Scott V. D. (1974) *J. Phys. D* **7**, 2142.
Love G., Cox M. G. C. & Scott V. D. (1975) *J. Phys. D* **8**, 1686.
Love G., Cox M. G. C. & Scott V. D. (1977) *J. Phys. D* **10**, 7.
Love G., Cox M. G. C. & Scott V. D. (1978a) *J. Phys. D* **11**, 7.

Love G., Cox M. G. C. & Scott V. D. (1978b) *J. Phys.* D **11**, 23.
Love G., Sewell D. A. & Scott V. D. (1984) *J. Physique* **45**, coll. C2, 21.
Lowe B. G. (1989) *Ultramicrosc.* **28**, 150.
Lurio A. & Reuter W. (1975) *Appl. Phys. Lett.* **27**, 704.
Lurio A. & Reuter W. (1977) *J. Phys.* D **10**, 2127.
Lurio A., Reuter W. & Keller J. (1977) *Adv. X-ray Anal.* **20**, 481.
Mackenzie A. (1990) *Microbeam Analysis – 1990*, p. 165.
Mackenzie A. (1991) Physica C **178**, 365.
Madden N. W., Hanepen G. & Clark B. C. (1986) *IEEE Trans. Nucl. Sci.* **33**, 303.
Madden N. W., Jaklevic J. M., Walton J. T. & Wiegand C. E. (1979) *Nucl. Instr. Meth.* **159**, 337.
Maenhaut W. & Raemdonck H. (1984) *Nucl. Instr. Meth.* B **1**, 123.
Mahesh K. (1976) *Nucl. Instr. Meth* **133**, 57.
Marinenko R. B., Myklebust R. L., Bright D. S. & Newbury D. E. (1989) *J. Microsc.* **155**, 183.
Mariscotti M. A. (1967) *Nucl. Instr. Meth.* **50**, 309.
Markowicz A., Storms H. & Van Grieken R. (1986) *X-ray Spectrom.* **15**, 115.
Marshall A. T. & Carde D. (1984) *J. Microsc.* **134**, 113.
Marshall A. T., Carde D. & Kent M. (1985) *J. Microsc.* **139**, 335.
Matsuya M., Fukuda H., Kawabe K., Sekiguchi H., Inagawa H. & Saito M. (1988) *Microbeam Analysis – 1988*, p. 329.
Maurice F., Seguin R. & Hénoc J. (1966) *ICXOM* 4, p. 357.
McAfee W. S. (1976) *J. Appl. Phys.* **47**, 1179.
McCoy D. D. & Gutmacher R. G. (1975) *Rev. Sci. Instr.* **46**, 460.
McMillan D. J., Baughman G. D. & Schamber F. H. (1985) *Microbeam Analysis – 1985*, p. 137.
Melford D. A. (1962) *J. Inst. Met.* **90**, 217.
Merlet C. & Bodinier J.-L. (1990) *Chem. Geol.* **83**, 55.
Moll S. H., Baumgarten N. & Donnelly W. (1980) *ICXOM* 8, p. 87.
Moseley H. G. J. (1913) *Phil. Mag.* **26**, 1024.
Moseley H. G. J. (1914) *Phil. Mag.* **27**, 703.
Mott N. F. & Massey H. S. W. (1949) *The Theory of Atomic Collisions* (Oxford: Clarendon Press).
Mulvey T. (1959) *J. Sci. Instr.* **36**, 350.
Mulvey T. (1967) In *Focusing of Charged Particles* (New York: Academic Press), p. 469.
Munden A. B. & Yeoman-Walker D. E. (1973) *J. Phys.* E **6**, 916.
Murata K., Matsukawa T. & Shimizu R. (1971) *Jap. J. Appl. Phys.* **10**, 678.
Musket R. G. (1986) *Nucl. Instr. Meth.* B **15**, 735.
Myklebust R. L., Fiori C. E. & Heinrich K. F. J. (1979) *Frame C: A Compact Procedure for Quantitative Energy-Dispersive Electron Probe X-ray Analysis*, NBS tech. note 1106 (Washington: U.S. Dept. Comm.).
Neubert G. & Rogaschewski S. (1980) *Phys. Stat. Sol.* A **59**, 35.
Neumann B., Reimer L. & Wellmans B. (1978) *Scanning* **1**, 130.
Nicholson J. B. & Hasler M. F. (1966) *Adv. X-ray Anal.* **9**, 420.
Nockolds C., Nasir M. J., Cliff G. & Lorimer G. W. (1980) In *Electron Microscopy and Analysis* 1979, ed. T. Mulvey (Bristol: Inst. Phys.), p. 417.
Nowlin C. H. & Blankenship J. L. (1965) *Rev. Sci. Instr.* **36**, 1830.
Nullens H., Van Espen P. & Adams F. (1979) *X-ray Spectrom.* **8**, 104.
O'Boyle D. (1965) *J. Appl. Phys.* **36**, 2849.
O'Brien L. P. (1963) *Adv. X-ray Anal.* **6**, 268.

Ordonez J. (1971) *Metallogr.* **4**, 575.
Packwood R. H. & Brown J. D. (1981) *X-ray Spectrom.* **10**, 138.
Paduch J. & Barszcz E. (1986) *J. Microsc. Spectrosc. Electr.* **11**, 81.
Page R. S. & Openshaw I. K. (1960) *ICXOM* 2, p. 385.
Parker B. A. (1980) *Micron* **11**, 331.
Pawley J. B., Hayes T. L. & Falk R. H. (1976) *Scanning Electron Microsc./1976/1*, 187.
Pell E. M. (1960) *J. Appl. Phys.* **31**, 291.
Philibert J. (1963) *ICXOM* 3, p. 379.
Philibert J. & Penot D. (1966) *ICXOM* 4, p. 365.
Philibert J. & Tixier R. (1968a) *J. Phys. D* **1**, 685.
Philibert J. & Tixier R. (1968b) In *Quantitative Electron Probe Microanalysis*, ed. K. F. J. Heinrich, N.B.S. spec. publ. 298 (Washington: U.S. Dept. Comm.), p. 13.
Philibert J. & Weinryb E. (1963) *ICXOM* 3, p. 451.
Poole D. M. & Thomas P. M. (1963) *ICXOM* 3, p. 411.
Poole D. M. & Thomas P. M. (1966) In *The Electron Microprobe*, ed. T. D. McKinley, K. F. J. Heinrich & D. B. Wittry (New York: Wiley) p. 269.
Pouchou J. L. & Pichoir F. (1984a) *Rech. Aérosp.* 1984–5, 13.
Pouchou J. L. & Pichoir F. (1984b) *Rech. Aérosp.* 1984–5, 47.
Pouchou J. L. & Pichoir F. (1985) *J. Microsc. Spectrosc. Electr.* **10**, 279.
Pouchou J. L. & Pichoir F. (1986) *J. Microsc. Spectrosc. Electr.* **11**, 229.
Pouchou J. L. & Pichoir F. (1987) *ICXOM* 11, p. 249.
Pouchou J. L. & Pichoir F. (1988) *Microbeam Analysis* – 1988, p. 319.
Pouchou J. L. & Pichoir F. (1991) In *Electron Probe Quantitation*, ed. K. F. J. Heinrich & D. E. Newbury (New York: Plenum), p. 31.
Pouchou J. L., Pichoir F. & Boivin D. (1990) *ICXOM* 12, p. 52.
Pouchou J. L., Pichoir F. & Girard F. (1980) *J. Microsc. Spectrosc. Electr.* **5**, 425.
Powell C. J. (1976) *Rev. Mod. Phys.* **48**, 33.
Powell C. J. (1990) *Microbeam Analysis* – 1990, p. 13.
Prins J. A. (1930) *Zeit. Phys.* **63**, 477.
Radzimski Z. J. (1987) *Scanning Microsc.* **1**, 975.
Ramachandran K. N. (1975) *Rev. Sci. Instr.* **46**, 1662.
Ranzetta G. V. T. & Scott V. D. (1967) *J. Sci. Instr.* **44**, 983.
Rao-Sahib T. S. & Wittry D. B. (1974) *J. Appl. Phys.* **45**, 5060.
Raznikov V. V., Dodonov A. F. & Lanin E. V. (1977) *Int. J. Mass. Spectrom. Ion Phys.* **25**, 295.
Reed S. J. B. (1964) Ph.D. thesis, Univ. Cambridge.
Reed S. J. B. (1965) *Brit. J. Appl. Phys.* **16**, 913.
Reed S. J. B. (1966) *ICXOM* 6, p. 339.
Reed S. J. B. (1972) *J. Phys. E* **5**, 997.
Reed S. J. B. (1975) *X-ray Spectrom.* **4**, 14.
Reed S. J. B. (1990a) *Microbeam Analysis* – 1990, p. 109.
Reed S. J. B. (1990b) *Microbeam Analysis* – 1990, p. 181.
Reed S. J. B. & Long J. V. P. (1963) *ICXOM* 3, p. 317.
Reed S. J. B. & Mason P. K. (1967) *EPASA* 2, paper no. 12.
Reed S. J. B. & Ware N. G. (1972) *J. Phys. E* **5**, 582.
Reed S. J. B. & Ware N. G. (1973) *X-ray Spectrom.* **2**, 69.
Rehbach W. & Karduck P. (1988) *Microbeam Analysis* – 1988, p. 285.
Rehbach W., Karduck P. & Burchard W.-G. (1985) *Mikrochim. Acta Suppl.* **11**, 309.

Reimer L. (1986) *Scanning Electron Microscopy – Physics of Image Formation and Microanalysis* (Glasgow: Blackie).
Reimer L. & Bernsen P. (1984) *Scanning Electron Microsc./1984/4*, 1707.
Reimer L. & Krefting E. R. (1975) In *Use of Monte-Carlo Calculations in Electron Probe Microanalysis and Scanning Electron Microscopy*, ed. K. F. J. Heinrich, D. E. Newbury & H. Yakowitz, N.B.S. spec. publ. 460 (Washington: U.S. Dept. Comm.), p. 45.
Reimer, L. & Lödding B. (1984) *Scanning* **6**, 128.
Reuter W. (1972) *ICXOM* 6, p. 121.
Ribbe P. H. & Smith J. V. (1966) *J. Geol.* **74**, 217.
Robertson, A., Prestwich W. V. & Kennett T. J. (1972) *Nucl. Instr. Meth.* **100**, 317.
Robinson V. N. E. (1974) *J. Phys. E* **7**, 650.
Robinson V. N. E. & Robinson B. W. (1978) *Scanning Electron Microsc./1978/1*, 595.
Robinson V. N. E., Cutmore N. G. & Burdon R. G. (1984) *Scanning Electron Microsc. 1984/2*, 483.
Roeder P. L. (1985) *Can. Mineral.* **23**, 263.
Rose A. (1948) *Adv. Electron.* **1**, 131.
Rosner B., Gur D. & Shabason L. (1975) *Nucl. Instr. Meth.* **131**, 81.
Rosseland S. (1923) *Phil. Mag.* **48**, 65.
Rouberol J. M., Tong M., Conty C. & Deschamps P. (1967) *Mikrochim. Acta Suppl.* **2**, 201.
Ruark A. & Brammer F. E. (1937) *Phys. Rev.* **52**, 322.
Russ J. C. (1972) *EPASA* 7, paper no. 76.
Russ J. C. (1974) *MAS-9*, paper no. 22.
Russ J. C., Sandborg A. O., Barnhart M. W., Soderquist C. E., Lichtinger R. W. & Walsh C. J. (1973) *Adv. X-ray Anal.* **16**, 284.
Ruste J. (1979) *J. Microsc. Spectrosc. Electr.* **4**, 123.
Ruste J. & Zeller C. (1977) *Comp. Rend. Acad. Sci. Paris B* **284**, 507.
Rutherford E. (1911) *Phil. Mag.* **21**, 669.
Sandstrom A. E. (1952) *Ark. Fys.* **4**, 519.
Sayce L. A. & Franks A. (1964) *Proc. Roy. Soc. A* **282**, 353.
Schamber F. (1973) *EPASA* 8, paper no. 85.
Schamber F. (1977) In *X-ray Fluorescence Analysis of Environmental Samples*, ed. T. G. Dzubay (Michigan: Ann Arbor Sci. Publ.), p. 241.
Schamber F. (1981) In *Energy Dispersive X-ray Spectrometry*. ed. K. F. J. Heinrich, D. E. Newbury, R. L. Myklebust & C. E. Fiori, N.B.S. spec. publ. 604 (Washington: U.S. Dept. Comm.), p. 193.
Schamber F., Wodke N. F. & McCarthy J. J. (1980) *ICXOM* 8, p. 124.
Schmitz U., Ryder P. L. & Pitsch W. (1969) *ICXOM* 5, p. 104.
Schwander H. & Gloor F. (1980) *X-ray Spectrom.* **9**, 134.
Scott V. D. (1985) *J. Microsc. Spectrosc. Electr.* **10**, 251.
Scott V. D. & Love G. (1991) In *Electron Probe Quantitation*, ed. K. F. J. Heinrich & D. E. Newbury (New York: Plenum), p. 19.
Self P. G., Norrish K., Milnes A. R., Graham J. & Robinson B. (1990) *X-ray Spectrom.* **19**, 59.
Sewell D. A., Love G. & Scott V. D. (1985a) *J. Phys. D* **18**, 1233.
Sewell D. A., Love G. & Scott V. D. (1985b) *J. Phys. D* **18**, 1245.
Sewell D. A., Love G. & Scott V. D. (1985c) *J. Phys. D* **18**, 1269.
Sewell D. A., Love G. & Scott V. D. (1987) *J. Phys. D* **20**, 1567.
Sharpe J. (1964) *Nuclear Radiation Detectors*, 2nd edn (London: Methuen).

Shimizu R., Ikuta T. & Murata K. (1972) *J. Appl. Phys.* **43**, 4233.
Shimizu R., Kataoka Y., Matsukawa T., Ikuta T., Murata K. & Hashimoto H. (1975) *J. Phys. D* **8**, 820.
Shimizu R., Kataoka Y., Ikuta T., Koshikawa T. & Hashimoto H. (1976) *J. Phys. D* **9**, 101.
Shimizu R., Murata K. & Shinoda G. (1966a) *Tech. Rep. Osaka Univ.* **16**, 415.
Shimizu R., Murata K. & Shinoda G. (1966b) *ICXOM* 4, p. 127.
Shimizu R., Murata K. & Shinoda G. (1967) *Tech. Rep. Osaka Univ.* **17**, 13.
Shinoda G. (1966) *Tech. Rep. Osaka Univ.* **4**, p. 97.
Shinoda G., Murata K. & Shimizu, R. (1968) In *Quantitative Electron Probe Microanalysis*, ed. K. F. J. Heinrich, N.B.S. spec. publ. 298 (Washington: U.S. Dept. Comm.), p. 155.
Short M. A. (1960) *Rev. Sci. Instr.* **38**, 291.
Short M. A. (1976) *X-ray Spectrom.* **5**, 169.
Smellie J. A. T. (1972) *Mineral. Mag.* **38**, 614.
Small J. A., Heinrich K. F. J., Fiori C. E., Myklebust R. L., Newbury D. E. & Dilmore M. P. (1978) *Scanning Electron Microsc./1978/1*, 445.
Small J. A., Leigh S. D., Newbury D. E. & Myklebust R. L. (1987) *J. Appl. Phys.* **61**, 459.
Small J. A., Newbury D. E., Myklebust R. L., Fiori C. E., Bell A. A. & Heinrich K. F. J. (1991) In *Electron Probe Quantitation*, ed. K. F. J. Heinrich & D. E. Newbury (New York: Plenum), p. 317.
Smith D. G. W. (1981) *X-ray Spectrom.* **10**, 78.
Smith D. G. W. & Gold C. M. (1979) *Microbeam Analysis – 1979*, p. 273.
Smith D. G. W. & Reed S. J. B. (1981) *X-ray Spectrom.* **10**, 198.
Smith D. G. W., Gold C. M. & Tomlinson D. A. (1975) *X-ray Spectrom.* **4**, 149.
Snetsinger K. G., Bunch T. E. & Keil K. (1968) *Amer. Mineral.* **53**, 1770.
Solosky L. F. & Beaman D. R. (1972) *Rev. Sci. Instr.* **43**, 1100.
Sousa C. A. (1975) *IEEE Trans. Nucl. Sci.* **22**, 109.
Spencer L. V. (1955) *Phys. Rev.* **98**, 1597.
Spendley W., Hext G. R. & Himsworth F. R. (1962) *Technometr.* **4**, 441.
Spielberg N. (1966) *Rev. Sci. Instr.* **37**, 1268.
Spielberg N. (1967a) *Adv. X-ray Anal.* **10**, 534.
Spielberg N. (1967b) *Rev. Sci. Instr.* **38**, 291.
Spielberg N. & Tsamas D. I. (1975) *Rev. Sci. Instr.* **46**, 1086.
Springer G. (1967a) *Fortschr. Mineral.* **45**, 103.
Springer G. (1967b) *Neues Jahrb. Mineral. Abhandl.* **106**, 241.
Springer G. (1972) *ICXOM* 6, p. 141.
Springer G. (1976) *X-ray Spectrom.* **5**, 88.
Springer G. & Nolan J. (1976) *Can. J. Spectrosc.* **21**, 134.
Springer G. & Rosner B. (1969) *ICXOM* 5, p. 170.
Statham P. J. (1976a) *J. Phys. E* **9**, 1023.
Statham P. J. (1976b) *X-ray Spectrom.* **5**, 154.
Statham P. J. (1977a) *X-ray Spectrom.* **6**, 94.
Statham P. J. (1977b) *Anal. Chem.* **49**, 2149.
Statham P. J. (1978) *X-ray Spectrom.* **7**, 132.
Statham P. J. (1979) *Mikrochim. Acta Suppl.* **8**, 229.
Statham P. J. (1980) *ICXOM* 9, 30.
Statham P. J. (1981) In *Energy Dispersive X-ray Spectrometry*, ed. K. F. J. Heinrich, D. E. Newbury, R. L. Myklebust & C. E. Fiori, N.B.S. spec. publ. 604 (Washington: U.S. Dept. Comm.), p. 141.

Statham P. J. & Nashishibi T. (1988) *Microbeam Analysis – 1988*, p. 50.
Statham P. J. & Pawley J. (1978) *Scanning Electron Microsc./1978/1*, 469.
Statham P. J., Long J. V. P., White G. & Kandiah K. (1973) *X-ray Spectrom.* **3**, 153.
Steel E. B. (1986) *Microbeam Analysis – 1986*, p. 439.
Steele I. M., Smith J. V., Pluth J. J. & Solberg T. N. (1975) *MAS-10*, paper no. 37.
Stephen J., Smith B. J., Marshall D. C. & Wittam E. M. (1975) *J. Phys. E* **8**, 607.
Stone R. E., Barkley V. A. & Fleming J. A. (1986) *IEEE Trans. Nucl. Sci.* **33**, 299.
Stone R. E., Walter F. J., Blackburn D. H., Pella P. & Kraner H. W. (1981) *X-ray Spectrom.* **10**, 81.
Sturrock P. A. (1951) *Phil. Trans. Roy. Soc. A* **243**, 387.
Sweatman T. R. & Long J. V. P. (1969a) *J. Petrol.* **10**, 332.
Sweatman T. R. & Long J. V. P. (1969b) *ICXOM* 5, p. 432.
Sweeney W. E., Seebold R. E. & Birks L. S. (1960) *J. Appl. Phys.* **31**, 1061.
Tanuma S. & Nagashima K. (1983) *Mikrochim. Acta 1983/1*, 299.
Tanuma S. & Nagashima K. (1984) *Mikrochim. Acta 1984/3*, 265.
Taylor P. G. & Burgess A. (1977) *J. Microsc.* **111**, 51.
Tenny H. (1968) *Metallogr.* **1**, 221.
Thomas P. M. (1963) *Brit. J. Appl. Phys.* **14**, 397.
Trammell R. C. (1978) *IEEE Trans. Nucl. Sci.* **25**, 910.
Tsukada M. (1962) *Nucl. Instr. Meth.* **14**, 241.
Van Amelsvoort J. M. M., Smits, H. T. J. & Stadhouders A. M. (1982) *Micron* **13**, 229.
Van der Mast K. D. (1983) *J. Microsc.* **130**, 309.
Van Espen P., Nullens H. & Adams F. (1980) *X-ray Spectrom.* **9**, 126.
Venuti G. S. (1983) *Scanning Electron Microsc./1983/4*, 1555.
Venuti G. S. (1987) *Scanning Microsc.* **1**, 939.
Vignes A. & Dez G. (1968) *J. Phys. D* **1**, 1309.
Vogel R. S. & Fergason L. A. (1966) *Rev. Sci. Instr.* **37**, 934.
Wakabayashi T., Miyake T., Date G. & Soezima H. (1972) *ICXOM* 6, p. 287.
Warburton W. K. & Iwanczyk J. S. (1987) *Scanning Microsc.* **1**, 135.
Ware N. G. (1981) *Comp. Geosci.* **7**, 167.
Ware N. G. (1991) *X-ray Spectrom.* **20**, 73.
Ware N. G. & Reed S. J. B. (1973) *J. Phys. E* **6**, 286.
Weisweiler W. (1969) *ICXOM* 5, p. 198.
Weisweiler W. (1972) *Mikrochim. Acta 1972*, 145.
Weisweiler W. (1974) *Arch. Eisenhütten.* **45**, 287.
Weisweiler W. (1975a) *Mikrochim. Acta 1975/1*, 365.
Weisweiler W. (1975b) *Mikrochim. Acta 1975/1*, 611.
Weisweiler W. (1975c) *Mikrochim. Acta 1975/2*, 179.
Weisweiler W. & Neff R. (1979) *Mikrochim. Acta Suppl.* **8**, 475.
Wendt M. & Schmidt A. (1978) *Phys. Stat. Sol. A* **46**, 179.
Wentzel G. (1927) *Zeit. Phys.* **43**, 524.
Wernisch J. (1985) *X-ray Spectrom.* **14**, 109.
Whiddington R. (1912) *Proc. Roy. Soc. A* **86**, 360.
White E. W. & Johnson G. G. (1970) *X-ray Emission and Absorption Wavelength and Two-Theta Tables*, ASTM Data Series DS37A, 2nd edn (Philadelphia: Amer. Soc. Test. & Mat.).

Williams C. W. (1968) *IEEE Trans. Nucl. Sci.* **15**, 297.
Williams K. L. (1987) *Introduction to X-ray Spectrometry* (London: Allen & Unwin).
Willich P. & Obertop D. (1990) *ICXOM* 12, p. 100.
Wilson R. R. (1941) *Phys. Rev.* **60**, 749.
Wittry D. B. (1958) *J. Appl. Phys.* **29**, 1543.
Wittry D. B. (1972) *J. Appl. Phys.* **6**, 471.
Wittry D. B. & Sun S. (1990a) *J. Appl. Phys.* **67**, 1633.
Wittry D. B. & Sun S. (1990b) *J. Appl. Phys.* **68**, 387.
Wittry D. B. & Sun S. (1991) *J. Appl. Phys.* **69**, 3886.
Worthington C. R. & Tomlin S. G. (1956) *Proc. Phys. Soc. A* **69**, 401.
Yakowitz H. & Heinrich K. F. J. (1969) *J. Res. NBS* **73A**, 113.
Yakowitz H., Myklebust R. L. & Heinrich K. F. J. (1973) *Frame: An On-line Correction Procedure for Quantitative Electron Probe Microanalysis*, NBS tech. note 796 (Washington: U.S. Dept. Comm.).
Zachariasen W. H. (1967) *Acta Cryst.* **23**, 558.
Ziebold T. O. & Ogilvie R. E. (1964) *Anal. Chem.* **36**, 322.
Ziebold T. O. & Ogilvie R. E. (1966) In *The Electron Microprobe*, ed. T. D. McKinley, K. F. J. Heinrich & D. B. Wittry (New York: Wiley), p. 378.

# Conference proceedings

Papers in conference proceedings are referred to in abbreviated form in the reference list; full details of these proceedings are given below.

### (1) EPASA–MAS series

The annual Electron Probe Analysis Society of America (EPASA) conferences held from 1966 to 1973 are listed below. In 1974 the name was changed to Microbeam Analysis Society (MAS), with annual conferences continuing to the present day. The only printed record of conferences up to and including 1978 consists of abstracts volumes. References to these are in the form: 'EPASA 6, paper no. 5', or from 1974 to 1978: 'MAS 11, paper no. 7', for example. Since 1979, proceedings have been published by San Francisco Press under the title: '*Microbeam Analysis* – ' followed by the year, with different editors, as listed below.

| Conf. | No. | Year | Location | Editors |
|---|---|---|---|---|
| EPASA | 1 | 1966 | College Park | |
| | 2 | 1967 | Boston | |
| | 3 | 1968 | Chicago | |
| | 4 | 1969 | Pasadena | |
| | 5 | 1970 | New York | |
| | 6 | 1971 | Pittsburgh | |
| | 7 | 1972 | San Francisco | |
| | 8 | 1973 | New Orleans | |
| MAS | 9 | 1974 | Ottawa | |
| | 10 | 1975 | Las Vegas | |
| | 11 | 1976 | Miami | |
| | 12 | 1977 | Boston | |
| | 13 | 1978 | Ann Arbor | |
| | 14 | 1979 | San Antonio | D. E. Newbury |
| | 15 | 1980 | San Francisco | D. B. Wittry |
| | 16 | 1981 | Vail | R. H. Geiss |
| | 17 | 1982 | Washington | K. F. J. Heinrich |
| | 18 | 1983 | Phoenix | R. Gooley |

| Conf. | No. | Year | Location | Editors |
|---|---|---|---|---|
| | 19 | 1984 | Bethlehem | A. D. Romig & J. I. Goldstein |
| | 20 | 1985 | Louisville | J. T. Armstrong |
| | 21 | 1986 | Albuquerque | A. D. Romig & W. F. Chambers |
| | 22 | 1987 | Hawaii | R. H. Geiss |
| | 23 | 1988 | Milwaukee | D. E. Newbury |
| | 24 | 1989 | Asheville | P. E. Russell |
| | 25 | 1990 | Seattle | D. B. Williams, P. Ingram & J. R. Michael |
| | 26 | 1991 | San Jose | D. G. Howitt |

**(2) ICXOM series**

The triennial international conferences which started in 1956 acquired the title 'International Congress on X-ray Optics and Microanalysis' (ICXOM) from 1968. The proceedings have been produced under various titles, with different editors and publishers, as listed below. References are given as '*ICXOM 9*, p. 123', for example.

| No. | Year | Location | Title, editors & publishers |
|---|---|---|---|
| 1 | 1956 | Cambridge | *X-ray Microscopy & Microradiography*<br>V. E. Cosslett, A. Engström & H. H. Pattee<br>New York: Academic Press (1957) |
| 2 | 1959 | Stockholm | *X-ray Microscopy & X-ray Microanalysis*<br>A. Engström, V. E. Cosslett & H. H. Pattee<br>Amsterdam: Elsevier (1960) |
| 3 | 1962 | Stanford | *X-ray Optics & X-ray Microanalysis*<br>H. H. Pattee, V. E. Cosslett & A. Engström<br>New York: Academic Press (1963) |
| 4 | 1965 | Orsay | *Optique des Rayons X et Microanalyse*<br>R. Castaing, P. Deschamps & J. Philibert<br>Paris: Hermann (1966) |
| 5 | 1968 | Tübingen | *5th Int. Congr. on X-ray Optics & Microanalysis*<br>G. Möllenstedt & K. Gaukler<br>Berlin: Springer (1969) |
| 6 | 1971 | Osaka | *6th I.C.X.O.M.*<br>G. Shinoda, K. Kohra & T. Ichinikowa<br>Tokyo: Tokyo Univ. Press (1972) |
| 7 | 1974 | Moscow-Kiev | *7th I.C.X.O.M.*<br>I. B. Borovsky & N. Komyak<br>Leningrad (1976) |
| 8 | 1977 | Boston | *8th I.C.X.O.M.*<br>D. R. Beaman, R. E. Ogilvie & D. B. Wittry<br>Midland: Pendell (1980) |
| 9 | 1980 | The Hague | *Proc. 7th Eur. Congr. on Electron Microsc. & Proc. 9th I.C.X.O.M.* – vol. 3 – *Analysis* |

| No. | Year | Location | Title, editors & publishers |
|---|---|---|---|
| | | | P. Brederoo & V. E. Cosslett |
| | | | Leiden: 7th Eur. Congr. Electr. Microsc. Found. (1980) |
| 10 | 1983 | Toulouse | 10ème Congr. Int. d'Optique de Rayons X et de Microanalyse, J. de Physique, Colloq. C2, suppl. 2, vol. 45. |
| 11 | 1986 | London (Ont.) | 11th I.C.X.O.M. |
| | | | J. D. Brown & R. H. Packwood |
| | | | London (Ont.): Univ. W. Ontario (1987) |
| 12 | 1989 | Cracow | 12th I.C.X.O.M. |
| | | | S. Jasiénska & L. J. Maksymowicz |
| | | | Cracow: Acad. Mining & Metallurgy (1990) |

# Index

aberrations, of electron lenses, 39–41
absorbed current image, *see* specimen current image
absorption, X-ray, 9–10
absorption correction, 12, 218–39, 263–4, 283–7
absorption edges, 237–9
absorption edge jump ratio, 244
accelerating voltage, 13, 26, 156
   choice of, 155
ADC, see analogue to digital converter
alignment,
   of column, 42
   of electron gun, 24–5
alpha coefficients, 262, 270–2
analogue to digital converter, 135–6
analytical electron microscope, 21
apertures, 41–2, 45
area/peak factor, 280
astigmatism, 33, 40
atomic number correction, 12
atomic structure, 4, 292–3
Auger effect, 7, 22, 301

background corrections,
   for e.d. analysis, 169–74
   for w.d. analysis, 149–52, 281
backscattered electron image, 55–6, 216–17
backscattering coefficient, 205–6
backscattering correction, 12, 212–14, 267

baseline restoration, 131–2
beam switching, 140–1
Bence–Albee coefficient, 272
beryllium window, 80, 108–9
Bethe range, 198
Bethe stopping power expression, 190–2
bias, grid, 26–30
Bloch's law, 190–1
Blodgett–Langmuir technique, 65
Bragg's law, 59
bremsstrahlung, *see* continuous spectrum
brightness, of electron beam, 28–30, 44–5
Burger–Dorgelo rule, 300

carbon coating, *see* conducting coating
cathodoluminescence, 16
cellulose nitrate, *see* collodion
characteristic fluorescence corrections, 243–51, 264
characteristic X-rays, 1, 4–6, 188–9, 193–6
charge restoration, in preamplifier, 126–8
chemical effects, on X-ray spectrum 275–6, 280
chromatic aberration, 33, 41
coherent scattering, *see* Rayleigh scattering
cold finger, 19, 282

# Index

collection efficiency, of Si(Li) detector, 120–1
collimator, for Si(Li) detector, 109
collodion counter window, 81
colour images, 53–4
Compton scattering, 9
computer control, of instrument, 48
computer programs, for corrections, 272–3
condenser lens, 33, 42–3
conducting coating, 156–8, 281–2
contamination, 19, 122–3, 158, 282
continuous slowing down approximation, 188
continuous X-ray spectrum, 8, 169–73
continuum *see* continuous X-ray spectrum
continuum fluorescence correction, 251–7, 265–7
continuum modelling, 151, 169–73
Coster–Kronig transitions, 197, 301–5
counter gas, 81–3
counter window, 80–1
counting statistics, 153
counting strategy, 153–4
counting system, for w.d. spectrometers, 91–102
critical excitation energy, 6
crossover, 24, 28, 33
cryogenic pump, 20
cryostat, 107–8
crystals, for w.d. spectrometers, 61–4

dead layer, in Si(Li) detector, 113–15
dead time, 100–2, 139
deflection system, 49
defocussing, of w.d. spectrometers, 76–8, 147–8
demagnification, of electron source, 42–3
depth distribution,
  of characteristic X-ray production, 218–22
  of continuum production, 169
derivative references, 179
detection limit, 154–5, 183–4
differentiation, of pulses, 94–5, 129–30

diffraction gratings, 66
diffusion depth, 207
diffusion pump, 18–19
digital filtering, 173–5
discriminator, 91, 96
display systems, 49–51
doubly-curved crystal geometry, 68
Duane–Hunt limit, 8, 164–5
Duncumb–Reed method,
  for backscattering correction, 212
  for stopping power correction, 194–5

efficiency,
  of Si(Li) detector, 118–23
  of w.d. spectrometer, 75–6
  of X-ray production, 7
electron backscattering, 202–17
electron detectors, 17, 56–8
electron gun, 13, 24–32
electron trap, for Si(Li) detector, 110
electronics,
  for e.d. spectrometers, 125–41
  for w.d. spectrometers, 91–102
energy, of X-rays, 5–6, 296–9
energy calibration, of e.d. spectrometers, 163–4
energy distribution, of backscattered electrons, 210–11
energy level diagram, 295
energy resolution,
  of proportional counter, 86–87
  of Si(Li) detector, 111, 132–3
escape peak,
  proportional counter, 89
  Si(Li) detector, 115–17, 180
exclusion principle, Pauli's, 292

Fano factor, 87, 111
Faraday cup, 47
FET, *see* field effect transistor
field effect transistor, 105, 125–8
field emission source, 32
filament, 13, 24, 26, 28–9
filter-fit method, 176–9
final lens, 33, 35–8, 42–3
flow counter, 80

fluorescence,
    X-ray, 10, 22–3, 242
    correction, 10, 12, 243–61
    near phase boundaries, 257–60
    uncertainty, 261
    yield, 7, 301–3
formvar counter window, 81

gas gain, in proportional counter, 83–4
gaussian distribution, 86–7
gaussian models, for phi-rho-z, 229–30
germanium detectors, 123–4
grid, 24–7
gun current, 25–7

Harwell pulse processor, 133–4
high brightness electron guns, 31–2
high order reflections, 63, 277

ice layer, on Si(Li) detector, 110
incident electron energy, 3
    measurement of, 156, 164–5
incomplete charge collection, in Si(Li) detector, 112–15, 279
integrated reflectivity, of diffracting crystal, 60–1
integration,
    of peaks, 165, 280
    of pulses, 94–5, 129–30
intensity, of X-ray lines, 300
interferences, in w.d. analysis, 152–3
internal fluorescence, in Si(Li) detector, 117–18
ion-implanted silicon detector, 124
ion pump, 18
ionisation, inner shell, 6–7
ionisation cross-section, 186–8
ionisation statistics,
    in proportional counter, 86–7
    in Si(Li) detector, 111
iteration, 12, 268–70

Johann geometry, 67
Johansson geometry, 67

$k$ ratio, 12
K spectrum, 4

KAP, *see* potassium acid phthalate
Kramers' law, 8, 169–71

L spectrum, 5
lanthanum hexaboride cathode, 31–2
layered samples, 240–1
least squares fitting, 175–9
Lenard coefficient, 199
lenses, electron, 13, 33–8
light element analysis, 274–91
limit of detection, *see* detection limit
line scan, 52, 55
lithium drifted silicon detector, 2, 103–24
    efficiency of, 118–23
    electronics for, 125–41
    energy calibration of, 163–4
    energy resolution of, 111
    windowless, 110
lithium fluoride crystal, 61–3
Love–Cox–Scott method, for stopping power correction, 195
Love–Scott method, for backscattering correction, 213–4

M spectrum, 5
magnetic lenses, 33–4
magnification, of scanning image, 17
main amplifier,
    for proportional counter, 93–6
    for Si(Li) detector, 129–33
mapping, elemental, 51–4, 161–2
mass absorption coefficient, *see* mass attenuation coefficient
m.a.c., *see* mass attenuation coefficient
mass attenuation coefficient, 9, 237–9, 287–91
matrix corrections, 3, 11–12, 262–73
mean excitation energy, 190–1, 195
mercuric iodide detectors, 124
mini-lenses, 37–8
Monte-Carlo method, 208–10, 286–7
mosaic crystals, 61
Moseley's law, 1
multilayers, 65–6, 276–7
multiple scattering, of electrons, 204
Mylar counter window, 81

# Index

noise, in Si(Li) detector preamplifier, 128–9, 279
non-diagram lines, 300
non-normal electron incidence, 17, 215–16, 234–36

objective lens, *see* final lens
optical microscope, 15–17, 35–6
opto-electronic charge restoration, 126–8
overlap, between peaks, 152–3, 166–8

PAP method, for matrix corrections, 196, 214, 231–3
parabolic model, for phi-rho-$z$, 231–2
peak broadening, in e.d. spectra, 132, 165–6
peak integration, in e.d. analysis, 165
peak shift, in e.d. spectra, 165–6
peak to background ratio, 75–6, 154
peak to background ratio method, 181–2
pentaerythritol crystal, 62–4
PET, *see* pentaerythritol
phi-rho-$z$ function, 218–32, 283–7
Philibert method, for absorption correction, 224–7
Philibert–Tixier method, for stopping power correction, 195
photoelectric effect, 9
pile-up, pulse, 137–9, 140, 141, 180–1
pinhole lens, 35
polepieces, 34, 36
pole-zero cancellation, 131
polycarbonate counter window, 81
polyethylene terephthalate, *see* Mylar
polypropylene counter window, 81
polyvinyl formaldehyde, *see* formvar
potassium acid phthalate crystal, 62
Pouchou–Pichoir method,
 for backscattering correction, 214
 for stopping power correction, 196
preamplifier,
 for proportional counter, 92–3
 for Si(Li) detector, 125–8
probe current, 13, 27–8, 44–6
 monitoring, 46–8
 stability, 30–1, 46–8

probe diameter, 13, 44–6
probe-forming system, 13, 33–48
profile factor, 280
proportional counters, 79–90
proton probe, 22
pulse height analyser, 91, 96
pulse height analysis, 14, 96–7, 148, 277–8
 multichannel, 135–6
pulse height depression, 85
pulse shaping, 93–5, 129–30

quadrilateral model, phi-rho-$z$, 227–9
qualitative analysis, 3, 142–3, 161
quantitative analysis, 3
 with w.d. spectrometers, 143–60
 with e.d. spectrometer, 162–85
quantum numbers, 292–3
quartz crystal film thickness monitor, 157

radiationless transitions, 7
range, of electrons, 197–8
RAP, *see* rubidium acid phthalate
ratemeter, 98–9
Rayleigh scattering, 9
reflection curve, 60–1
reflection efficiency, of w.d. spectrometer crystal, 75
refraction, of X-rays, 63–4
resolution, of w.d. spectrometer, 71–4
rocking curve, *see* reflection curve
rotary pump, 18
Rowland circle, 67, 69
rubidium acid phthalate crystal, 62
Rutherford scattering formula, 202–3

satellite lines, 300–1
saturation, of probe current, 27–8
scaler, 99–100
scanning, 2, 17–18, 49–58
scanning coils, 49
scanning electron microscope, 2, 20–21
scanning image,
 electron, 17–18, 54–6
 X-ray, 17–18, 51–2, 78